Carbohydrates in Drug Design and Discovery

RSC Drug Discovery Series

Titles in the Series:

1: Metabolism, Pharmacokinetics and Toxicity of Functional Groups
2: Emerging Drugs and Targets for Alzheimer's Disease; Volume 1
3: Emerging Drugs and Targets for Alzheimer's Disease; Volume 2
4: Accounts in Drug Discovery
5: New Frontiers in Chemical Biology
6: Animal Models for Neurodegenerative Disease
7: Neurodegeneration
8: G Protein-Coupled Receptors
9: Pharmaceutical Process Development
10: Extracellular and Intracellular Signaling
11: New Synthetic Technologies in Medicinal Chemistry
12: New Horizons in Predictive Toxicology
13: Drug Design Strategies: Quantitative Approaches
14: Neglected Diseases and Drug Discovery
15: Biomedical Imaging
16: Pharmaceutical Salts and Cocrystals
17: Polyamine Drug Discovery
18: Proteinases as Drug Targets
19: Kinase Drug Discovery
20: Drug Design Strategies: Computational Techniques and Applications
21: Designing Multi-Target Drugs
22: Nanostructured Biomaterials for Overcoming Biological Barriers
23: Physico-Chemical and Computational Approaches to Drug Discovery
24: Biomarkers for Traumatic Brain Injury
25: Drug Discovery from Natural Products
26: Anti-Inflammatory Drug Discovery
27: New Therapeutic Strategies for Type 2 Diabetes: Small Molecules
28: Drug Discovery for Psychiatric Disorders
29: Organic Chemistry of Drug Degradation
30: Computational Approaches to Nuclear Receptors
31: Traditional Chinese Medicine

32: Successful Strategies for the Discovery of Antiviral Drugs
33: Comprehensive Biomarker Discovery and Validation for Clinical Application
34: Emerging Drugs and Targets for Parkinson's Disease
35: Pain Therapeutics; Current and Future Treatment Paradigms
36: Biotherapeutics: Recent Developments using Chemical and Molecular Biology
37: Inhibitors of Molecular Chaperones as Therapeutic Agents
38: Orphan Drugs and Rare Diseases
39: Ion Channel Drug Discovery
40: Macrocycles in Drug Discovery
41: Human-based Systems for Translational Research
42: Venoms to Drugs: Venom as a Source for the Development of Human Therapeutics
43: Carbohydrates in Drug Design and Discovery

How to obtain future titles on publication:
A standing order plan is available for this series. A standing order will bring delivery of each new volume immediately on publication.

For further information please contact:
Book Sales Department, Royal Society of Chemistry, Thomas Graham House, Science Park, Milton Road, Cambridge, CB4 0WF, UK
Telephone: +44 (0)1223 420066, Fax: +44 (0)1223 420247
Email: booksales@rsc.org
Visit our website at www.rsc.org/books

Carbohydrates in Drug Design and Discovery

Edited by

Jesús Jiménez-Barbero
CIB-CSIC, Madrid
Email: jjbarbero@cicbiogune.es

F. Javier Cañada
CIB-CSIC, Madrid
Email: jcanada@cib.csic.es

Sonsoles Martín-Santamaría
CIB-CSIC, Madrid
Email: smsantamaria@cib.csic.es

RSC Drug Discovery Series No. 43

Print ISBN: 978-1-84973-939-9
PDF eISBN: 978-1-84973-999-3
ISSN: 2041-3203

A catalogue record for this book is available from the British Library

Published by The Royal Society of Chemistry,
Thomas Graham House, Science Park, Milton Road,
Cambridge CB4 0WF, UK

Registered Charity Number 207890

For further information see our web site at www.rsc.org

Printed and bound by CPI Group (UK) Ltd, Croydon, CR0 4YY

Preface

Carbohydrate interactions are widely spread through nature and play essential roles in diverse biological functions. Without being exhaustive, we can mention gamete–gamete interactions that initiate fertilization, recognition by (and attachment to) host cells by pathogenic organisms, or leukocyte rolling during the course of inflammation. Many other processes of biomedical relevance also involve glycan recognition. During the last three decades, key developments have focused on attempting to comprehend the chemical and molecular basis of saccharide interactions and how they manifest in medicine and biology. Many research fields interconnect herein. Indeed, the progress in glycosciences is probably due to advances in the areas of: synthetic carbohydrate chemistry methods as well as in the development of powerful tools that allow the rapid, robust, and diverse screening of protein–carbohydrate systems, such as carbohydrate or protein-based microarrays; in the instrumentation used to make biophysical measurements; in the advances in high-resolution structural analysis by X-ray crystallography and NMR; in the development of carbohydrate-based therapeutics, *etc.*

Many studies of protein–carbohydrate interactions arise from their involvement in diseases. Indeed, the outer surfaces of the cells of most pathogens have a distinct set of complex carbohydrate structures present as glycolipids or glycoproteins, together with their partners: the carbohydrate-binding proteins (lectins or toxins) that specifically recognize particular carbohydrate motifs. In fact, the initial contact between two cells or a cell and pathogen almost certainly comes down to protein–glycan recognition events. It is now clear that carbohydrates play a main role in the highly specific molecular recognitions of lectins, enzymes, and antibodies, mediating the key cellular activities mentioned above, as cell recognition, growth, and apoptosis.

RSC Drug Discovery Series No. 43
Carbohydrates in Drug Design and Discovery
Edited by Jesús Jiménez-Barbero, F. Javier Cañada and Sonsoles Martín-Santamaría

Published by the Royal Society of Chemistry, www.rsc.org

Given the interdisciplinary nature of the glycoscience field, the study of carbohydrate interactions in drug research and discovery is widely open for exploration. This book gathers different efforts in this area. It covers basic aspects for exploring carbohydrate interactions, especially employing NMR spectroscopy, passing to the description of cutting edge research on the development of carbohydrate-based drugs and vaccines. Using chemical glycobiology methods and drug discovery approaches, the importance of multivalency for increasing recognition and the development of glyconanotechnology tools are also highlighted. Incursions in different key areas, such as Alzheimer's disease, cancer, bacterial infection, and immune response are also presented.

Jesús Jiménez-Barbero, F. Javier Cañada and Sonsoles Martín-Santamaría

Contents

RSC Drug Discovery Series No. 43
Carbohydrates in Drug Design and Discovery
Edited by Jesús Jiménez-Barbero, F. Javier Cañada and Sonsoles Martín-Santamaría

Published by the Royal Society of Chemistry, www.rsc.org

CHAPTER 1

Carbohydrate–Protein Interactions: A 3D View by NMR

ANA ARDÁ,[a] ANGELES CANALES,[a] F. JAVIER CAÑADA[a] AND JESÚS JIMÉNEZ-BARBERO*[a,b,c]

[a] Chemical and Physical Biology, CIB-CSIC, Ramiro de Maeztu 9, 28040 Madrid, Spain; [b] CIC bioGUNE, Parque Tecnológico de Bizkaia, Edif. 801A-1, 48160 Derio-Bizkaia, Spain; [c] Ikerbasque, Basque Foundation for Science, Bilbao, Spain
*Email: jjbarbero@cicbiogune.es

1.1 Introduction

Nowadays it is well established that carbohydrates have exceptional properties for coding information.[1] This also holds true for carbohydrate receptors, lectins, antibodies and enzymes, which translate the sugar-based signals into cellular effects.[2] In fact, one of the important roles of carbohydrates in Nature is to serve as recognition points for molecular receptors (which can be grouped into enzymes, lectins and antibodies), giving rise to a specific molecular recognition process that triggers a given biological response. The knowledge of the structural elements that govern such molecular recognition events is fundamental, and furthermore opens the possibilities to intervene in them with therapeutic purposes.

Oligosaccharides are involved in a plethora of regulatory processes, such as bacterial/viral infection, angiogenesis, inflammation, cell growth and

RSC Drug Discovery Series No. 43
Carbohydrates in Drug Design and Discovery
Edited by Jesús Jiménez-Barbero, F. Javier Cañada and Sonsoles Martín-Santamaría

Published by the Royal Society of Chemistry, www.rsc.org

development.[3] Understanding the chemical basis of carbohydrate–receptor interactions not only gives a functional meaning to structures and changes occurring in diseases but also helps devise innovative therapeutic approaches. Therefore, the comprehension of the conformational, dynamics and spatial presentation features of saccharides is of paramount importance.

NMR spectroscopy has been demonstrated to be a robust tool for carbohydrate research. In fact, different NMR approaches are widely employed to study the interactions of carbohydrates[4] and chemical analogues (glycomimetics) with their receptors up to the level of atomic resolution, both from the perspective of the carbohydrate ligand[5] and from the receptor.[6] The most accessible observables related to the structure are chemical shifts (δ), scalar couplings (J) and nuclear Overhauser effects (NOEs). However, in studies with complex oligosaccharides there are limits to the amount of relevant structural information provided by these observables, due to problems of signal overlapping, strong coupling and/or the scarcity of the key NOE information. In this sense, there is increasing use of additional parameters with structural information, such as residual dipolar couplings (RDCs), paramagnetic relaxation enhancements (PREs) or pseudo contact shifts (PCSs) induced by a paramagnetic ion. We will discuss all these parameters in this chapter.

Carbohydrates are rather flexible molecules. Therefore, NMR observables do not always correlate with a single conformer but with an ensemble of low free-energy conformers that can be accessed by thermal fluctuations. In this regard, NMR parameters should be complemented by computational methods in attempts to unravel the structural and conformational features of the molecular recognition process unambiguously.

Depending on the system under study, different NMR approaches can be followed to characterize protein–carbohydrate interactions; the standard methodologies can usually be classified as "ligand-based" or "receptor-based". The selection of the proper methodology is usually determined by the size of the receptor, the dissociation constant of the complex (K_D), the availability of labelled protein (^{15}N, ^{13}C) and the access to soluble receptors at enough concentration for NMR measurements.

1.2 Ligand-Based Approach

This is the most frequently employed methodology in NMR-based screening applied to drug discovery programs, both in academia and in industry. As the "ligand-observed detection" name suggests, detection takes place on the resonances of the free ligand. Ligand recognition can be identified thanks to the different motions of the receptor and ligand molecules (Figure 1.1): upon carbohydrate recognition the motional properties of the ligand are similar to the receptor and this change in mobility can be detected by different NMR experiments.

Ligand-based methods require complexes with relatively fast kinetics. This means dissociation constants in the range $K_D \geq 100$ μM. If k_{on} is well

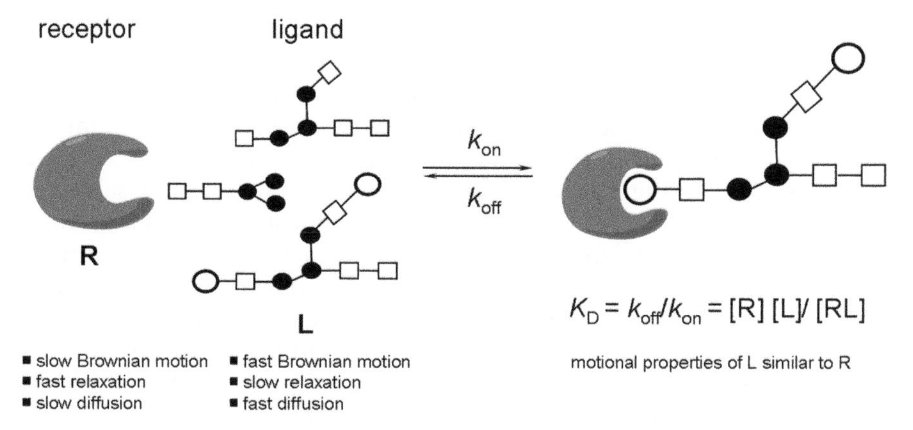

Figure 1.1 Illustration of the different physical properties of ligand and receptor molecules.

approximated by a diffusion-limited value (107–109 M^{-1} s^{-1}), then the slowest exchange rate (k_{off}) values lie in the range $1000 < k_{off} < 100\,000$ s^{-1}. Even though, in the biological context, carbohydrates are generally attached to proteins or lipids (glycoproteins and glycolipids), most of the studies of the interactions between carbohydrates and their receptors by NMR in solution are carried out by using free sugars. Thus, in the absence of multivalent effects, their binding strengths to proteins are usually rather weak. Thereby, in neutral carbohydrates the dissociation constants are usually in the mM to low μM range and the ligand-based approach can be successfully used.

In such a situation, we profit from the fact that, in most cases, carbohydrate ligands dissociate relatively fast from the receptor's binding site. If the receptor is much larger than the ligand, its larger correlation time will dominate the relaxation properties of the ligand, provided that an interaction takes place. These are the basis of two of the most commonly used methods for obtaining structural information from the ligand's perspective in carbohydrate–protein interactions by NMR: saturation transfer difference (STD) and transferred NOESY (trNOESY). The main advantages of these methods are the small amount of receptor required (experiments are acquired with an excess of the ligand with respect to the receptor). In addition, there is no requirement of the labelled protein, no limitations in size for the target molecule, or any requirement of knowledge about the target structure.

1.2.1 Saturation Transfer Difference

STD is a powerful technique; it allows us not only to identify binders to a receptor, discriminating from non-binders, but also to define the epitope in the case of an intermolecular interaction. Opposite to the usually folded and packed structures of other biomolecules, such as proteins and nucleic acids, carbohydrates present extended and flexible structures with large hydrodynamic radii. Defining the ligand regions in closer contact with the protein provides an important piece of information in order to deduce the 3D shape of the ligand in the binding site, and to identify the structural elements that provide specificity to the recognition process. The simplest version of the methodology implies acquiring two sets of 1D ^{1}H-based spectra: in one of them, the protein's proton resonances are saturated (on-resonance spectrum) and in the other one they are not (off-resonance spectrum). In the first experiment (on-resonance) the saturation of magnetization (protein) is transferred to the protons of the bound ligand, whose resonances will thus suffer a decrease in intensity with respect to the second spectrum (off-resonance), in which no protein saturation occurs. By subtraction (on-resonance and off-resonance), the STD spectrum is obtained, in which only resonances of the protons close to the protein show up. Furthermore, their intensity will be dependent on the proximity to the protein, yielding epitope information.

Based on this simple experiment, different strategies have been developed with the aim of overcoming the problem of proton overlapping, typical of carbohydrates. One of them has been to extend the ^{1}H STD information into a second dimension, whether homonuclear (STD-TOCSY and STD-NOESY)[7] or heteronuclear (STD-HSQC and STD-HMQC).[8] These experiments have allowed us to define epitopes in larger glycans, such as undeca- and nonasaccharide N-glycans[5] interacting with lectins or the hepta-xyloglucans recognised by a glycoside hydrolase.[9] A different strategy has consisted of applying the STD strategy to heteronuclei, such as ^{19}F.[10] In this case we fully avoid the resonance overlapping, but fluorine atoms have to be introduced through chemical synthesis.

These strategies have been applied thoroughly to study the recognition of glycans and glycomimetics by different carbohydrate recognising elements of therapeutic interest, such as lectins, antibodies, enzymes, viruses or even cells.[11] Some of these examples are gathered together below.

The recognition of carbohydrates on the host cell surface is an important step in viral entry. Taking advantage of the large size of viruses and virus-like particles (VLPs), STD is a well-suited method for studding these recognition processes. Even though contradictory results had been reported about the relevance of sialic acid in certain human viral infections,[12] much evidence is related to infectivity by rotavirus or influenza virus with the recognition of sialic acid on host cell surfaces. STD experiments demonstrated in 2009[13,14] that the previously thought sialidase-insensitive spike protein VP8* from the human rotavirus strain Wa is actually a sialic acid-dependent protein that recognises GD1 and GM1 and GM3-associated subterminal sialic acid residues. Later on, this recognition process was confirmed by performing STD experiments with native viruses, instead of the isolated viral protein. Furthermore, these experiments provided strong evidence that it is VP8* on the virus surface that is responsible for the recognition of sialic acids at the initial virus/host cell contacts.[15] Indeed, the application of STD to study the interaction between ligands and native viruses had been previously proposed by Peters *et al.*,[16] a strategy that requires very low amounts of virus and a high throughput of ligands. It was demonstrated to discriminate between binding (antiviral) from non-binding ligands to human rhinovirus serotype 2 (HRV2), providing, besides, complete binding epitope information. STD with viral particles has also been applied to gain insights into the infection process by avian influenza virus.[17] Increasing evidence had proposed that human infectivity by the avian influenza virus H5N1 involved a switch in preference of the viral protein hemagglutinin (HA) from α-(2–3)-linked Neu5Ac (major form in avian intestinal cells in birds where the infection is enteric) to α-(2–6)-linked Neu5Ac (abundant in lung and airway epithelial cells in humans in which the infection is respiratory) that could be due to a single amino acid mutation of Asp190 to Glu190. STD experiments carried out with virus-like particles derived from the H5N1 avian influenza containing the hemagglutinin (HA) protein demonstrated that HA from non-humans is indeed able to discriminate between 3′SLN and 6′SLN, binding preferentially to the first one. A more recent work has used STD to relate HA of different subtypes with preference to bind either 3′SLN or 6′SLN.[18] The interaction of noroviruses, the major cause of nonbacterial gastroenteritis, with host attachment factors, like human blood group antigens (HBGA)[19] and inhibitors,[20] has also been studied through STD-NMR.

STD has also been applied to living cells, by overexpressing the corresponding receptor on the cell surface. It was first applied to demonstrate the recognition of the *S. cerevisiae* mannan by cells containing the lectin DC-SIGN (dendritic cell-specific ICAM-3 grabbing nonintegrin).[11a] One of the drawbacks of this approach is to maintain the living cells suspended in solution. More recently, the combination of this strategy with high-resolution magic angle spinning (HR-MAS) was successfully applied to the recognition of the glycoside natural product phlorizin by the Na^+/glucose co-transporter hSGLT1, for which it is known to be a potent inhibitor.[21]

In all the above-mentioned examples, the carbohydrate binding protein attached to the surface of either cells, viruses or virus-like particles is a lectin. Lectins are proteins of non-enzymatic and non-immune origin in charge of recognising carbohydrates, playing a wide variety of roles in different biological events. Indeed, lectins can be potential targets for drug development in different pathologies.[22] In particular, the role of certain lectins in cancer development and apoptosis has focused the attention on them as possible therapeutic anti-tumour targets.[23]

A typical strategy in drug discovery is the development of enzyme inhibitors with the aim of blocking their activity. Many of them are involved in carbohydrate processing. STD has also been applied in order to understand how substrates and products are recognised by their processing enzymes, an important piece of information for designing inhibitors. Interesting targets are the enzymes involved somehow in the synthesis of the precursors of the glycoconjugates on the parasite's cell surface, like UDP-glucose pyrophosphorylase from *Leishmania major*[24] or trans-sialidase from *Trypanosoma cruzi*[25] or in bacterial cell walls, like UDP-galactopyranose mutase.[26] In this last case, STD not only provided epitopic information but STD competition experiments were used in order to compare binding affinities of fluorinated inhibitors.[27]

STD was initially proposed for screening ligand libraries, permitting the use of mixtures of compounds to speed up the identification of ligands for a given target.[7] An interesting proposal has been to apply STD to dynamic libraries. The method was applied to the identification of the best β-galactosidase inhibitors among a pool of hemithioacetals, formed through the fast and reversible reaction between thiomonosaccharides with small chemical fragments containing aldehyde functionalities. By using eight different monosaccharides and three different fragments, a dynamic library of 18 possible inhibitors was created, from which the best one was identified by giving the strongest STD response.[28]

Carbohydrates act also as antigens, being recognised by antibodies and generating an immune response. Even though the details that confer immunogenicity to carbohydrates are not yet fully understood, their therapeutic use is promising and STD is being used as an important tool for describing epitope selectivity in these recognition processes. Some recent examples involve the use of antibodies as targets in autoimmune diseases like Guillain–Barre syndrome,[29] in cancer,[30] in infectious diseases like HIV[31] or in bacterial infections like those produced by *Y. pestis*,[32] group A streptococcus (GAS),[33] *Shigella flexneri*[34] or *Bacillus anthracis.*[35]

STD can also provide an estimation of the binding affinity. The traditional strategy has been based on following the ligand STD changes upon the presence of a competitor, whose K_D is known.[36] Alternatively, single ligand titration experiments have been proposed to directly extract K_D from STD measurements.[37]

In spite of the wide possibilities of STD, this method is applicable to a limited range of binding affinities, between μM and mM for K_D, due to the requirements for fast dissociation rate complexes. In cases where the affinity is too high (in fact, when k_{off} is too slow), no STD signal will be detected, because proton ligand relaxation will occur before detection. In such cases the absence of STD signal might be misinterpreted as a nonbinding event. On the other hand, nonspecific interactions can give rise to a STD response, especially when dealing with aromatic patches prone to establish hydrophobic interactions, particularly in water, and whose long relaxation rates can bias the STD intensities.[38]

1.2.2 Transferred Nuclear Overhauser Effect Spectroscopy

The transferred NOESY[39] is in this sense a more robust and reliable technique. It is based on the different sign of the NOE depending on molecular correlation time, and thus on molecular size: positive for small molecules, but negative for large ones.[40] As stated above, if a small ligand is in the presence of an interacting large protein, and in fast exchange between the free and the protein-bound states, the large correlation time of the protein–ligand complex will dominate, even if only a small fraction of the ligand is in the bound state. As a result, while the NOEs of the free ligand are positive, the NOEs of the ligand in the presence of the protein will become negative. This NOE sign change is known as transferred NOE. This method permits us to prove that one small ligand binds to a large receptor, although the reverse is not necessarily true.

The first applications of trNOE permitted us to identify different scenarios that might take place in the recognition of carbohydrates. One of them is a conformational selection process, in which a single conformation of the ligand, among the ones present when it is free, is recognised, as in the recognition of the disaccharide melibiose [Gal-α-(1→6)-Glc] by the lectin domain of ricin, where only the major ligand's conformation found in the free state is recognised by the lectin.[41] In a few cases the conformational selection proceeds in such a way that a ligand conformation with minor contributions in the free state is the one recognised, like the case of the Glc*p*NAc-β-(1→6)-α-Man*p* disaccharide bound to the lectin WGA.[42]

The study of the recognition of sialyl Lewis X tetrasaccharide by E-selectin, a key recognition event in inflammatory processes, allowed identifying a conformational selection process in which the major conformer in solution is the one being recognised.[43] In parallel, the binding mode of a family of E-selectin antagonists with higher affinity than the natural ligand was also studied by trNOE. Interestingly, the comparison between the natural ligand's and antagonist's binding modes allowed the researchers to relate conformational preorganization with increased binding affinity.[44] Many other examples of lectin/carbohydrate recognition processes have been studied by trNOE.[45]

Glycosaminoglycans (GAGs) have well-known therapeutic applications, such as anticoagulants or antithrombotic drugs. The advances in structure–function relationship knowledge are providing them with new possible therapeutic applications.[46] trNOESY has contributed to gain insights into the recognition of GAGs by their receptors, where glycan flexibility involves not only dihedral angles[47] but also pyranose ring conformation, like in the case of the synthetic 6-nonsulfated tetrasaccharide ANS-I2S-ANS-I2s, where the IdoA residues adopt a $^{1}C_{4}$ conformation when bound to FGF2.[48]

One the main difficulties in trNOESY is to achieve the proper experimental conditions to detect such an effect, the protein/ligand ratio being a critical parameter. The most important bias in this experiment comes from spin-diffusion in the ligand mediated by the protein.[49] This can give rise to the misinterpretation of spin-diffusion crosspeaks as trNOESY crosspeaks, which would lead to a virtual bound conformation. The way to identify such effects is through the co-acquisition of trROESY experiments.[50]

1.2.3 Combinations

In many cases, STD and trNOESY experiments are combined together in order to have a more complete view of the interaction from the ligand perspective: the epitope mapping and bound conformation. This strategy has significantly contributed to gain insights into biological processes with therapeutic applications. For instance, it provided confirmation that the different binding modes found by X-ray[51] for the recognition of the terminal Man-α-(1–2)-Man fragment by DC-SIGN also takes place in solution.[52] DC-SIGN is a C-type lectin found on the surface of dendritic cells and macrophages, known to recognise mannose and Lewis X-based pathogen-associated molecular patterns (PAMPs) to subsequently activate phagocytosis. It has attracted much attention since it was found to be the target to which some viruses (*e.g.* VIH,[53] Ebola,[54] hepatitis C[55]), bacteria (*e.g. H. pylori*, *K. pneumoniae*, *M. tuberculosis*[56]), parasites (*e.g. L. pifanoi*) and fungi (*Candida albicans*)[57] bind, being essential for the infection. Thus, the development of inhibitors with the aim of blocking its carbohydrate binding properties has been proposed as a potential therapeutic strategy.[58] In certain cases the combination of trNOESY and STD has allowed the identification of their different binding modes.[59]

1.2.4 When Tight Binding Occurs

Despite the advantages of the STD and trNOESY experiments, there are systems in which this methodology cannot be applied due to strong binding. In the context of carbohydrates, this is usually the case in charged systems such as glycosaminoglycans that establish electrostatic interactions with their protein receptors.[3] In these systems the carbohydrate conformation can be inferred by using filtered experiments with ^{13}C-labelled protein samples. The protons of the receptor (attached to ^{13}C) can be selectively removed from the spectrum, allowing the analysis of the carbohydrate signals (protons attached to ^{12}C) without interference from the protein signals. The ^{13}C double-filtered 2D NOESY method has been successfully applied to obtain the conformation of a hexasaccharide bound to acidic fibroblast growth factor.[60]

These experiments can also be acquired in the 3D version. For instance, the 3D ^{13}C F1-edited F3-filtered HSQC-NOESY experiment has been applied in the structure calculation of the complex between *Coprinopsis cinerea* lectin 2 (CCL2) and a fucosylated chitobiose. In this case, not only intramolecular NOEs but also 82 intermolecular NOEs could be identified.[61]

1.2.5 Residual Dipolar Couplings

Residual dipolar couplings (RDCs) originate from the dipolar interaction occurring between a pair of nuclei (*e.g.* ^{1}H–^{13}C, ^{1}H–^{1}H). RDCs are orientational in nature since they depend on the angle θ that forms the vector joining two nuclei with the spectrometer magnetic field and are independent of the distance between protons. This feature makes RDCs highly complementary to the conventional distance constraints derived from NOE data and are especially useful in systems with small numbers of NOEs due to carbohydrate flexibility or extended shapes.

The dipolar coupling between a pair of nuclei (*e.g.* $^{1}H–^{13}C$) displays the angular dependence shown in eqn (1.1), where θ is the angle that forms the vector joining the two nuclei with the direction of the B_0 spectrometer magnetic field and r_{HC} is the distance between interacting spins:

$$D_{(^{1}H-^{13}C)} = \frac{\mu_0 \gamma_H \gamma_C h}{\langle 2\pi r_{HC} \rangle^3} \left\langle \frac{3\cos^2\theta - 1}{2} \right\rangle \tag{1.1}$$

The terms enclosed within brackets in eqn (1.1) correspond to the time average distance $\langle r_{HC}{}^{-3} \rangle$ and time angular averaging $\langle \cos^2\theta \rangle$. In liquid-state NMR the function goes to zero due to the isotropic distribution of the $^{1}H–^{13}C$ vectors and no dipolar couplings are observed. Therefore, RDC measurement requires a weak alignment of the carbohydrate molecules in order to obtain some orientations of the interatomic vector connecting the two coupled nuclei slightly more favourably than others.[62]

The calculation of the dipolar interaction between a pair of dipolar-coupled nuclei requires determination of $\langle \cos^2\theta \rangle$ in eqn (1.1). This term can be determined by using the alignment tensor methodology.[63] In the so-called principal axis frame, in which the three axes of the alignment tensor are diagonal, eqn (1.2) provides the theoretical value of the RDC for any $^{1}H–^{13}C$ vector of the carbohydrate, where A_{ax} and A_{rh} are the axial and rhombic components, respectively, θ is the angle that forms vector H–C with the *z*-axis in the molecular frame and ϕ is the angle that describes the projection of vector H–C over the plane *xy* of the molecular frame (see Figure 1.2):

$$D_{(^{1}H-^{13}C)} = \frac{-\mu_0}{16\pi^2} \frac{\gamma_H \gamma_C h}{r_{HC}^3} \left[A_{ax}(3\cos^2\theta - 1) + 3/2A_{rh}(\sin^2\theta\cos^2 2\phi)\right] \tag{1.2}$$

For a given signal in the spectrum, the magnitude of the RDC is determined from the difference in splitting between the sample oriented in the aligned media (RDC $+J$) and the isotropic non-aligned sample (J).

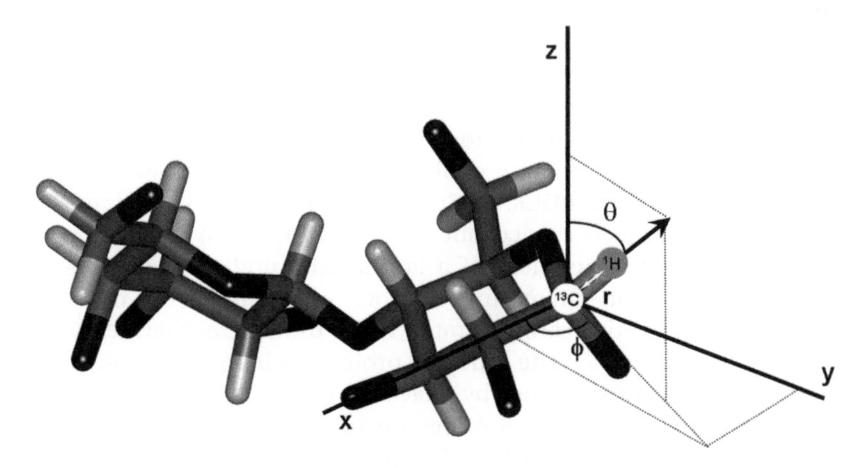

Figure 1.2 Definition of the angular parameters θ and ϕ used in eqn (1.2).

RDCs have been efficiently used to derive the conformation of carbohydrates in solution.[64] However, fewer examples have been described on the use of this parameter to study the carbohydrate conformation in the bound state with their receptors. In contrast to transferred NOEs that are dominated by NOE bound contributions even with high ligand/receptor ratios, RDCs of exchanging ligands can be dominated by free-state contributions. In order to overcome this drawback, it is possible to enhance the alignment of the protein under study by addition of a hydrophobic alkyl tail that anchors to the bicelles of the ordering medium.[65] However, this approach is not generally applicable because of the nature of the required protein modifications. The use of a His-tagged protein with a nickel chelate-carrying lipid inserted into the lipid bilayer-like alignment media has been proposed as an alternative approach.[66]

1.2.6 Pseudo Contact Shifts

Pseudo contact shifts (PCSs) also arise from a dipolar interaction, but in this case the effect is due to the dipolar interaction between the unpaired electrons of a paramagnetic ion and the nuclei in the vicinity of this ion. PCSs give rise to large changes in chemical shifts that decrease with the distance between the metal ion and the nuclear spin with a $1/r^3$ dependence. Therefore, PCSs provide long-range structural information. The equation that governs PCSs is given in eqn (1.3), where $\Delta\delta_{PCS}$ is the difference in chemical shifts measured between diamagnetic and paramagnetic samples, r is the distance between the metal ion and the nuclear spin and θ and ϕ are the polar coordinates describing the position of the nuclear spin with respect to the principal axes of the $\Delta\chi$ tensor:

$$\Delta\delta_{PCS} = \frac{1}{12\pi r^3}\left[\Delta\chi_{ax}(3\cos^2\theta - 1) + 3/2\Delta\chi_{rh}\sin^2\theta\cos^2\phi\right] \tag{1.3}$$

The effect is proportional to the anisotropy of the magnetic susceptibility tensor ($\Delta\chi$) and independent of the isotropic component of the χ tensor. It is therefore sufficient to describe the PCS as a function of the axial and rhombic components of the $\Delta\chi$ tensor, $\Delta\chi_{ax}$ and $\Delta\chi_{rh}$.

In addition, all lanthanide ions with non-vanishing susceptibility anisotropy tensors ($\Delta\chi$) generate weak molecular alignment with the magnetic field and therefore RDCs.[67a] Conformational studies of diamagnetic carbohydrates have been carried out by converting them to paramagnetic compounds through attachment of a lanthanide binding tag by chemical synthesis. There is an increasing number of examples of the use of PCS methodology in the carbohydrates field, from disaccharides to nonasaccharide structures.[67,68] The standard protocol is based on the generation of an ensemble of conformations by molecular dynamics simulations and to back-calculate the expected PCS values for each conformation. The fitting experimental *versus* calculated PCSs allow defining the carbohydrate conformation.

PCSs have been recently used to derive the conformation of a complex type N-glycan. The characterization of this molecule is challenging since the molecular pseudo-symmetry leads to isochronous NMR in both branches (see Figure 1.3).[68] N-Glycans are common modifications of membrane/secreted proteins and immunoglobulin G (IgG) antibodies. For instance, the Fc domain of IgG from healthy humans has a complex-type biantennary N-glycan attached to Asn297. While the Fc glycans are quite heterogeneous, a high percentage is normally terminated with galactose. Deviations from this pattern correlate well with certain diseases, as well as differential effects on inflammatory response.[69] Despite their importance, there is a need for

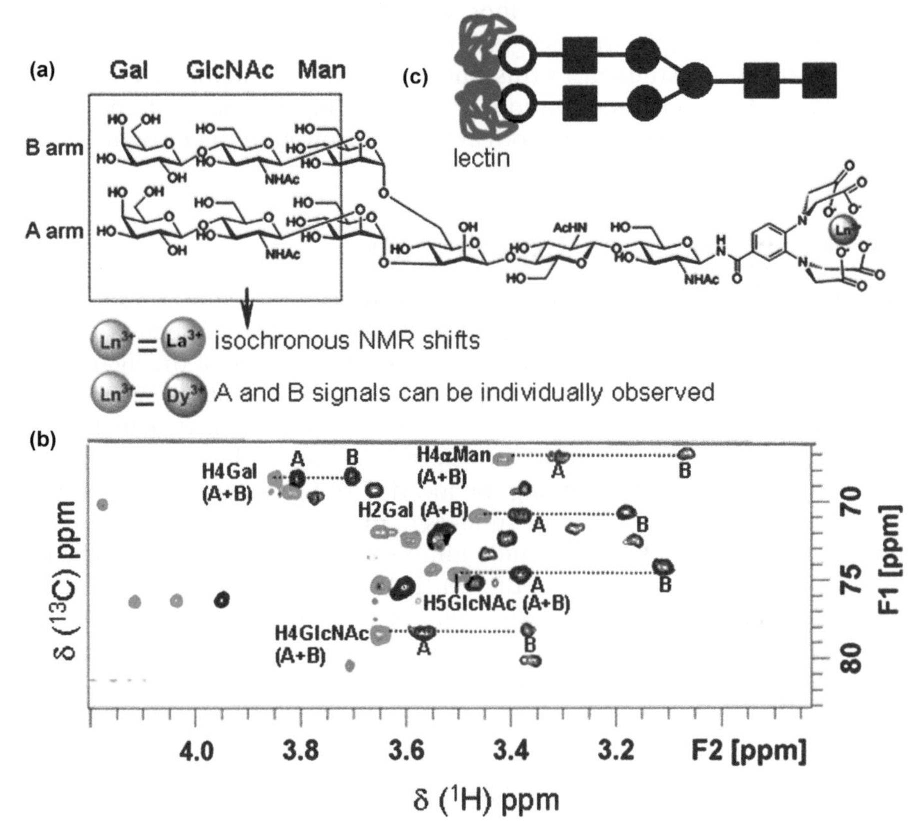

Figure 1.3 (a) Complex-type N-glycan derivative synthesized bearing a lanthanide binding tag. ^{1}H and ^{13}C chemical shifts of the terminal Gal and GlcNAc moieties as well as those of the C3, C4, and C6 atom pairs of Man residues A and B are isochronous in the natural compound and also in the reference spectrum acquired with the diamagnetic ion La^{3+} [*red spectrum* in (b) panel]. (b) Superimposition of the ^{1}H–^{13}C HSQC acquired in diamagnetic (*red*) and paramagnetic (*black*) conditions. Some of the signals that can be individually assigned in the paramagnetic sample are labelled. (c) Schematic representation of the recognition of the N-glycan by galectin-3.

development of new methods to characterize these flexible carbohydrates. The conformational features of the different classes of N-glycans (from complex to high-mannose type) have been investigated using different techniques, especially NMR spectroscopy,[70] since the flexibility of the glycosidic linkages has usually hampered their study by X-ray crystallography. However, typical NMR parameters (*J* couplings and NOEs) provide only short-range structural information. Therefore, it is not easy to define the global shape of these nonglobular molecules.

In this context, the use of the lanthanide tag has permitted breakage of the inherent pseudo-symmetry of the N-glycan as shown in Figure 1.3, revealing that the T-shaped *gg* rotamer at the Man-α-(1–6)-Man junction is the major one in solution, with minor contributions of other backfolded geometries. In addition, the recognition of this nonasaccharide by human galectin-3 has been studied. In this line, the novel methodology employed has permitted the characterization of the binding epitopes of the symmetrical N-glycan, showing that both arms are involved in the recognition of human galectin-3.[68]

1.3 Receptor-Based Experiments

The monitoring of molecular recognition processes using these NMR methods is based on the comparison of different NMR parameters of the receptor molecule resonances in the presence and absence of the ligand (or mixtures of putative ligands). Owing to the direct observation of receptor signals, this method cannot be applied to large proteins (it is usually employed in receptors below 40 kDa). The advantage of the receptor-based approach is that it allows defining the location of the protein binding site. This is especially important in the understanding of the action mechanism of a drug, since it is possible to find additional binding sites (as happens in allosteric modulators). The main drawback of receptor-based techniques is the requirement of a previous assignment of the protein NMR resonances. This assignment should be ideally combined with the *a priori* knowledge of the receptor's 3D structure (either from X-ray or NMR) to drive lead generation.

1.3.1 Chemical Shift Perturbations

One of the most frequently used receptor-based methods in drug discovery programs relies on the perturbation of the receptor's chemical shifts upon binding of the ligand. The chemical shift is an extremely sensitive parameter to the environmental changes and therefore can be used to detect interaction processes. Chemical shift mapping obtained *via* NMR titration experiments is a straightforward NMR technique to define those protein residues that are involved in the binding to a partner molecule. Usually, a series of NMR experiments (generally ^{1}H–^{15}N HSQC or ^{1}H–^{13}C HSQC spectra) are recorded with increasing amounts of the ligand (Figure 1.4).

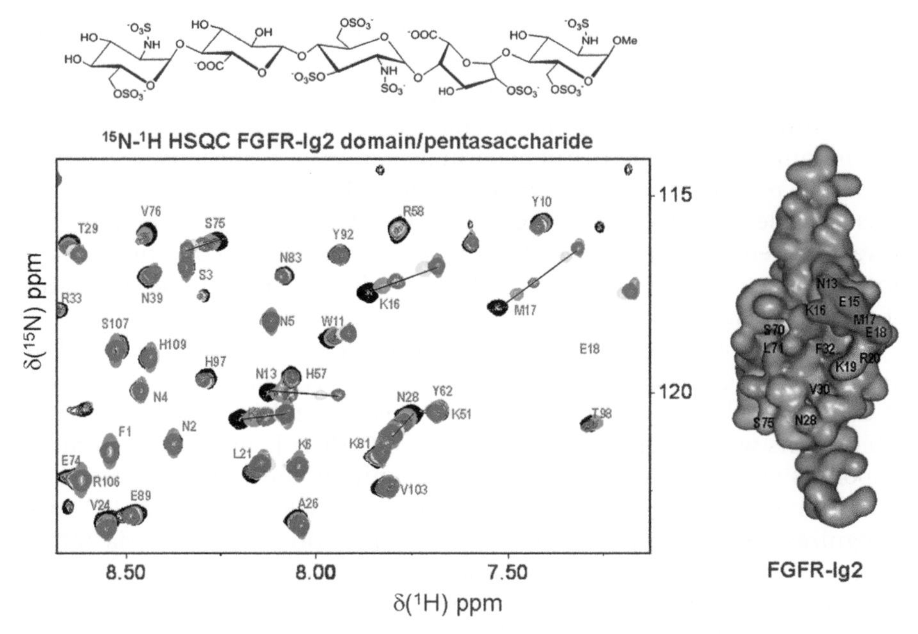

Figure 1.4 $^{1}H-^{15}N$ HSQC spectra of FGFR-Ig2 acquired in the presence of a heparin pentasaccharide. Titration carried out with 0.25, 0.5, 1.0, 1.5 and 2.0 molar equivalents of the pentasaccharide. The expansion shows significant changes in the protein signals (up to 0.15 ppm) upon additions of the carbohydrate.

This approach has been used to unravel the role of heparin in the activation of fibroblast growth factors (FGFs) and fibroblast growth factor receptors (FGFRs).[6,71] FGFs are key pharmacological targets since they are involved in cell replication, angiogenesis, differentiation, cell adhesion, migration and wound healing. Down- and up-regulation of FGFs are associated with many pathologies.[72]

The interactions of heparin fragments with both acidic FGFs and FGFRs were characterized using NMR.[69,71] As an example, in Figure 1.4 is shown the superimposition of the $^{1}H-^{15}N$ HSQC spectra of the immunoglobulin-like 2 domain (Ig2) of FGFR in the presence of increasing amount of a heparin pentasaccharide.[73] As a first step, assignment of the backbone HN correlations of the protein was carried out by combination of the 3D triple resonance HNCO, HN(CA)CO, HNCACB and CBCA(CO)NH experiments. $^{1}H-^{15}N$ HSQC experiments were collected with different ligand/protein ratios from 1 : 0.25 to 1 : 2. The residues with the largest chemical shift perturbations are clustered on a well-defined surface of the protein, indicating that there is a specific interaction between the protein and the oligosaccharide at this particular region.

In the heparin field, NMR receptor approaches also have been proved to be useful in determining the role of heparin in the cytotoxicity of eosinophil cationic protein (ECP).[74] ECP is secreted by activated eosinophil granulocytes during infections, acting as a mediator in human immune host defence. ECP's cytotoxicity is the subject of intense research due to the tissue damage associated with eosinophil degranulation at the inflammation site.[75] Novel eosinophil targeting therapies are under development and the anti-inflammatory properties of heparin derivatives are a promising research field.[76] In recent work, the structure of the ECP in complex with a heparin trisaccharide was obtained using NMR. ^{1}H–^{15}N HSQC experiments of ECP were acquired in the presence of increasing trisaccharide concentrations to obtain both the location of the carbohydrate binding site and the estimation of the dissociation constant.[74]

NMR receptor-based methods have also been used to study DC-SIGN inhibitors. DC-SIGN is a C-type (Ca^{2+} dependent) lectin present in the surface of dendritic cells. DC-SIGN binds to mannosides and fucosides to mediate interactions with other cells or pathogens. DC-SIGN is involved in immune responses since it mediates pathogen recognition and cellular interactions that lead to pathogen neutralization. As a pathogen receptor, DC-SIGN recognizes several viruses (HIV,[53] ebola,[54] hepatitis C[55]), bacteria (*Mycobacterium tuberculosis*[56]) and fungi (*Candida albicans*[57]). However, numerous studies have demonstrated that some of these pathogens subvert DC's function to escape immune surveillance by targeting DC-SIGN, such as HIV-1 that exploits the DC-SIGN internalization pathway to transinfect T cells.[53] DC-SIGN binds to the mannosylated surface glycoprotein gp120 on HIV, and this recognition event is the initial entry port of the human immunodeficiency virus to the host.

In this context, it is important to design inhibitors of the carbohydrate–lectin interaction. Oligomannose and oligofucose derivatives could be envisioned as potential inhibitors of DC-SIGN interactions. However, the presence of additional C-type lectins with similar specificities precludes their application. In order to overcome this problem, glycomimetics with higher affinity for DC-SIGN have been designed. Addition of the glycomimetic to DC-SIGN carbohydrate recognition domain resulted in ^{1}H–^{15}N HSQC chemical shift perturbations similar to those obtained with *N*-acetylmannosamine or fucose.[77] These data demonstrate that the glycomimetic occupies the same carbohydrate binding site and interacts with the same side-chain residues on DC-SIGN as the natural ligands.

^{1}H–^{15}N HSQC experiments were also used to study bifunctional inhibitors, designed for targeting both matrix metalloproteinases (MMPs) and galectins.[78] There is strong evidence supporting the fact that MMPs are involved in tumour progression.[79] In addition, galectins are overexpressed in several tumours.[80] The presence of galectins in the zone of tumour invasion and the role in tumour progression of up-regulating MMPs reveal a functional connection between these two effector proteins. In this context, a dual inhibitor was synthesized and its ability to bind both receptors was explored using ^{15}N-labelled galectin-3 and ^{15}N-labelled metalloproteinase 12.[78] The simultaneous binding of the dual inhibitor to the two proteins was demonstrated using a wise combination of ligand-based and receptor-based NMR experiments, employing the proper combination of labelled and non-labelled proteins to monitor the interaction process.

1.3.2 Paramagnetic Relaxation Enhancements

In the context of carbohydrate–protein interactions, it is also worth mentioning the application of nitroxide spin-labelled carbohydrates for mapping the carbohydrate binding sites on the protein surface. Nitroxide spin-labelled molecules, such as 2,2,6,6-tetramethylpiperidine-1-oxyl (TEMPO), give rise to intensity losses in the NMR peaks due to the paramagnetic relaxation enhancement effect (PRE) of the nitroxide.

Nitroxide spin labels have been incorporated by chemical synthesis in *N*-acetyllactosamine,[81] cellotriose and cellotetraose to carry out binding studies.[82]

The presence of the spin label allows defining the interaction surface in the receptor by analysing the perturbation of the intensities of cross-peaks in the protein ^{15}N HSQC spectrum. In this context, the perturbation of the intensities of cross-peaks in the ^{15}N HSQC spectrum of full-length galectin-3 has been monitored and used to identify protein residues proximate to the binding site for *N*-acetyllactosamine.[81] In addition, it is possible to convert intensity measurements into distances between discrete protein amide protons and the bound spin label.[81]

1.4 Perspectives

Studying carbohydrate interactions by using NMR with the drug design perspective is a growing field. We foresee many developments in the coming years. The standardization of novel expression systems for biomedically relevant glycoproteins will allow us to generate these entities in the proper amounts for the required detailed generation and analysis of NMR parameters. Current NMR methodologies, based on those described herein, will also find their application in the derivation of the key factors influencing the existence of induced fitting or conformational selection processes. The combination with advanced modelling procedures will, with no doubt, generate novel methodologies, novel scientific achievements and also molecules that can be employed as drugs or probes.

References

1. D. Solís, N. V. Bovin, A. P. Davis, J. Jiménez-Barbero, A. Romero, R. Roy, K. Smetana Jr. and H.-J. Gabius, *Biochim. Biophys. Acta*, 2015, **1850**, 186.
2. H.-J. Gabius, S. André, J. Jiménez-Barbero, A. Romero and L. Solís, *Trends Biochem. Sci.*, 2011, **36**, 298.
3. I. Capila and R. J. Linhardt, *Angew. Chem., Int. Ed.*, 2002, **41**, 391.
4. *NMR Spectroscopy of Glycoconjugates*, ed. J. Jiménez-Barbero and T. Peters, Wiley-VCH, Weinheim, 2003.

5. A. Ardá, P. Blasco, D. Varón Silva, V. Schubert, S. André, M. Bruix, F. J. Cañada, H.-J. Gabius, C. Unverzagt and J. Jiménez-Barbero, *J. Am. Chem. Soc.*, 2013, **135**, 2667.
6. A. Canales, R. Lozano, B. López-Méndez, J. Angulo, R. Ojeda, P. M. Nieto, M. Martín-Lomas, G. Giménez-Gallego and J. Jiménez-Barbero, *FEBS J.*, 2006, **273**, 4716.
7. M. Mayer and B. Meyer, *Angew. Chem., Int. Ed.*, 1999, **38**, 1784.
8. M. Vogtherr and T. Peters, *J. Am. Chem. Soc.*, 2000, **122**, 6093–6099.
9. A. Silipo, J. Larsbrink, R. Marchetti, R. Lanzetta, H. Brumer and A. Molinaro, *Chem. – Eur. J.*, 2012, **18**, 13395.
10. T. Diercks, J. P. Ribeiro, F. J. Cañada, S. André, J. Jiménez-Barbero and H.-J. Gabius, *Chem. – Eur. J.*, 2009, **15**, 5666.
11. (a) S. Mari, D. Serrano-Gómez, F. J. Cañada, A. L. Corbí and J. Jiménez-Barbero, *Angew. Chem., Int. Ed.*, 2004, **44**, 296; (b) B. Claasen, M. Axmann, R. Meinecke and B. Meyer, *J. Am. Chem. Soc.*, 2005, **127**(3), 916–919.
12. (a) K. Fukudome, O. Yoshie and T. Konno, *Virology*, 1989, **172**, 196; (b) M. Ciarlet, S. E. Crawford and M. K. Estes, *J. Virol.*, 2001, **75**, 11834; (c) J. M. Nicholls, M. C. W. Chan, W. Y. Chan, H. K. Wong, C. Y. Cheung, D. L. W. Kwong, M. P. Wong, W. H. Chui, L. L. M. Poon, S. W. Tsao, Y. Guan and J. S. M. Peiris, *Nat. Med.*, 2007, **13**, 147; (d) Y. Ha, D. J. Stevens, J. J. Skehel and D. C. Wiley, *Proc. Natl. Acad. Sci. U. S. A.*, 2001, **98**, 11181.
13. T. Haselhorst, F. E. Fleming, J. C. Dyason, R. D. Hartnell, X. Yu, G. Holloway, K. Santegoets, M. J. Kiefel, H. Blanchard, B. S. Coulson and M. von Itzstein, *Nat. Chem. Biol.*, 2008, **5**, 91.
14. T. Haselhorst, H. Blanchard, M. Frank, M. J. Kraschnefski, M. J. Kiefel, A. J. Szyczew, J. C. Dyason, F. Fleming, G. Holloway, B. S. Coulson and M. von Itzstein, *Glycobiology*, 2007, **17**, 68.
15. T. Haselhorst, T. Fiebig, J. C. Dyason, F. E. Fleming, H. Blanchard, B. S. Coulson and M. von Itzstein, *Angew. Chem., Int. Ed.*, 2011, **50**, 1055.
16. A. J. Benie, R. Moser, E. Bäuml, D. Blaas and T. Peters, *J. Am. Chem. Soc.*, 2003, **125**, 14.
17. T. Haselhorst, J. M. Garcia, T. Islam, J. C. Lai, F. J. Rose, J. M. Nicholls, J. S. Peiris and M. von Itzstein, *Angew. Chem., Int. Ed.*, 2008, **47**, 1910.
18. C. McCullough, M. Wang, L. Rong and M. Caffrey, *PLoS One*, 2012, 7, e33958.
19. (a) B. Fiege, C. Rademacher, J. Cartmell, P. I. Kitov, F. Parra and T. Peters, *Angew. Chem., Int. Ed.*, 2012, **51**, 928; (b) M. Zakhour, N. Ruvoën-Clouet, A. Charpilienne, B. Langpap, D. Poncet, T. Peters, N. Bovin and J. Le Pendu, *PLoS Pathogens*, 2009, **5**, e1000504.
20. (a) G. S. Hansman, S. Shahzad-ul-Hussan, J. S. McLellan, G.-Y. Chuang, I. Georgiev, T. Shimoike, K. Katayama, C. A. Bewley and P. D. Kwong, *J. Virol.*, 2012, **86**, 284; (b) C. Rademacher, J. Guiard, P. I. Kitov, B. Fiege, K. P. Dalton, F. Parra, D. R. Bundle and T. Peters, *Chem. – Eur. J.*, 2011, **17**, 7442.
21. C. Airoldi, S. Giovannardi, B. La Ferla, J. Jiménez-Barbero and F. Nicotra, *Chem. – Eur. J.*, 2011, **17**, 13395.

22. *Animal Lectins: Form, Function and Clinical Applications*, ed. G. S. Gupta, A. Gupta and R. K. Gupta, Springer, Vienna, 2012.
23. Z. Liu, Q. Zhang, H. Peng and W. Z. Zhang, *Appl. Biochem. Biotechnol.*, 2012, **168**, 629.
24. A.-C. Lamerz, T. Haselhorst, A. K. Bergfeld, M. von Itzstein and R. Gerardy-Schahn, *J. Biol. Chem.*, 2006, **281**, 16314.
25. T. Haselhorst, J. C. Wilson, A. Liakatos, M. J. Kiefel, J. C. Dyason and M. von Itzstein, *Glycobiology*, 2004, **14**, 895.
26. Y. Yuan, X. Wen, D. A. Sanders and B. M. Pinto, *Biochemistry*, 2005, **44**, 14080.
27. (a) I. N'Go, S. Golten, A. Ardá, J. Cañada, J. Jiménez-Barbero, B. Linclau and S. P. Vincent, *Chem. – Eur. J.*, 2014, **20**, 106; (b) Y. Yuan, D. W. Bleile, X. Wen, D. A. Sanders, K. Itoh, H. W. Liu and B. M. Pinto, *J. Am. Chem. Soc.*, 2008, **130**, 3157.
28. R. Caraballo, H. Dong, J. P. Ribeiro, J. Jiménez-Barbero and O. Ramström, *Angew. Chem., Int. Ed.*, 2010, **49**, 589.
29. Y. E. Tsvetkov, M. Burg-Roderfeld, G. Loers, A. Ardá, E. V. Sukhova, E. A. Khatuntseva, A. A. Grachev, A. O. Chizhov, H. C. Siebert, M. Schachner, J. Jiménez-Barbero and N. E. Nifantiev, *J. Am. Chem. Soc.*, 2012, **134**, 426.
30. (a) N. Yuasa, T. Koyama, G. P. Subedi, Y. Yamaguchi, M. Matsushita and Y. Fujita-Yamaguchi, *J. Biochem.*, 2013, **154**, 521; (b) H. Möller, N. Serttas, H. Paulsen, J. M. Burchell, J. Taylor-Papadimitriou and B. Meyer, *Eur. J. Biochem.*, 2002, **269**, 1444; (c) M. B. Tessier, O. C. Grant, J. Heimburg-Molinaro, D. Smith, S. Jadey, A. M. Gulick, J. Glushka, S. L. Deutscher, K. Rittenhouse-Olson and R. J. Woods, *PLoS One*, 2013, **8**, e54874.
31. (a) P. Di Gianvincenzo, F. Chiodo, M. Marradi and S. Penadés, *Methods Enzymol.*, 2012, **509**, 21; (b) P. M. Enríquez-Navas, M. Marradi, D. Padro, J. Angulo and S. Penadés, *Chem. – Eur. J.*, 2011, **17**, 1547; (c) P. M. Enríquez-Navas, F. Chiodo, M. Marradi, J. Angulo and S. Penadés, *ChemBioChem*, 2012, **13**, 1357.
32. F. Broecker, J. Aretz, Y. Yang, J. Hanske, X. Guo, A. Reinhardt, A. Wahlbrink, C. Rademacher, C. Anish and P. H. Seeberger, *ACS Chem. Biol.*, 2011, **6**, 252.
33. M. A. Johnson and B. M. Pinto, *J. Am. Chem. Soc.*, 2002, **124**, 15368.
34. F. X. Theillet, C. Simenel, C. Guerreiro, A. Phalipon, L. A. Mulard and M. Delepierre, *Glycobiology*, 2011, **21**, 109.
35. M. A. Oberli, M. Tamborrini, Y. H. Tsai, D. W. Werz, T. Horlacher, A. Adibekian, D. Gauss, H. M. Möller, G. Pluschke and P. H. Seeberger, *J. Am. Chem. Soc.*, 2010, **132**, 10239.
36. (a) M. Mayer and B. Meyer, *J. Am. Chem. Soc.*, 2001, **123**, 6108; (b) M. G. Szczepina, R. B. Zheng, G. C. Completo, T. L. Lowary and B. M. Pinto, *ChemBioChem*, 2009, **10**, 2052.
37. J. Angulo, P. M. Enríquez-Navas and P. M. Nieto, *Chem. – Eur. J.*, 2010, **16**, 7803.

38. J. Yan, A. D. Kline, H. Mo, M. J. Shapiro and E. R. Zartler, *J. Magn. Reson.*, 2003, **163**, 270.
39. (a) G. M. Clore and A. Gronenborn, *J. Mag. Reson.*, 1982, **48**, 402; (b) G. M. Clore and A. Gronenborn, *J. Mag. Reson.*, 1983, **52**, 423.
40. *The Nuclear Overhauser Effect in Structural and Conformational Analysis*, ed. W. Neuhaus, Wiley-VCH, Weinheim, 2000.
41. V. L. Bevilacqua, Y. Kim and J. H. Prestegard, *Biochemistry*, 1992, **31**, 9339.
42. K. Lycknert, M. Edblad, A. Imberty and G. Widmalm, *Biochemistry*, 2004, **43**, 9647.
43. T. Peters, K. Scheffler, B. Ernst, A. Katopodis, J. L. Magnani, W. T. Wang and R. Weisemann, *Angew. Chem., Int. Ed. Engl.*, 1995, **34**, 1841.
44. (a) G. Thoma, J. L. Magnani, J. T. Patton, B. Ernst and W. Jahnke, *Angew, Chem., Int. Ed.*, 2001, **40**, 1941; (b) W. Jahnke, C. K. Hartmuth, J. J. Marcel, J. L. Blommers, J. L. Magnani and B. Ernst, *Angew. Chem., Int. Ed. Engl.*, 1997, **36**, 2603.
45. T. Haselhorst, T. Weimar and T. Peters, *J. Am. Chem. Soc.*, 2001, **123**, 10705.
46. N. Volpi, *Curr. Med. Chem.*, 2006, **13**, 1799.
47. M. Hricovíni, M. Guerrini and A. Bisio, *Eur. J. Biochem.*, 1999, **261**, 789.
48. S. Guglieri, M. Hricovıni, R. Raman, L. Polito, G. Torri, B. Casu, R. Sasisekharan and M. Guerrini, *Biochemistry*, 2008, **47**, 13862.
49. (a) T. Haselhorst, J. F. Espinosa, J. Jiménez-Barbero, T. Sokolowski, P. Kosma, H. Brade, L. Brade and T. Peters, *Biochemistry*, 1999, **38**, 6449; (b) T. Sokolowski, T. Haselhorst, K. Scheffler, R. Weisemann, P. Kosma, H. Brade, L. Brade and T. Peters, *J. Biomol. NMR*, 1998, **12**, 123.
50. (a) J. L. Asensio, F. J. Cañada and J. Jiménez-Barbero, *Eur. J. Biochem.*, 1995, **233**, 618; (b) S. R. Arepalli, P. J. Glaudemans, G. D. Daves, P. Kovac and A. Bax, *J. Magn. Reson.*, 1995, **B106**, 195.
51. H. Feinberg, R. Castelli, K. Drickamer, P. H. Seeberger and W. I. Weis, *J. Biol. Chem.*, 2007, **282**, 4202.
52. J. Angulo, I. Díaz, J. J. Reina, G. Tabarani, F. Fieschi, J. Rojo and P. M. Nieto, *ChemBioChem*, 2008, **9**, 2225.
53. T. B. Geijtenbeek, D. S. Kwon, R. Torensma, S. J. van Vliet, G. C. van Duijnhoven, J. Middel, I. L. Cornelissen, H. S. Nottet, V. N. KewalRamani, D. R. Littman, C. G. Figdor and Y. van Kooyk, *Cell*, 2000, **100**, 587.
54. C. P. Alvarez, F. Lasala, J. Carrillo, O. Muñiz, A. L. Corbí and R. Delgado, *J. Virol.*, 2002, **76**, 684.
55. S. Pöhlmann, J. Zhang, F. Baribaud, Z. Chen, G. J. Leslie, G. Lin, A. Granelli-Piperno, R. W. Doms, C. M. Rice and J. A. McKeating, *J. Virol.*, 2003, 77, 4070.
56. L. Tailleux, O. Schwartz, J. L. Herrmann, E. Pivert, M. Jackson, A. Amara, L. Legres, D. Dreher, L. P. Nicod, J. C. Gluckman, P. H. Lagrange, B. Gicquel and O. Neyrolles, *J. Exp. Med.*, 2003, **197**, 121.
57. A. Cambi, K. Gijzen, I. J. de Vries, R. Torensma, B. Joosten, G. J. Adema, M. G. Netea, B. J. Kullberg, L. Romani and C. G. Figdor, *Eur. J. Immunol.*, 2003, **33**, 532.

58. M. J. Borrok and L. L. Kiessling, *J. Am. Chem. Soc.*, 2007, **129**, 12780.
59. (a) C. Guzzi, J. Angulo, F. Doro, J. J. Reina, M. Thépaut, F. Fieschi, A. Bernardi, J. Rojo and P. M. Nieto, *Org. Biomol. Chem.*, 2011, **9**, 7705; (b) M. Thépaut, C. Guzzi, I. Sutkeviciute, S. Sattin, R. Ribeiro-Viana, N. Varga, E. Chabrol, J. Rojo, A. Bernardi, J. Angulo, P. M. Nieto and F. Fieschi, *J. Am. Chem. Soc.*, 2013, **135**, 2518.
60. A. Canales, J. Angulo, R. Ojeda, M. Bruix, R. Fayos, R. Lozano, G. Giménez-Gallego, M. Martín-Lomas, P. M. Nieto and J. Jiménez-Barbero, *J. Am. Chem. Soc.*, 2005, **127**, 5778.
61. M. Schubert, S. Bleuler-Martinez, A. Butschi, M. A. Wälti, P. Egloff, K. Stutz, S. Yan, I. B. H. Wilson, M. O. Hengartner, M. Aebi, F. H. T. Allain and M. Künzler, *PLoS Pathog.*, 2012, **8**, e1002706.
62. A. Annila and P. Permi, *Concepts Magn. Reson.*, 2004, **23A**, 22.
63. (a) A. Saupe, *Angew. Chem., Int. Ed. Engl.*, 1968, **7**, 97; (b) J. A. Losonczi, M. Andrec, M. W. F. Fischer and J. H. Prestegard, *J. Magn. Reson.*, 1999, **138**, 334; (c) F. Kramer, M. V. Deshmukh, H. Kessler and S. J. Glaser, *Concepts Nucl. Magn. Reson.*, 2004, **21A**, 10.
64. (a) M. Martín-Pastor and C. A. Bush, *J. Biomol. NMR*, 2001, **19**, 125; (b) H. F. Azurmendi, M. Martín-Pastor and C. A. Bush, *Biopolymers*, 2002, **63**, 89.
65. T. Zhuang, H. Leffler and J. H. Prestegard, *Protein Sci.*, 2006, **15**, 1780.
66. R. D. Seidel III, T. Zhuang and J. H. Prestegard, *J. Am. Chem. Soc.*, 2007, **129**, 4834.
67. (a) M. Erdélyi, E. d'Auvergne, A. Navarro-Vázquez, A. Leonov and C. Griesinger, *Chem. – Eur. J.*, 2011, **17**, 9368; (b) A. Mallagaray, A. Canales, G. Domínguez, J. Jiménez-Barbero and J. Pérez-Castells, *Chem. Commun.*, 2011, **47**, 7179; (c) S. Yamamoto, T. Yamaguchi, M. Erdélyi, C. Griesinger and K. Kato, *Chem. – Eur. J.*, 2011, **17**, 9280; (d) Y. Zhang, S. Yamamoto, T. Yamaguchi and K. Koichi, *Molecules*, 2012, **17**, 6658.
68. A. Canales, A. Mallagaray, J. Pérez-Castells, I. Boos, C. Unverzagt, S. André, H. J. Gabius, F. J. Cañada and J. Jiménez-Barbero, *Angew. Chem., Int. Ed.*, 2013, **52**, 13789.
69. (a) J. N. Arnold, M. R. Wormald, R. B. Sim, P. M. Rudd and R. A. Dwek, *Annu. Rev. Immunol.*, 2007, **25**, 21; (b) A. Alavi and J. S. Axford, *Rheumatology*, 2008, **47**, 760; (c) A. W. Barb and J. H. Prestegard, *Nat. Chem. Biol.*, 2011, **7**, 147.
70. S. W. Homans, R. A. Dwek and T. W. Rademacher, *Biochemistry*, 1987, **26**, 6571.
71. L. Nieto, A. Canales, I. S. Fernández, E. Santillana, R. González-Corrochano, M. Redondo-Horcajo, F. J. Cañada, P. Nieto, M. Martín-Lomas, G. Giménez-Gallego and J. Jiménez-Barbero, *ChemBioChem*, 2013, **14**, 1732.
72. C. J. Powers, S. W. McLeskey and A. Wellstein, *Endocr. – Relat. Cancer*, 2000, **7**, 165.
73. L. Nieto, A. Canales, G. Giménez-Gallego, P. M. Nieto and J. Jiménez-Barbero, *Chem. – Eur. J.*, 2011, **17**, 11204.

74. M. F. García-Mayoral, A. Canales, D. Díaz, J. López-Prados, M. Moussaoui, J. L. de Paz, J. Angulo, P. M. Nieto, J. Jiménez-Barbero, E. Boix and M. Bruix, *ACS Chem. Biol.*, 2013, **8**, 144.
75. E. A. Jacobsen, R. A. Helmers, J. J. Lee and N. A. Lee, *Blood*, 2012, **120**, 3882.
76. (a) M. E. Wechsler, P. C. Fulkerson, B. S. Bochner, G. M. Gauvreau, G. J. Gleich, T. Henkel, R. Kolbeck, S. K. Mathur, H. Ortega, J. Patel, C. Prussin, P. Renzi, M. E. Rothenberg, F. Roufosse, D. Simon, H. U. Simon, A. Wardlaw, P. F. Weller and A. D. Klion, *J. Allergy Clin. Immunol.*, 2012, **130**, 563; (b) T. Ahmed, G. Smith, I. Vlahov and W. M. Abraham, *Respir. Res.*, 2012, **13**, 6.
77. L. R. Prost, J. C. Grim, M. Tonelli and L. L. Kiessling, *ACS Chem. Biol.*, 2012, 7, 1603.
78. M. Bartoloni, B. E. Domínguez, E. Dragoni, B. Richichi, M. Fragai, S. André, H.-J. Gabius, A. Ardá, C. Luchinat, J. Jiménez-Barbero and C. Nativi, *Chem. - Eur. J.*, 2013, **19**, 1896.
79. A. Noël, M. Jost and E. Maquoi, *Semin. Cell Dev. Biol.*, 2008, **19**, 52.
80. G. Radosavljevic, V. Volarevic, I. Jovanovic, M. Milovanovic, N. Pejnovic, N. Arsenijevic, D. K. Hsu and M. L. Lukic, *Immunol. Res.*, 2012, **52**, 100.
81. N. U. Jain, A. Venot, K. Umemoto, H. Leffler and J. H. Prestegard, *Protein Sci.*, 2001, **10**, 2393.
82. P. E. Johnson, E. Brun, L. F. MacKenzie, S. G. Withers and L. P. McIntosh, *J. Mol. Biol.*, 1999, **287**, 609.

CHAPTER 2

NMR as a Tool to Unveil the Molecular Basis of Glycan-mediated Host–Pathogen Interactions

ROBERTA MARCHETTI,* ANTONIO MOLINARO AND ALBA SILIPO*

Dipartimento di Scienze Chimiche, Università di Napoli Federico II, Complesso Universitario Monte Sant'Angelo, Via Cintia 4, I-80126 Napoli, Italy
*Email: roberta.marchetti@unina.it; silipo@unina.it

2.1 Introduction

Perception of pathogens by their hosts is the outcome of a highly conserved and sophisticated defense system, tightly regulated by both pathogen and host signaling molecules exposed on their cell surfaces. Glycoconjugates and glycan-binding receptors, which cover human, plant, fungal, bacterial and viral cell membranes, provide the first point of contact for any kind of host–guest interaction and influence, to an important degree, all stages of infection, from early colonization of host epithelial surfaces to the induction of immune response and inflammation.[1] Molecular recognition studies of glycans at the host–pathogen surfaces are pivotal to deeply comprehend essential biological mechanisms and offer many possibilities to control disease states. Furthermore, understanding how host proteins recognize and

RSC Drug Discovery Series No. 43
Carbohydrates in Drug Design and Discovery
Edited by Jesús Jiménez-Barbero, F. Javier Cañada and Sonsoles Martín-Santamaría

Published by the Royal Society of Chemistry, www.rsc.org

bind microbial patterns plays a key role in the rational design of glycomimetics with improved drug-like properties, as well as in the development of carbohydrate-based vaccines.[2]

A large variety of techniques, such as crystallographic analysis, titration microcalorimetry (ITC), surface plasmon resonance (SPR) and fluorescence spectroscopy, among others, allows the elucidation of molecular recognition events. However, several interactions involving glycans are weak and transient and therefore difficult to study given that, for instance, the complexes do not easily crystallize. NMR spectroscopy has steadily caught on in the last decade, since NMR-based methods exhibit the unique ability to retrieve information, in solution, of sugar–receptor complexes, providing a dynamic picture of the binding system. Thus modern techniques, such as saturation transfer difference NMR and transferred NOE, have become among the most powerful and versatile methods for the detection and characterization of binding processes between a ligand and its receptor.[3]

The present chapter focuses on the applications of novel NMR methodologies employed by our research team, as well as by other groups, in the study of glycoconjugate–protein interactions, deeply implicated in human and plant disease.

2.2 Host–Pathogen Interactions in Human Infections

Innate immunity is an evolutionarily conserved system of protection against invading microbes. It relies, across the animal kingdom, on specific host proteins termed Pattern Recognition Receptors (PRRs), possessing the ability to detect peculiar microbial signatures known as Pathogen Associated Molecular Patterns (PAMPs).[4] The analysis, at a molecular level, of the complex interplay between pathogen-specific glycoconjugates and host proteins on immune cells permits us to gain detailed insight into the basis of the host immune defense, as shown in the following examples.

2.2.1 NMR Molecular Recognition Studies on Human C-Type Lectins

In recent years, several NMR interaction studies have been carried out to investigate the binding mode of C-type lectin receptors, a class of PRRs that has been suggested to play a key role in antimicrobial, antiviral and antifungal immunity.[5]

Among the components of the C-type lectin receptors family, DC-SIGN (dendritic cell-specific ICAM-3 grabbing non-integrin) has attracted considerable attention, since it is involved in the recognition of a great variety of pathogens including some viruses, bacteria and fungi, through mannosylated or fucosylated epitopes.[6] In addition, it is involved in the transmission of human immunodeficiency virus (HIV) to T cells and in the modulation of the immune response. Many studies have been carried out to develop good strategies to inhibit the sugar binding site of DC-SIGN since it is considered a potential target for the design of anti-infective agents.[7,8]

Noteworthy, *in vivo* saturation transfer difference (STD) NMR experiments, applied to living cells carrying a DC-SIGN receptor, were performed taking advantage to prevent protein isolation and purification. This protocol permitted the detection of the direct interaction of the transmembrane protein with *S. cerevisiae* mannan,[9] which has the ability to compete with many DC-SIGN dependent pathogen recognitions.[10]

Recently, a combinatorial approach based on transferred NOESY (tr-NOESY), STD NMR and the complete relaxation and conformational exchange matrix analysis of saturation transfer (CORCEMA-ST) protocol has provided key information on binding mode and interaction properties of model glycomimetic compounds, improving the design of selective DC-SIGN binders.[8a,11] Indeed, the described exclusive interaction of the pseudo-1,2-mannobioside with DC-SIGN *versus* langerin, a C-type lectin able to block HIV transmission, makes this glycomimetic an attractive lead compound for the rational improvement of binding selectivity.

Finally, a solution NMR analysis of the carbohydrate recognition domain (CRD) of DC-SIGN has increased the current knowledge of its interaction with long glycan chains exposed on dangerous pathogens.[12] NMR data allowed the first assignment of the active CRD of DC-SIGN and provided deep insight into the unique binding modes adopted by the receptor in the interaction with different virus-associated ligands. In detail, NMR titration experiments were performed on different substrates, with special focus on the long glycan $Man_9GlcNAc$, which exhibits a structure very similar to the main oligosaccharide of the HIV envelope glycoprotein gp120. The chemical shift perturbations of the CRD of DC-SIGN, upon titration of different glycans, suggested that the protein interacted in a different manner with oligosaccharides on the viral coat compared with the smaller glycans previously investigated by crystallography. These findings open a new avenue for the rational design of selective inhibitors for high-mannose oligosaccharides, like those on HIV.

We have recently applied NMR methods to a study of the binding mode of the human mannose binding lectin (MBL),[13] a soluble C-type lectin, able to recognize invariant exposed structures adorning pathogens, such as carbohydrates and acetylated compounds, among others. Advanced NMR techniques allowed characterization of the interaction of rhMBL with hexose carbohydrate ligands, coating the cell surface of numerous pathogens. Tr-NOESY experiments were performed in order to analyze ligand conformational changes upon binding (Figure 2.1). The interaction of a substrate to its receptor can be easily distinguished by looking at the sign and size of the observed NOEs. Briefly, according to their molecular weight, molecules with a short correlation time exhibit high positive NOEs, whereas molecules with a slow rotation in solution undergo negative NOEs. As expected, positive NOEs were observed for the small ligand in the free

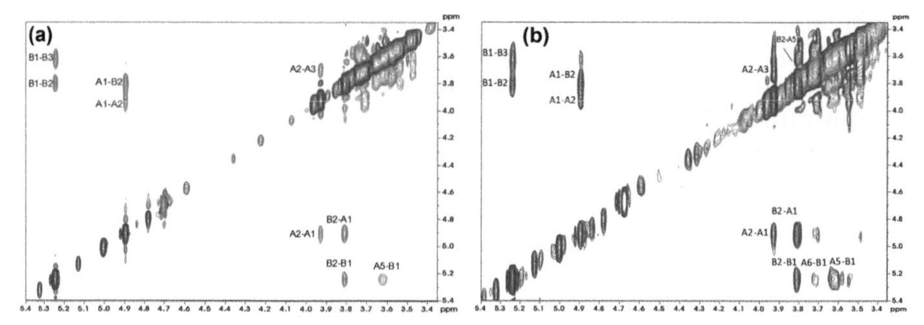

Figure 2.1 NOESY (a) and tr-NOESY (b) on α-(1–2)-mannose disaccharide and on MBL : α-(1–2)-mannose disaccharide (1 : 20). When the ligand bound to the receptor protein, the NOEs undergo changes in the sign demonstrating the binding. In addition, key NOEs, going from the free to the bound state, showed changes, increasing (B1-A5, B2-A5) or decreasing (B1-A6) in intensity. The corresponding inter-proton distances were compared with distances calculated for both energetic minima predicted for α-(1–2)-mannose disaccharide ($\Phi = -33°$, $\Psi = 57°$ or $\Phi = -33°$, $\Psi = 57°$), indicating that the bound conformational state was characterized from negative values of the inter-glycosidic dihedral angles.

state, while upon binding it behaved as a part of the sugar–protein complex and showed negative NOEs. In addition, the bioactive conformation of the ligand was determined from a detailed analysis of inter-residue NOEs (Figure 2.1), demonstrating that there were conformational changes upon binding.

Given the great therapeutic potential of MBL against glycosylated enveloped viruses, such as lethal ebola as well as influenza A, the binding dynamics and affinity also of the fusion protein L-FCN/MBL76 have been investigated. Its less complex quaternary structure, with respect to that of MBL, permits overcoming limits related to the cost and the difficulty of industrial scale production. NMR data showed that L-FCN/MBL76 had good functional activity, according to previous findings, indicating that the chimeric molecule exhibits a superior effector activity, probably mediated by greater structural flexibility.

An NMR approach has been employed also to characterize the recognition of opportunistic fungal pathogens by host cells. Among other receptors which have the ability to detect non-self 1,3-linked β-glucans, dectin-1 has been shown to be one of the main host proteins involved in the fungal recognition pathways.[13] It amplifies the production of cytokines induced by Toll-like receptors (TLRs) and, at the same time, works independently for the production of interleukins. This is the reason why mice deficient in dectin-1 exhibit an increased susceptibility to yeasts such as *C. albicans*[14] and *P. carinii*.[15] NMR titration and STD NMR experiments were performed on the extracellular C-type lectin domain of dectin-1 and a synthetic β-(1,3)-hexadecasaccharide.[16] NMR results allowed the delineation of the ligand portion more involved in the binding, suggesting that the protons H-3 and

H-5 at the α-face of the inner glucose established the strongest interaction with the receptor. These data are in agreement with previous results using a laminarine polysaccharides as a probe.[17]

2.2.2 The Role of Sialic Acid in Bacterial and Viral Adhesion

Many host–pathogens recognition processes are affected by the presence of sialylated glycoconjugates exposed on the host cell surfaces. Several papers have been published with the aim to describe more precisely the interaction of different receptors with sialylated ligands at atomic resolution.

It is worth noting that, since *Vibrio cholerae* is responsible for several million cases of diarrheal diseases and is a common cause of death among children, a lot of attention has been focused on the molecular basis of its recognition by the host. In this context, the mode of action of *V. cholerae* neuraminidase (VCNA) has been investigated and a model for the role of the enzyme in the bacterial pathogenesis has been suggested.[18] It is known that VCNA is responsible for the removal of sialic acid from higher order gangliosides to unmask the receptor for cholera toxin. The molecular recognition details of sialylated ligands bound to the N-terminal lectin domain were explored by STD NMR experiments on a non-hydrolyzable thiosialoside [Neu5,9Ac$_2$-2-S-(α-2,6)-GlcNAcβ1Me]. NMR results, in combination with X-ray and isothermal titration calorimetry (ITC) analysis, allowed mapping of the ligand epitope, indicating that the Neu5,9Ac residue was the ligand portion more involved in the binding with the lectin, which instead gave a minor interaction with the β1Me aglycon moiety (Figure 2.2).

Thanks to the presence of a sialic acid binding domain, VCNA plays a key role in cholera acute bacterial infection; this binding region, indeed, promotes the catalytic efficiency of the neuraminidase, locating it in close proximity to the gangliosides, thus improving the enzyme's ability to reveal the receptor for cholera toxin.

Recently, the severity of cholera infections caused by *V. cholerae* El Tor biotype has been investigated by NMR and ITC.[19] The ability of the El Tor biotype to infect preferentially blood group O individuals, unlike classical *V. cholerae*, has been confirmed by experiments on different oligosaccharides mimicking the blood groups B, A or O. The results allowed attribution of the blood group dependence of El Tor to a mutation in its CTB toxin. Thus in El Tor toxin the threonine 47 is mutated to isoleucine, whereas it is retained in enterotoxigenic *E. coli* and *V. cholera* O1 classical biotypes. This mutation represents a hindrance for the binding to blood group B oligosaccharides and explains the enhanced ability of the El Tor biotype to distinguish between different blood groups.

Sialylated cell-surface oligosaccharides are the major target not only for bacteria but also for several viruses. Within this frame, NMR-based methods have been employed to investigate the role of sialic acids in bird and human infections caused by the influenza virus. It is well established that the initial step of virus entry into host cell membranes is mediated by the envelope

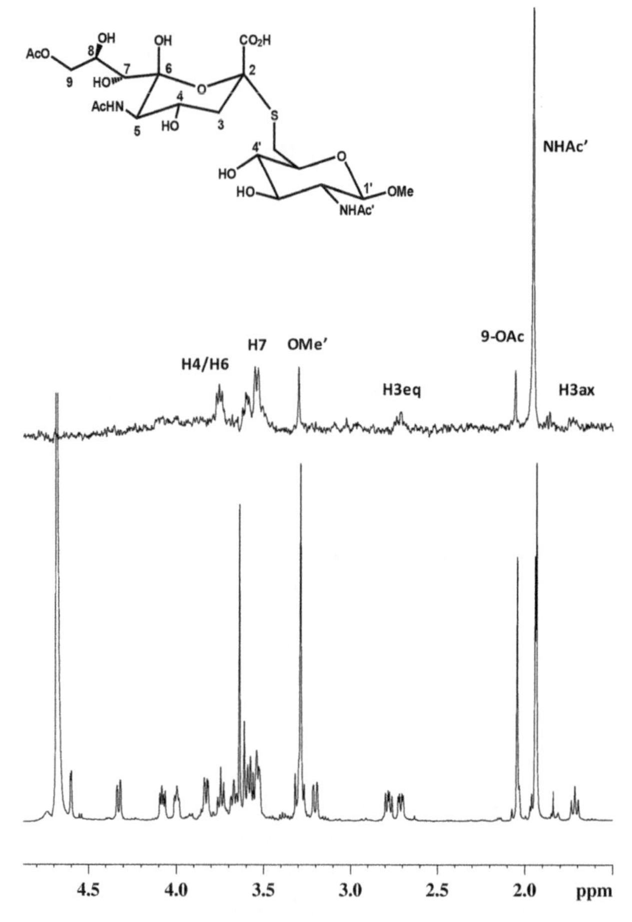

Figure 2.2 Reference (*upper*) and 1D STD NMR (*lower*) spectra of Neu5,9Ac$_2$-2-S-(α-2,6)-GlcNAcβ1Me in complex with VCNA.
(Adapted from Moustafa *et al.*[18]).

glycoprotein hemagglutinin (HA), which binds sialic acid on target cells and allows the fusion of viral and host membranes. Thus HA binds α-2,3-linked *N*-acetylneuraminic acid containing glycans (most abundant in the digestive tract of avian species) and α-2,6-linked sialic receptors (most abundant in the human upper respiratory tract).[20] A more accurate comprehension of the interaction between HA and receptor analogues has been provided by STD NMR, initially applied to purified proteins from different subtypes of influenza A.[21] First, STD NMR competition experiments permitted establishment of the relative affinities of HAs for two different sialylmimetics, α-2,3- and α-2,6-sialyllactose. More recently, the binding epitope of small molecules, such as *tert*-butylhydroquinone (TBHQ)[22] or glycopeptide mimetics and analogues,[23] bound to HA has been investigated by NMR with the aim to find potential inhibitors for different influenza viruses.

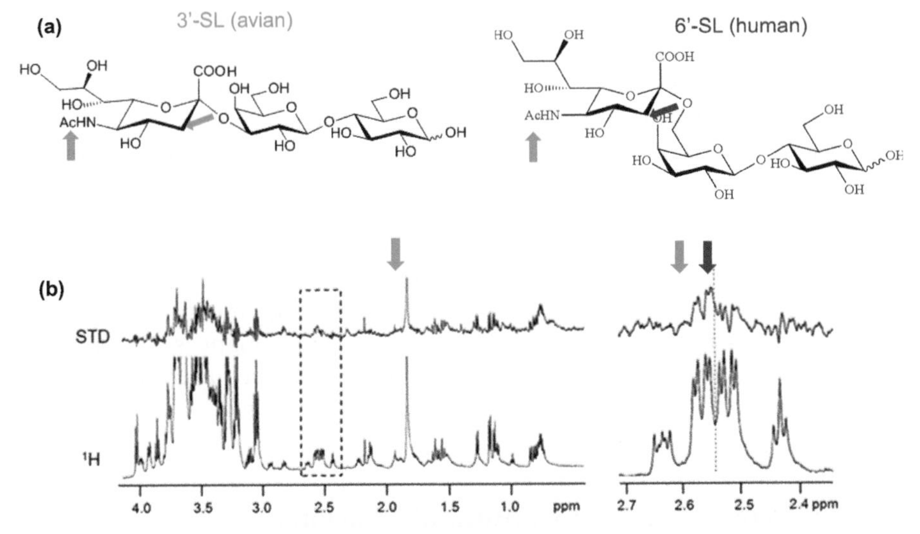

Figure 2.3 Application of STD NMR to the study of the interaction between HA-VLPs and sialylated ligands, α-2,3- and α-2,6-sialyllactose (3′-SL and 6′-SL). (a) Chemical structure of 3′-SL and 6′-SL; signals easily distinguishable by NMR are indicated with arrows. (b) STD NMR spectrum of H5-VPLs in the presence of an equimolar mix of α-2,3-SL and α-2,6-SL. A strong signal belonging to the methyl protons of the *N*-acetamido group indicates a clear binding between H5-VLPs and sialyllactose; a distinction between the two ligands was impaired because of the overlapping of $NHCH_3$ signals. However, differences in the binding mode of sialic acid containing glycans with the HA subtype H5 could be detected, given the diagnostic chemical shifts of the $H3_{eq}$ protons of 3′-SL and 6′-SL. As expected, the differences in the relative intensities of the $H3_{eq}$ STD signals showed a preferential binding of H5-VLPs to α-2,3-SL. (Adapted from Haselhorst *et al.*[24b])

Interesting experiments have been performed also on rather complex systems by using non-infectious virus-like particles (VLPs) bearing HA proteins (Figure 2.3).[24] This approach allows the biosafety problems related to the handling of live viruses to be overcome, maintaining the advantage of working in near-physiological environment.

This last strategy has been successfully applied to the study of rabbit hemorrhagic disease virus (RHDV)[25] and human norovirus (NV),[26] that attach histo blood group antigens (HBGAs), implicated in the viral life cycle. STD NMR experiments, performed on virus-like particles, added to the understanding of the minimal structural requirements for the binding of histo blood group antigens to the viruses. In addition, tr-NOESY spectra provided crucial information about the bioactive conformation of synthetic HBGA fragments, in particular of sLe^x, fundamental for the design of potential inhibitors.[26a]

In some cases, NMR has been used to study the interaction of small ligands with entire native viruses. In 2003, Benie and colleagues demonstrated, for the first time, that STD NMR was a valuable tool to reveal the binding epitope of low molecular weight molecules when bound to a known binding site on the viral coat; the system studied was the human rhinovirus serotype 2 (HRV-2) in the interaction with the synthetic entry inhibitor Repla 394.[27] Later, the VP8 subunit of the rotavirus, a cause of acute gastroenteritis in young children worldwide, has been studied in the interaction with selected *N*-acetylneuraminic acid derivatives using STD NMR and molecular modeling.[28] Although all these examples underline the great potential of NMR methodologies for the investigation of complex systems of interaction, it is worth noting that the setting of experimental parameters is very tricky. A careful choice of the off- and on-resonances frequencies should be done and several control spectra have to be acquired in order to reduce background signals and avoid artifacts belonged to non-specific binding.

2.2.3 Glycoconjugates as Candidate Vaccines

It is well known that host immune defense against several diseases caused by bacteria and viruses is related to the presence of serum antibodies able to recognize surface carbohydrate antigens;[29] therefore, the potential use of glycoconjugates as vaccine candidates against bacterial and viral infections is a topic that focuses a lot of attention. NMR and molecular modeling methods, which provide key structural details, such as the conformational behavior of glycoconjugates in solution, supply a solid help to the design of novel compounds which stimulate the immune system. The knowledge of the glycan's shape and the comprehension of the molecular basis of the recognition and binding between glycoconjugates and antibodies[30] is a crucial step in the conceiving of structure-based design of carbohydrate antigens.

Among all the bacterial virulence factors, amphiphilic molecules anchored to the outer membrane of Gram-negative microorganisms, known as lipopolysaccharides (LPSs), are of particular interest since the diversity of LPS structures and their differential recognition by host receptors have been associated with several bacterial diseases.[31] The emergence of drug-resistant strains and the increase in the number of immunocompromised individuals, particularly susceptible to infection,[32] have led to heightened interest in the identification of new drugs and vaccines for the treatment and prevention of these diseases.

The knowledge of virulence mechanisms of species belonging to the *Burkholderia cepacia* complex (BCC) in patients affected by cystic fibrosis (CF) is still limited. Within this frame, we used NMR and molecular recognition techniques to investigate the interaction between the O-polysaccharide chain of the Gram-negative bacterium *Burkholderia anthina*, isolated from a CF patient from the UK, and a monoclonal antibody (mAb), designated as 5D8, specific for *B. cepacia* LPS.[33] A multidisciplinary approach permitted us to reveal the sugar residues of the O-antigen more involved in the interaction (Figure 2.4). Key structural details were provided by STD NMR and tr-NOESY

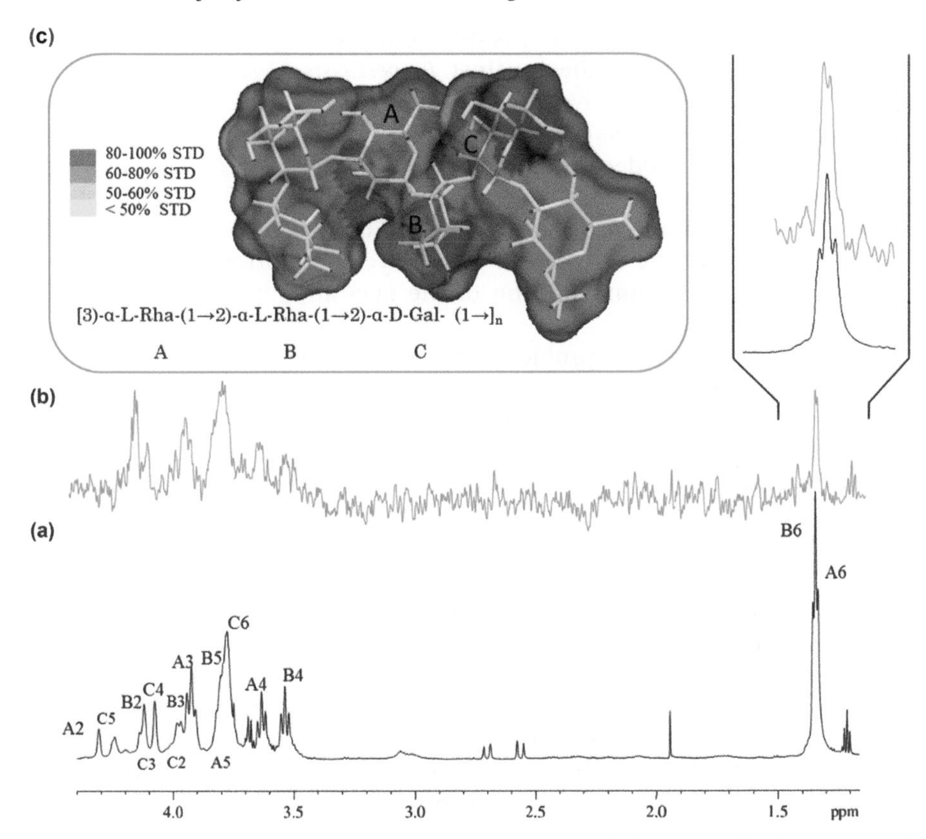

Figure 2.4 Reference 1H NMR spectrum (a) and 1D STD NMR spectrum (b) of a mixture of mAb 5D8 : O-chain from *B. anthina* (1 : 25). (c) STD-derived epitope mapping on the molecular envelope of synthetic hexasaccharide (containing the repeating unit of the O-chain from *B. anthina*) with color coding from the highest (*red*) to lowest (*yellow*) observed STD effect. The quantitative analysis of the STD NMR spectra, based on the STD build-up curves,[34] underlined the key role of the galactose residue in the binding event, since highest STD effects were observed for H3, H5 and H6 of the Gal moiety (above 80%). According to the STD data, the two rhamnose residues also contributed to the molecular recognition process, although to a minor extent.

experiments performed on both purified O-antigen and synthetic compounds containing the repeating unit.

A similar approach has been used to unveil the molecular recognition of a *B. anthracis* tetrasaccharide and the monoclonal antibody MTA 1-3.[35] NMR studies demonstrated that the entire ligand participated in the antibody binding, although the tightest binding site was located within the anthrose-β-(1–3)-rhamnose substructure.[36]

LPS epitopes as a target for antibodies against the *Bordetella pertussis* have been evaluated[37] by using a combinatorial approach including immunochemical, biological and NMR methods. The reactivity of specific immunoglobulin G antibodies against *B. pertussis* LPSs were thoroughly analyzed. This Gram-negative bacterium produces two different types of lipooligosaccharides (LOSs) possessing a lipid A moiety linked to a core nonasaccharide or dodecasaccharide. STD NMR spectroscopy provided the epitope mapping of the LOS-derived pentasaccharide [composed of α-D-GlcNAc-(1→4)-β-D-ManNAc3NAcA-(1→3)-β-L-FucNAc4NMe-(1→6)-α-D-GlcN-(4→1)-α-L,D-Hep] in the interaction with polyclonal antibodies, suggesting that the immunodominant portion of the LOS antigen is situated in the distal trisaccharide.

In order to overcome problems related to the weak immunogenic effects of carbohydrates that limit their use as vaccines, a mimetic approach has steadily increased in recent years. This alternative strategy is based on the use of peptides that mimic bacterial sugar moieties, with the aim to increase the host immune response. In this context, a carbohydrate-mimetic peptide (MDWNMHAA) of the O-antigen from *Shigella flexneri* Y has been identified to be cross-reactive with the monoclonal antibody SYA/J6.[38] In detail, the knowledge of the conformational behavior of *S. flexnery* O-antigen, obtained by NMR and molecular modeling, made up of a linear chain [→2)-α-L-Rha-(1→2)-α-L-Rha-(1→3)-α-L-Rha-(1→3)-β-D-GlcNAc-(1→2)-α-L-Rha-(1→], was necessary for a successful mimic approach.[39] Later, the structure of the mimic octapeptide, when free or bound to mAb SYA/J6, was investigated by a combinatorial approach including a STD-NMR, tr-NOESY and CORCEMA protocol.[38] The resulting findings put down the basis for the conceiving of novel higher-affinity binders to be used as vaccine candidates.

The HIV-1 infection is the cause of the acquired immunodeficiency syndrome (AIDS) that represents a significant worldwide health threat; thus, the development of a HIV-1 vaccine is needed critically. Different studies have been carried out with the aim to define the structural and affinity details of the interaction between human antibodies and glycoconjugates which cover the viral envelope. In particular, the HIV-1 surface component gp120 is heavily glycosylated, with N-linked glycans accounting for about half the mass of the glycoprotein surface. Despite the dense glycan arrangement that surrounds the viral envelope, antibodies from HIV-1 infected patients exhibit low N-glycan reactivity.[40] However, some HIV-1 neutralizing antibodies, such as 2G12, PG9 and PG16, reactive against N-glycans, have been identified and their binding affinities among different oligomannose ligands have been investigated.[41] A model work in this field was performed by Navas and Angulo, who analyzed the interaction of anti-HIV 2G12 with linear and branched oligomannosides by means of STD NMR and tr-NOESY experiments.[34b,42] The binding epitope of the synthetic compounds, motifs of the natural high-mannose $Man_9(GlcNAc)_2$, were obtained by a fine and detailed NMR analysis, providing an accurate description of the molecular recognition of these ligands in solution. The results revealed different binding modes and affinities among linear and branched ligands, underlying that the increase of affinity is not related to the complexity of the glycan structure.

A combination of NMR, mutagenesis and structure analyses was used to describe the recognition of both high mannose and complex-type N-linked glycans by the clonally related PG9 and PG16 human antibodies.[41,43] First of all, the structure of both PG9 and the antigen-binding fragment (Fab) of PG16 in complex with gp120 variable regions V1–V2 was solved. Furthermore, the binding of each antibody with $Man_5GlcNAc_2$ and biantennary complex-type glycans was analyzed using STD NMR (Figure 2.5). The comparison of STD NMR enhancements showed that PG-16 binds complex-type glycans more tightly than $Man_5GlcNAc_2$; on the other hand, PG9 prefers $Man_5GlcNAc_2$. In order to go deeper into the analysis of glycan molecular recognition, mutated and chimeric forms of PG9–PG16 antibodies were created and their binding specificities were studied at atomic level, providing key structural information.

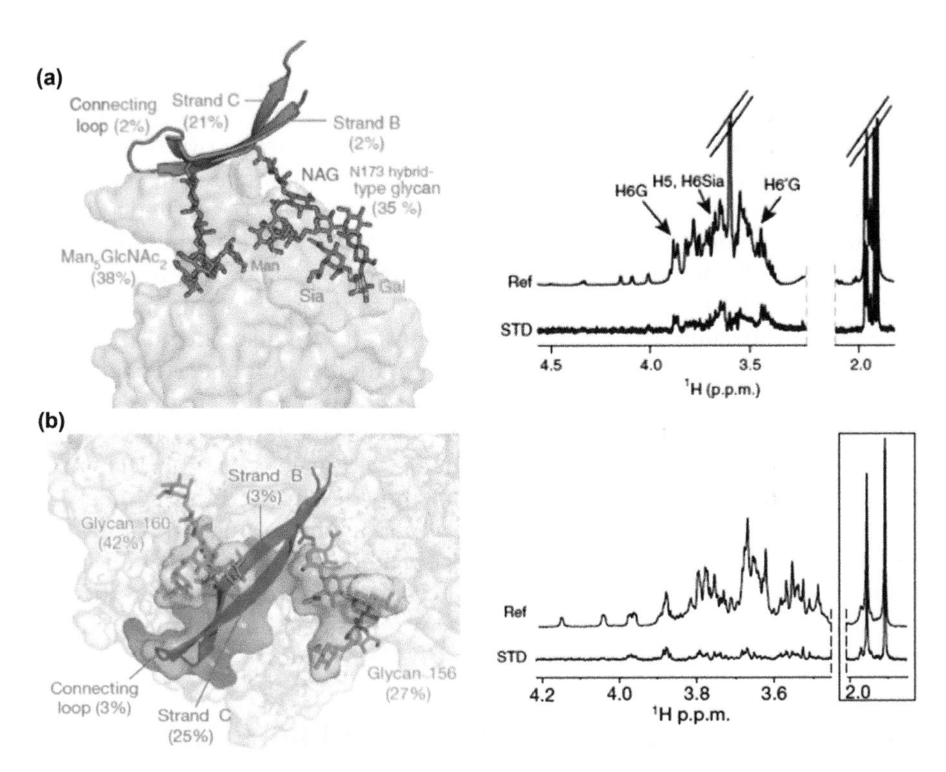

Figure 2.5 (a) PG16 interactions with hybrid-type N-linked glycan at residues 156 or 173 on gp120. On the *left*, the paratope–epitope interface; PG16 is shown as a *gray* surface. On the *right*, reference and STD NMR spectra of the mixture Fab PG16:complex-type carbohydrate. (b) PG9 interaction with $Man_5GlcNAc_2$. On the *left*, the paratope–epitope interface; PG9 is shown as a *gray* surface. On the *right*, reference and STD NMR spectra for $Man_5GlcNAc_2$-Asn binding to PG9.
(Adapted from McLellan *et al.*[41] and Pancera *et al.*[43])

All these significant findings are of crucial importance in the context of vaccine development strategy, especially considering that active drugs could be delivered by the conjugation of weak binders to nano compounds including polymers, clusters, dendrimers and gold particles.[44]

Despite the development of potentially curative chemotherapy, tuberculosis (TB) is still a global emergence, being a leading cause of human mortality in the developing world due to the continuous appearance of drug resistance. *Mycobacterium tuberculosis* is the causal agent of the TB pathological condition and shares with the *Mycobacterium* genus a unique cell wall, which forms an efficient permeability barrier and plays a crucial role in intrinsic drug resistance as well as in macrophage survival under stress conditions. The pathogenicity of *M. tuberculosis* is attributed to the mycolyl-arabinogalactan-peptidoglycan (mAGP) complex, and to another important cell wall component, a lipoarabinomannan, which encapsulates the microorganism. Since the enzymes responsible for mycobacterium cell wall biosynthesis and modeling are absent in humans, they have attracted a lot of attention as a potential treatment of TB disease. For instance, deep insights into the mechanism by which the bacterial galactan is synthesized by two glycosyltransferases have been acquired by means of STD NMR.[45] Moreover, a multidisciplinary approach involving mass spectrometry, titration calorimetry and NMR spectroscopy has been used to investigate the binding affinity of the monoclonal antibody CS-35 for a series of 17 synthetic oligosaccharide fragments of mycobacterial arabinans.[46] NMR experiments have been performed on both strong and weak binders, showing the ligand moieties essential for the interaction that are the non-reducing end β-Ara*f* residue and especially the reducing end α-Ara*f*. It was also shown that the affinity of the binding increased for more highly branched structures. Furthermore, two different binding modes were detected, the "native" one in which the two terminal residues, β- and α-Ara*f*, are in close proximity to the protein, and the "alternative" binding mode, with a lower affinity, characterized by a different orientation of ligand moieties in the binding pocket. Finally, insights into the contribution of aliphatic and aromatic residues to the interaction with arabinan fragments were made. All the results delivered by the detailed refinement of CS-35 ligand specificity represent significant information, given that antibodies reactive against lipoarabinomannan motifs have been presented as potential anti-TB vaccines.[47]

2.3 Host–Pathogen Interactions in Plant Disease

The innate immunity system is a well-described phenomenon in vertebrates and insects but is less studied in plant systems.[48] However, as with mammals, plants exhibit the ability to recognize, as non-self, invariant microbial-associated molecular patterns, characteristic of microbial organisms but completely absent in the host. Several plant receptor proteins, indeed, serve as biochemical modules implicated in the recognition of pathogen glycoconjugates improving the establishment of bacterial and/or fungal resistance.

In this context, we investigated the molecular basis of recognition and interaction of cell-wall fungal chitooligosaccharides by a rice receptor binding protein, CEBiP, carrying extracellular lysine motif (LysM) domains.[49] Advanced STD NMR, tr-NOE spectroscopic techniques and molecular dynamics, taken together with biological data, allowed for the delineation of the structural requirements of the aforementioned plant–fungus interaction. A detailed analysis of the binding of a plethora of chitooligosaccharides of different length to the central LysM domain of CEBiP indicated a minor binding for short-chain oligosaccharides. A strong interaction was instead detected for longer chitin fragments, namely $(GlcNAc)_7$ and $(GlcNAc)_8$. Moreover, NMR results highlighted the pivotal role of the *N*-acetyl groups in the recognition process (Figure 2.6). A 3D model of the interaction between CEBiP and chitin oligosaccharides, based on both biochemical and NMR findings, has been produced (Figure 2.6). According to that sandwich-like model, a hydrophobic cleft, with an isoleucine residue in the central region, accommodates four residues of GlcNAc. In detail, two CEBiP molecules simultaneously bind the chitooligosaccharide from opposite sides to produce a dimer, which is most likely involved in the activation of the rice receptor kinase, CERK1, triggering the immune response.

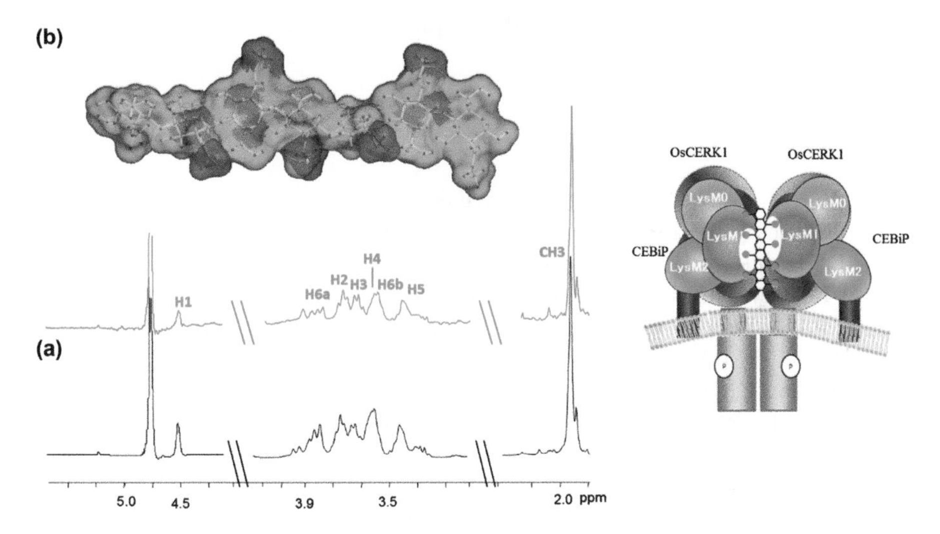

Figure 2.6 On the *left*, the STD NMR spectrum[50] of $(GlcNAc)_7$ bound to trx-LysM1-2 of CEBiP and the corresponding ligand epitope mapping as derived by the STD quantitative analysis. STD effects are color coded from the highest (*red*) to the lowest (*yellow*). On the *right*, a hypothetical sandwich-like model of the activation of the CEBiP-OsCERK1 complex by $(GlcNAc)_8$. Two LysM1 domains, located on two CEBiP molecules, anchor *N*-acetyl moieties from opposite sides to produce a dimer. The dimerization of CEBiP is probably involved in the activation of CERK1, crucial in the chitin signaling in rice.

In several plants, a protective role against various pathogenic attacks is carried out by many lectins possessing a common structural motif of 30–43 residues, termed the "hevein domain", which is able to recognize and bind chitin, a key polysaccharide component found in fungi, insects and parasites. Over the years, much work has been performed using NMR spectroscopy to investigate the affinity and geometry of the hevein binding site.[51] The specific interaction of the "chitin binding" domain belonging to different plant lectins with several $(GlcNAc)_n$ oligomers has been studied, providing a detailed structural model for the hevein–chitin complex.

2.4 Conclusions

Major roles of glycoconjugates in health and disease are mediated by binding proteins that decode the information content of the glycome through their recognition of glycans as ligands. Therefore, understanding at a molecular level the basis of the complex cross-talk between microbial glycoconjugates and their cognate receptors is mandatory to design efficient and novel antimicrobial compounds, as well as carbohydrate mimic-based vaccines. In this context, NMR represents a valuable tool to supply a molecular three-dimensional view of receptor–ligand complexes and to enhance the knowledge on structural requisites at the basis of fundamental molecular recognition processes.

References

1. A. Varki, R. D. Cummings, J. D. Esko, *et al. Essentials of Glycobiology*, Cold Spring Harbor Laboratory Press, 2nd edn, 2009, Cap 39.
2. J. E. Hudak and C. R. Bertozzi, *Chem. Biol.*, 2014, **21**, 16.
3. (a) B. Meyer and T. Peters, *Angew. Chem., Int. Ed.*, 2003, **42**, 864–890; (b) A. C. Lepre, M. J. Moore and W. J. Peng, *Chem. Rev.*, 2004, **104**, 3641–3675; (c) J. Angulo and P. M. Nieto, *Eur. Biophys. J.*, 2011, **40**, 1357–1369; (d) L. P. Calle, F. J. Canada and J. Jimenez-Barbero, *Nat. Prod. Rep.*, 2011, **28**, 1118–1125; (e) M. A. Johnson and B. M. Pinto, *Carbohydr. Res.*, 2004, **339**, 907–928; (f) A. Bhunia, S. Bhattacharjya and S. Chetterjee, *Drug Discovery Today*, 2012; (g) T. Carlomagno, *Annu. Rev. Biophys. Biomol. Struct.*, 2005, **34**, 245–66; (h) J. L. Wagstaff, S. L. Taylor and M. J. Howard, *Mol. Biosyst.*, 2013, **9**, 571–577; (i) L. Unione, S. Galante, D. Díaz, F. J. Cañada and J. J. Barbero, *Med. Chem. Commun.*, 2014, **5**, 1280–1289.
4. S. Akira, K. Takeda and T. Kaisho, *Nat. Immunol.*, 2001, **2**, 675–680.
5. *The Sugar Code. Fundamentals of Glycosciences*, ed. H.-J. Gabius, Wiley-VCH, Weinheim, 2009.
6. Y. van Kooyk and T. B. H. Geijtenbeek, *Nat. Rev. Immunol.*, 2003, **3**, 697–709.

7. J. J. Reina, S. Sattin, D. Invernizzi, S. Mari, L. M.-P. G. Tabarani, F. Fieschi, R. Delgado, P. M. Nieto, J. Rojo and A. Bernardi, *ChemMedChem*, 2007, **2**, 1030–1036.
8. (a) G. Tabarani, J. J. Reina, C. Ebel, C. Vivès, H. Lortat-Jacob, J. Rojo and F. Fieschi, *FEBS Lett.*, 2006, **580**, 2402–2408; (b) M. Ciobanu, K.-T. Huang, J.-P. Daguer, S. Barluenga, O. Chaloin, E. Schaeffer, C. G. Mueller, D. A. Mitchell and N. Winssinger, *Chem. Commun.*, 2011, **47**, 9321–9323.
9. S. Mari, D. Serrano-Gomez, J. F. Canada, A. L. Corbo and J. Jimenez-Barbero, *Angew. Chem., Int. Ed.*, 2005, **44**, 296–298.
10. M. Relloso, A. Puig-Kroger, O. M. Pello, J. L. Rodriguez-Fernandez, G. de la Rosa, N. Longo, J. Navarro, M. A. Munoz-Fernandez, P. Sanchez-Mateos and A. L. Corbi, *J. Immunol.*, 2002, **168**, 2634.
11. M. Thépaut, C. Guzzi, I. Sutkeviciute, S. Sattin, R. Ribeiro-Viana, N. Varga, E. Chabrol, J. Rojo, A. Bernardi, J. Angulo, P. M. Nieto and F. Fieschi, *J. Am. Chem. Soc.*, 2013, **135**, 2518–2529.
12. F. Probert, S. B. Whittaker, M. Crispin, D. A. Mitchell and A. M. Dixon, *J. Biol. Chem.*, 2013, **288**, 22745–22757.
13. R. Marchetti, R. Lanzetta, C. J. Michelow, A. Molinaro and A. Silipo, *Eur. J. Org. Chem.*, 2012, **27**, 5275–5281.
14. P. R. Taylor, S. V. Tsoni, J. A. Willment, *et al.*, *Nat. Immunol.*, 2007, **8**, 31–38.
15. S. Saijo, N. Fujikado, T. Furuta, *et al.*, *Nat. Immunol.*, 2007, **8**, 39–46.
16. H. Tanaka, T. Kawai, Y. Adachi, S. Hanashimaì, Y. Yamaguchi, N. Ohno and T. Takahashi, *Bioorg. Med. Chem.*, 2012, **20**, 3898–3914.
17. B. Sylla, J.-P. Guégan, J.-M. Wieruszeski, C. Nugier-Chauvin, L. Legentil, R. Daniellou and V. Ferrières, *Carbohydr. Res.*, 2011, **346**, 1490.
18. I. Moustafa, H. Connaris, M. Taylor, V. Zaitsev, J. C. Wilson, M. J. Kiefel, M. von Itzstein and G. Taylor, *J. Biol. Chem.*, 2004, **279**, 40819–40826.
19. P. K. Mandal, T. R. Branson, E. D. Hayes, J. F. Ross, J. A. Gavín, A. H. Daranas and W. B. Turnbull, *Angew. Chem., Int. Ed.*, 2012, **51**, 5143–5146.
20. R. J. Connor, Y. Kawaoka, R. G. Webster and J. C. Paulson, *Virology*, 1994, **205**, 17–23.
21. C. McCullough, M. Wang, L. Rong and M. Caffrey, *PLoS One*, 2012, 7, e33958.
22. A. Antanasijevic1, H. Cheng, D. J. Wardrop, L. Rong and M. Caffrey, *PLoS One*, 2013, **8**(10), e76363.
23. M. Waldmann, R. Jirmann, K. Hoelscher, M. Wienke, F. C. Niemeyer, D. Rehders and B. Meyer, *J. Am. Chem. Soc.*, 2014, **136**, 783–788.
24. (a) *Influenza Virus Sialidase – A Drug Discovery Target*, ed. J. M. von Itzstein, 2012, Springer-Verlag/Wien; (b) T. Haselhorst, J. M. Garcia, T. Islam, J. C. Lai, F. J. Rose, J. M. Nicholls, J. S. Peiris and M. von Itzstein, *Angew. Chem., Int. Ed.*, 2008, **47**, 1910–1912.
25. C. Rademacher, N. R. Krishna, M. Palcic, F. Parra and T. Peters, *J. Am. Chem. Soc.*, 2008, **130**, 11.

26. (a) B. Fiege, C. Rademacher, J. Cartmell, P. I. Kitov, F. Parra and T. Peters, *Angew. Chem., Int. Ed.*, 2012, **51**, 928–932; (b) C. Rademacher, J. Guiard, P. I. Kitov, B. Fiege, K. P. Dalton, F. Parra, D. R. Bundle and T. Peters, *Chem. - Eur. J.*, 2011, **17**, 7442–7453.
27. A. J. Benie, *et al.*, *J. Am. Chem. Soc.*, 2003, **125**, 14–15.
28. T. Haselhorst, *et al.*, *Glycobiology*, 2006, **17**(1), 68–81.
29. Z. Kossaczka, J. Shiloach, V. Johnson, D. N. Taylor, R. A. Finkelstein, J. B. Robbins and S. C. Szu, *Infect. Immun.*, 2000, **68**, 5037–5043.
30. F. Marcelo, F. J. Cañada, J. Jiménez-Barbero, in *Anticarbohydrate Antibodies*, 2012, ed. P. Kosma and S. Müller-Loennies, Springer-Verlag/Wien.
31. S. I. Miller, R. K. Ernst and M. W. Bader, *Nat. Rev. Microbiol.*, 2005, **3**, 36–46, DOI: 10.1038/nrmicro1068.
32. B. C. De Jong, D. M. Israelski, E. L. Corbett and P. M. Small, *Annu. Rev. Med.*, 2004, **55**, 283–301.
33. R. Marchetti, A. Canales, R. Lanzetta, I. Nilsson, C. Vogel, D. E. Reed, D. P. Aucoin, J. Jiménez-Barbero, A. Molinaro and A. Silipo, *ChemBioChem*, 2013, **14**(12), 1485–1493.
34. (a) J. Angulo, I. Díaz, J. J. Reina, G. Tabarani, F. Fieschi, J. Rojo and P. M. Nieto, *ChemBioChem*, 2008, **9**, 2225–2227; (b) P. M. Enríquez-Navas, M. Marradi, D. Padro, J. Angulo and S. Penadés, *Chem. - Eur. J.*, 2011, **17**, 1547–1560.
35. M. Tamborrini, D. B. Werz, J. Frey, G. Pluschke and P. H. Seeberger, *Angew. Chem., Int. Ed.*, 2006, **45**, 6581–6582.
36. M. A. Oberli, M. Tamborrini, Y.-H. Tsai, D. B. Werz, T. Horlacher, A. Adibekian, D. Gauss, H. M. Möller, G. Pluschke and P. H. Seeberger, *J. Am. Chem. Soc.*, 2010, **132**, 10239–10241.
37. T. Niedziela, I. Letowska, J. Lukasiewicz, M. Kaszowska, A. Czarnecka, L. Kenne and C. Lugowski, *Infect. Immun.*, 2005, **3**(11), 7381–7389.
38. M. G. Szczepina, D. W. Bleile and B. M. Pinto, *Chem. - Eur. J.*, 2011, **17**, 11446–11455.
39. M.-J. Clément, A. Imberty, A. Phalipon, S. Pérez, C. Simenel, L. A. Mulard and M. Delepierre, *J. Biol. Chem.*, 2003, **278**(48), 47928–47936.
40. (a) L. M. Walker, *et al.*, *PLoS Pathog.*, 2010, **6**, e1001028; (b) L. M. Walker, *et al.*, *Nature*, 2011, **477**, 466–e1001470.
41. J. S. McLellan, M. Pancera, *et al.*, *Nature*, 2011, **480**, 336–345.
42. P. M. Enriquez-Navas, M. Marradi, D. Padro, J. Angulo and S. Penades, *Chem. - Eur. J.*, 2011, **17**, 1547–1560.
43. M. Pancera, *et al.*, *Nat. Struct. Mol. Biol.*, 2013, **20**, 804–812.
44. A. Bernardi, *et al.*, *Chem. Soc. Rev.*, 2013, **42**, 4709–4727.
45. (a) M. G. Szczepina, R. B. Zheng, G. C. Completo, T. L. Lowary and B. M. Pinto, *ChemBioChem*, 2009, **10**, 2052–2059; (b) M. G. Szczepina, R. B. Zheng, G. C. Completo, T. L. Lowary and B. M. Pinto, *Bioorg. Med. Chem.*, 2010, **18**, 5123–5128.
46. C. Rademacher, G. K. Shoemaker, H.-S. Kim, R. B. Zheng, H. Taha, C. Liu, R. C. Nacario, D. C. Schriemer, J. S. Klassen, T. Peters and T. L. Lowary, *J. Am. Chem. Soc.*, 2007, **129**, 10489–10502.

47. B. Hamasur, M. Haile, A. Pawlowski, U. Schröder, A. Williams, G. Hatch, G. Hall, P. Marsh, G. Källenius and S. B. Svenson, *Vaccine*, 2003, **21**, 4081–4093.
48. T. Nurnberger, F. Brunner, B. Kemmerling and L. Piater, *Immunol. Rev.*, 2004, **198**, 249–266.
49. M. Hayafune, R. Berisio, R. Marchetti, A. Silipo, M. Kayama, Y. Desaki, S. Arima, F. Squeglia, A. Ruggiero, K. Tokuyasu, A. Molinaro, H. Kaku and N. Shibuya, *Proc. Natl. Acad. Sci. U. S. A.*, 2014, **111**, E404–E413.
50. B. Claasen, M. Axmann, R. Meinecke and B. Meyer, *J. Am. Chem. Soc.*, 2005, **127**(3), 916–919.
51. (a) J. Jiménez-Barbero, F. J. Cañada, J. L. Asensio, N. Aboitiz, P. Vidal, A. Canales, P. Groves, H. J. Gabius and H. Siebert, *Adv. Carbohydr. Chem. Biochem.*, 2006, **60**, 303–354; (b) J. L. Asensio, F. J. Cañada, M. Bruix, A. Rodríguez-Romero and J. Jiménez-Barbero, *Eur. J. Biochem.*, 1995, **230**, 621–633; (c) M. I. Chávez, M. Vila-Perelló, F. J. Cañada, D. Andreu and J. Jiménez-Barbero, *Carbohydr. Res.*, 2010, **345**, 1461–1468; (d) N. Aboitiz, M. Vila-Perelló, P. Groves, J. L. Asensio, D. Andreu, F. J. Cañada and J. Jiménez-Barbero, *ChemBioChem*, 2004, **5**, 1245–1255.

CHAPTER 3

Lipopolysaccharides as Microbe-associated Molecular Patterns: A Structural Perspective

FLAVIANA DI LORENZO, CRISTINA DE CASTRO, ROSA LANZETTA, MICHELANGELO PARRILLI, ALBA SILIPO AND ANTONIO MOLINARO*

Department of Chemical Sciences, University of Naples Federico II, Via Cinthia 4, 80126, Naples, Italy
*Email: molinaro@unina.it

3.1 Introduction

The history of the lipopolysaccharide (LPS) macromolecule, also known as "endotoxin", starts in the year 1892 from a work on *Vibrio cholerae*[1,2] executed by Richard Pfeiffer, a disciple of Robert Koch, who showed that heat-killed cholera bacteria were able to cause toxic shock reactions in guinea pigs, thus demonstrating that they were themselves toxic, unlike the previous dogma regarding the toxicity belonging to secreted products from living microorganisms, known as "exotoxins".[1,2] Later, the endo-bacterial nature of this toxic material was further confirmed by Florence Seibert in 1923,[2] identifying the bioactive material contaminating infusion fluids and pharmaceutical drugs.[2] In 1943, the pharmacologist Murray J. Shear introduced the current term "lipopolysaccharide", highlighting the glycolipid

RSC Drug Discovery Series No. 43
Carbohydrates in Drug Design and Discovery
Edited by Jesús Jiménez-Barbero, F. Javier Cañada and Sonsoles Martín-Santamaría

Published by the Royal Society of Chemistry, www.rsc.org

nature of the endotoxins,[2] although it took several years to reach the complete elucidation of the structure of the macromolecule; this was obtained in 1952 by Otto Lüderitz and Otto Westphal, who were the first to isolate the endotoxic material in sufficiently pure amounts required for structural studies.[3] Their protocol of LPS extraction from bacterial dried cells is currently used in the lipopolysaccharide research field. The possibility to use efficient protocols of LPS purification propelled an intensive and highly productive research in all fields of life science, including chemistry, biology, genetics, biophysics, medicine and immunology. In particular, in the 1960s it was demonstrated that the lipidic portion of LPS is responsible for the endotoxicity of the entire macromolecule, and this was later also confirmed by Tetsuo Shiba in the 1980s, who prepared synthetically the first free lipid A.[4] Moreover, in the same years, Westphal and Lüderitz in collaboration with Anne-Marie Staub conducted experiments on several animals which were inoculated with Gram-negative bacteria, showing the production of antibodies specifically directed against the O-side chain of bacterial LPSs.[2] In this context, the definition of the molecule moieties responsible for LPS endotoxic and antigenic properties also opened a parallel enormous branch of research aimed to elucidate its immunostimulant properties and its structure-dependent bioactivity. Indeed, since their discovery, LPSs have been isolated from a large number of Gram-negative bacteria and the structural details of each have been characterised with advanced techniques such as NMR spectroscopy, HPLC and mass spectrometry; moreover, the majority of the publications regarding the LPS structure also report extensive *in vivo* and *in vitro* studies to test the capability and the biochemical aspects of the purified macromolecule to interact and activate the host innate and adaptive immune response cell components.

The purpose of the present chapter is to provide the main information related to the structural profile of the LPS molecule, illustrating the common and uncommon features found in several microorganisms and showing the most used procedure to reach the structural characterisation of LPSs. A section of this chapter will be also dedicated to description of the biosynthetic pathway and the biological role of LPS and its structure-dependent activity.

3.2 General Aspects of the LPS Macromolecule

LPSs are heat-stable amphiphilic molecules comprising approximately 75% of the outer surface of almost all Gram-negative bacteria and of some cyanobacteria.[5–7] They are indispensable for viability and survival of such bacteria, as they heavily contribute to the structural integrity of the outer membrane (OM) and to the protection of the bacterial cell envelope.[8] The asymmetric bilayer of the Gram-negative OM resembles that of typical membranes composed of phospholipids resulting from the back-to-back coupling of a LPS leaflet with an analogue structure formed by phospholipids. On the other hand, most of the OM physicochemical properties depend on the LPS self-aggregating behaviour, which in turn is finely tuned by its molecular structure. For instance, the negative charges of the phosphate groups present on the saccharide moiety establish electrostatic interactions with divalent cations (mainly Ca^{2+} and Mg^{2+}), contributing to the ordered structure and low fluidity of the LPS monolayer.[9] This is reflected in their low permeability and high resistance to dangerous environmental compounds as well as to harsh conditions, as in case of extremophile bacteria.[10,11] Indeed, most of the commonly used antibiotics directed against Gram-negative bacteria, such as polymyxin B and mellitin, are able to destabilise the above-mentioned interactions, leading to the disruption of membrane integrity.[12] Furthermore, LPSs are involved in a plethora of host–bacterium interactions as colonisation, symbiosis, adhesion, tolerance in commensal bacteria and, in the case of pathogens, virulence.[13] The pathogenicity of the toxic LPS molecule is strictly related to its capability, once released, to trigger the activation of both innate and adaptive immune systems in a wide range of eukaryotic hosts, ranging from insects to humans.[8] Concerning their structure, LPSs belonging to different bacterial species possessing different structures, but also the LPS of an individual bacterial strain is not a single molecule possessing a specific chemical structure but rather a blend of various molecules characterised by an intrinsic size and structural heterogeneity.[14,15] Furthermore, bacteria of the same species producing diverse LPS molecules under different growth conditions have been also found.[15] Despite this high structural heterogeneity, three common basic regions (Figure 3.1) can be usually identified in all LPSs, since they are encoded by different gene clusters. Indeed, a glycolipid domain, termed lipid A, anchors LPSs in the outer leaflet of the OM, while an oligosaccharide

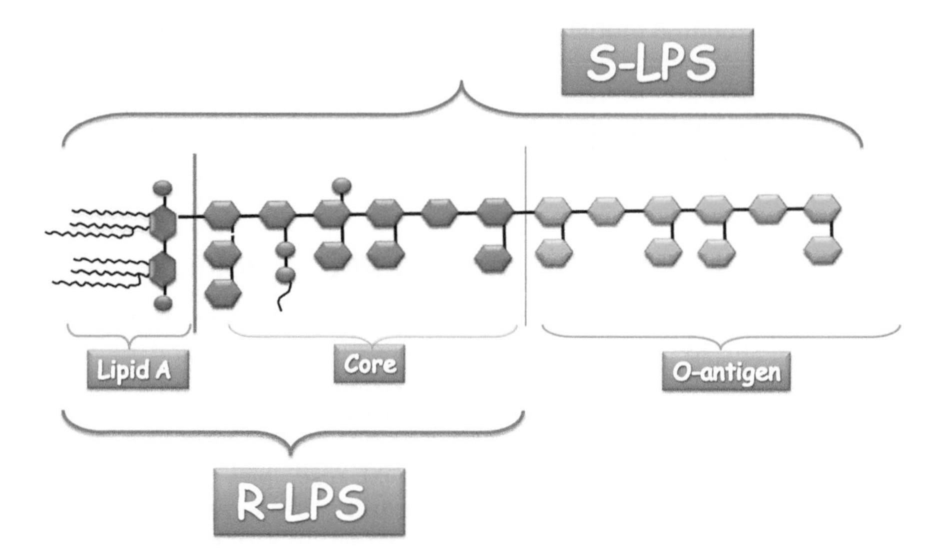

Figure 3.1 Schematic representation of a lipopolysaccharide macromolecule.

named the core region, covalently linked to lipid A, is oriented outwards.[15] Complete LPSs, termed S-LPS or *smooth*-type LPS, comprise a third region which may be the O-polysaccharide (or O-antigen or O-side chain), a capsular polysaccharide or, only in *Enterobacteriaceae*, the enterobacterial common antigen.[16] The O-polysaccharide chain is not ubiquitous, as it can be lacking or partly truncated in some Gram-negative strains, whose colonies assume a rough morphology; thus the terminology currently used to designate this LPS type, namely without the O-polysaccharide portion, is R-LPS (or *rough*-type LPS or lipooligosaccharide, LOS).[17]

3.3 The Lipid A Moiety: The Endotoxic Heart of LPSs

Lipid A possesses a high conservative structure consisting of a β-(1→6)-linked D-glucosamine disaccharide backbone substituted at positions 2 and 3 of both glucosamine residues by amide- and ester-linked 3-hydroxyl fatty acids, respectively. The acyl groups that are directly linked to the sugar backbone are defined as primary. Some of the primary fatty acids are further acylated at the hydroxyl groups by secondary acyl chains. Furthermore, the sugar backbone is generally α-phosphorylated at position O-1 of the reducing glucosamine (Glc*p*N I) and at position O-4′ of the non-reducing glucosamine (Glc*p*N II).[18] Since its first characterisation, *E. coli* LPS lipid A (Figure 3.2) is considered to date the prototype to which other LPS structures are often compared; it is built up of a bis-phosphorylated disaccharide backbone of D-Glc*p*Ns [P→4-β-D-Glc*p*N-(1→6)-α-D-Glc*p*N-1→P] acylated at positions 2 and 3 of both Glc*p*Ns by four C14:0 (3-OH) (Figure 3.2). The primary fatty acids located on the Glc*p*N II are both esterified at their hydroxyl group by two secondary fatty acids: the N-linked C14:0 (3-OH) is esterified by a C12:0 while the O-linked C14:0 (3-OH) by a C14:0 (Figure 3.2).

In spite of its conserved general structure, the lipid A moiety also presents a sort of microheterogeneity due to subtle chemical differences arising from bacterial adaptation, incomplete biosynthesis, change of environment, presence of external stimuli and chemical modifications resulting from the procedures used for lipid A extraction from bacterial cells. Such microheterogeneity has been observed in the acylation (number, type and distribution of acyl chains) and phosphorylation patterns, and less commonly, also in the disaccharide backbone, as in the case of a number of bacterial species, such as *Aquifex pyrophilus*[19] (Figure 3.2), *Brucella abortus*,[20] *Bacteriovorax stolpii*,[21] *Caulobacter crescentus*,[22] *Mesorhizobium huakuii*,[23] *Bradyrhizobium elkanii*[24] (Figure 3.2), *Bartonella henselae*[25] and *Legionella pneumophila*,[26] whose GlcN residues may be replaced with 2,3-diamino-2,3-dideoxy-D-glucopyranose (Glc*p*N3N) residues. Moreover, phosphate groups can be substituted by further phosphate groups, producing a pyrophosphate, but also by other polar substituents such as 4-amino-4-deoxy-L-arabinopyranose (arabinosamine, Ara*p*4N) and a 2-aminoethanol group (EtN), or by acid residues such as galacturonic acid (Gal*p*A) (Figure 3.2). Phosphate groups can be absent, as in the case of *Bdellovibrio bacteriovorus* LPS lipid A that is characterised by the replacement of phosphate groups with two mannose residues, generating a totally neutral lipid A.[27] Exceptionally, the lipid A of the marine bacterium *Loktanella rosea* has a very unusual structure as the molecule is non-phosphorylated, both Glc*p*N residues are β-linked, and the Glc*p*N I forms with α-Gal*p*A a unique mixed trehalose-like structure.[28] Concerning the acylation pattern, fatty acids can be attached to the disaccharide backbone either symmetrically (3 + 3, *e.g. Neisseria meningitides*) or asymmetrically (4 + 2, *e.g. Escherichia coli*). Finally, lipid A fatty acids less frequently present further structural features such as a methyl branch, different functional and hydroxyl groups, length of the chains up to 28 carbon atoms and odd-numbered carbon chains. In this context, *L. pneumophila* possesses a lipid A with a branched 2,3-dihydroxy fatty acid, 2,3-di-OH-*i*-14:0,[29] while 3-keto fatty acids are retrieved in *Rhodobacter sphaeroides* and *R. capsulatus*.[30,31] Furthermore, in *B. elkanii* (Figure 3.2), two long-chain acyl groups are present, which is unusual, and either of them may be non-stoichiometrically *O*-acylated.[32]

Environmental- or growth conditions-dependent changes in lipid A structure have been found in several bacterial species;[33] the most known

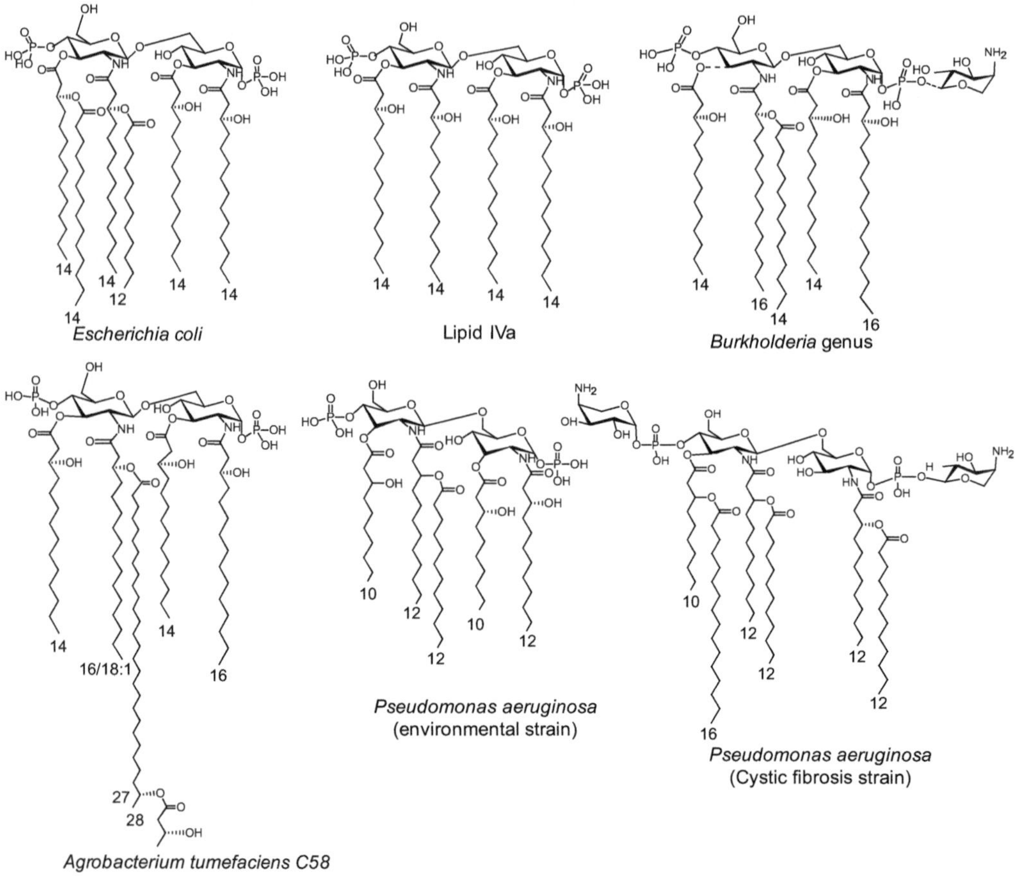

Escherichia coli

Lipid IVa

Burkholderia genus

Agrobacterium tumefaciens C58

Pseudomonas aeruginosa (environmental strain)

Pseudomonas aeruginosa (Cystic fibrosis strain)

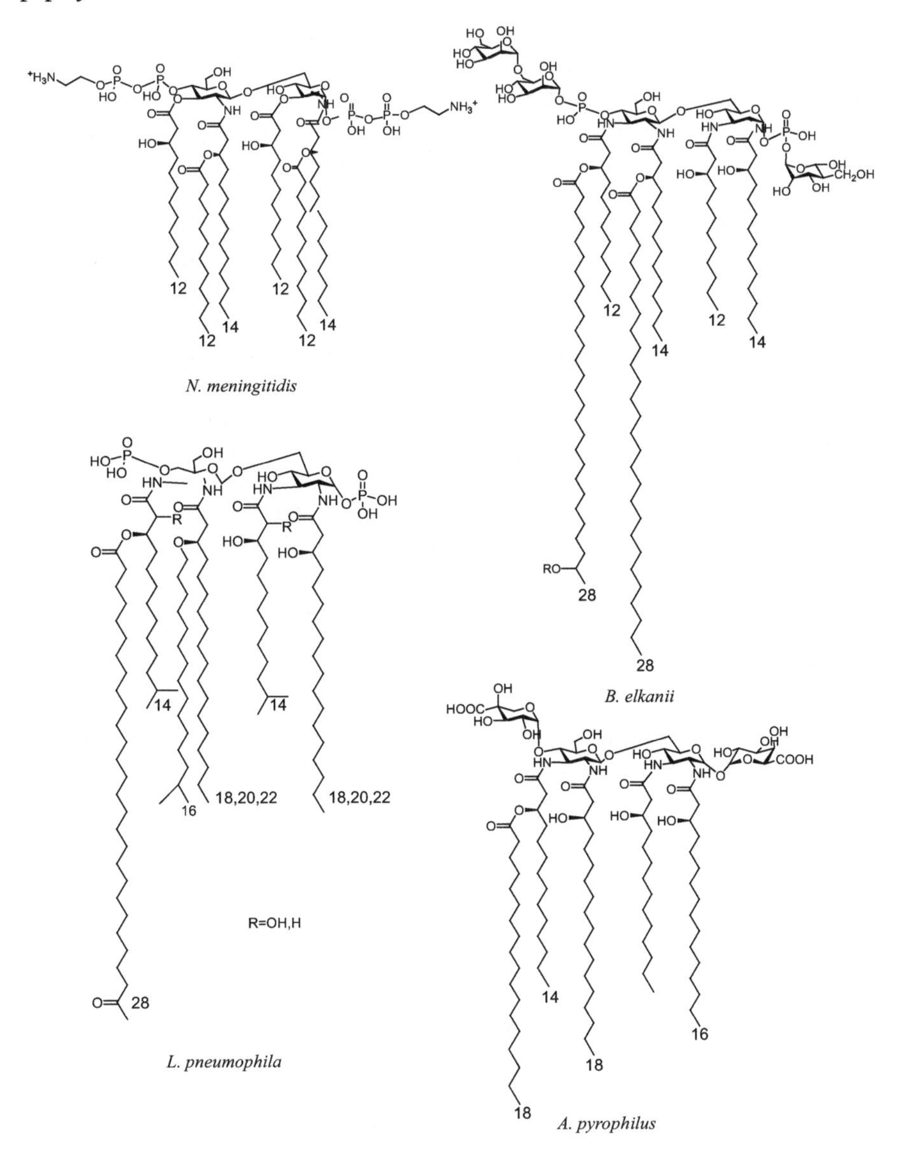

Figure 3.2 Structural heterogeneity in lipid A structure.

case regards lipid A from *E. coli*, which can undergo structural modifications such as the addition of palmitate residues in the presence of antimicrobial peptides and low concentrations of Mg^{2+}. Moreover, the presence of late functioning phosphatases in *Rhizobium etli*, *R. leguminosarum* and *Francisella tularensis* induces the synthesis of partially or completely dephosphorylated lipid A species.[34] It is worth noting that some bacterial species, such as the cystic fibrosis opportunistic pathogens *Burkholderia* (Figure 3.2),[35] were found to constitutively synthesize a lipid A substituted by

L-Ara4N residues, that are, in physiological conditions, positively charged elements whose presence reduces the net charge surface of the OM inferred by LPS molecules, increasing bacterial resistance against cationic antimicrobial compounds such as polymyxin B or cationic antimicrobial peptides (CAMP).[35] A further example of environmental-dependent changes in lipid A structure concerns extremophiles that are bacteria able to live in extreme habitats such as high temperature or high pressure; indeed, most of these particular microorganisms were found to produce LPS substituted by galacturonic acid residues either on the Kdo moiety or on the lipid A glucosamine backbone.[28] The presence of galacturonic acid residues provides additional anionic charges in the lipid A core region, helping the connection of LPS molecules to each other through interaction with environmental divalent cations, thus contributing to the stability and resistance of the OM, which is pivotal for these bacteria always exposed to potentially killer conditions.

Even the most subtle variation in chemical structure of lipid A can affect the LPS molecule bioactivity, since this latter is strictly structure related; indeed, the lipid A intrinsic conformation is responsible for its agonistic and antagonistic activity[36–42] (see below). The exposition to tiny and balanced amounts of agonistic lipid A prepares the immune system to protect the host against further infection, while uncontrolled and excessive exposure (or excessive responses) can lead to an uncontrolled inflammatory response culminating in septic shock and cellular death. Given these premises and this being a structure-dependent phenomenon, it is needless to say that it is pivotal to elucidate the structure of the lipid A moiety, as well as of the entire LPS molecule, in order to understand the molecular mechanisms underlying the innate and acquired immune response and the inflammatory process.

3.4 The Core Oligosaccharide: Structure and Diversity

The core oligosaccharide is a complex component of the LPS molecule since it can be characterised by up to 15 monosaccharides which can be often phosphorylated and organised to give either a linear or a branched structure.[43] In such cases, in the core LPS portion, two different regions can be distinguished on the basis of the monosaccharide composition, termed inner core and outer core.[43] The inner core region, linked to lipid A, is less variable and consists of uncommon sugar residues such as heptoses (L-*glycero*-D-*manno*-heptose and D-*glycero*-D-*manno*-heptose) and Kdo (3-deoxy-D-*manno*-octulosonic acid);[44] this latter, which connects the core oligosaccharide to the lipid A backbone with an α-configured ketosidic linkage in almost every LPS investigated to date,[45] is considered a diagnostic marker for all Gram-negative bacteria, since it is clear that all LPS molecules contain at least one Kdo residue. The only two exceptions in which Kdo is not the first residue of the inner core are *Acinetobacter haemolyticus*[46] and *Shewanella algae*;[47] in the former, Kdo is replaced by its C-3 hydroxy derivative, the

D-*glycero*-D-*talo*-oct-2-ulosonic acid (Ko),[46] whereas in the latter it is replaced by its C-8 amino derivative, the 8-amino-8-deoxy-*manno*-oct-2-ulosonic acid (Kdo8N).[47] The main core oligosaccharide usually presents at O-5 of the Kdo residue a *manno*-configured monosaccharide such as the L-*glycero*-D-*manno*-heptose (L,D-Hep) or, less frequently, its biosynthetic precursor the D-*glycero*-D-*manno*-heptose residue (D,D-Hep).[48] For example, *Helicobacter pylori* LPS presents an abundance of D,D-Hep in its core moiety; indeed, in some isolates the linking region that connects the outer core to the O-polysaccharide is characterised by a long D,D-heptoglycan polymer. Some bacteria produce core oligosaccharides lacking heptose residues, which are mainly substituted by mannose, glucose, galactose or galacturonic acid residues.[49,50] In enterobacterial LPS this first Kdo unit bears at its O-5 position a heptose trisaccharide fragment: α-L,D-HepIII-(1→7)-α-L,D-HepII-(1→3)-α-L,D-HepI; Kdo always carries another negatively charged substituent at its O-4, generally a second Kdo unit, a phosphate group or a further Kdo residue.[51] An interesting core oligosaccharide has been found in aerobic halotolerant marine bacteria belonging to *Loktanella rosea* strain KMM 6003 and isolated from the Sea of Japan; indeed, *L. rosea* LPS turned out to have a core oligosaccharide characterised by a novel trisaccharide fragment composed exclusively of ulosonic sugars and containing neuraminic acid. Whereas in a few cases the first Kdo residue has been found carrying two ulosonic acid residues, as in the case of *Acinetobacter lwoffii* F78, on the contrary a neuraminic acid has never been found directly linked to the Kdo residue.

Residues comprising the inner core region are often decorated with charged substituents like phosphate, pyrophosphate, Ara*p*4N or uronic acids, as found in bacteria belonging to the *Pseudomonas* genus and many other species. Particularly, the Ara*p*4N residue is always present as a terminal non-reducing monosaccharide of the trisaccharide Ara*p*4N-(1→8)-α-Ko-(2→4)-α-Kdo in *Burkholderia* LPSs (Figure 3.3). It is speculated that all the substituents bearing a positively charged free amino group, as Ara*p*4N residues, might play a role in pathogenesis since they reduce the negatively charged surface on the OM, rendering it positively charged or in an isoelectric state that, in turn, confers resistance to antibiotic compounds and antimicrobial peptides.[15,31] Moreover, all the negatively charged residues found in both inner and outer core regions seem to maintain an association with divalent cations required for membrane integrity.[37]

The outer core region is the most exposed portion, often branched, and is characterised by a higher structural variability than the inner core region. It is typically composed of common hexoses such as glucose, galactose, *N*-acetylglucosamine or *N*-acetylgalactosamine and it may also contain residues such as 6-deoxy-L-mannose (L-Rha) or *N*-acetyl-2,6-dideoxy-D-glucosamine (D-QuiNAc). It is possible to find peculiarity also in this region, although less commonly; for example, in *Shewanella* and *Proteus* LPSs a new kind of glycosydic linkage has been found that involves an open-chain acetal linkage of a glucosamine residue that is present as a non-cyclic carbonyl form.[52] The plant pathogen *Xanthomonas campestris* pv. *campestris* was shown to possess a

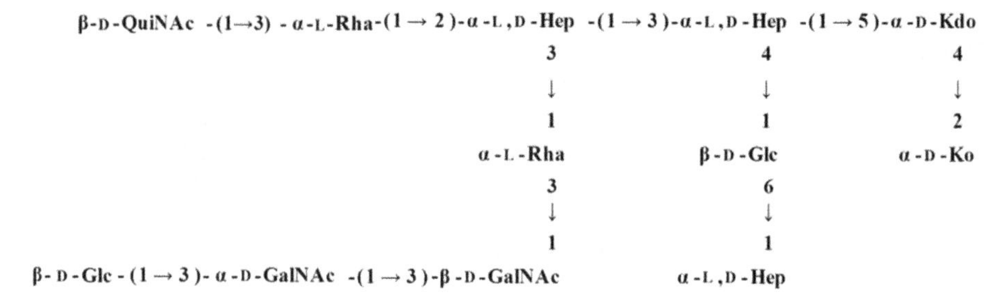

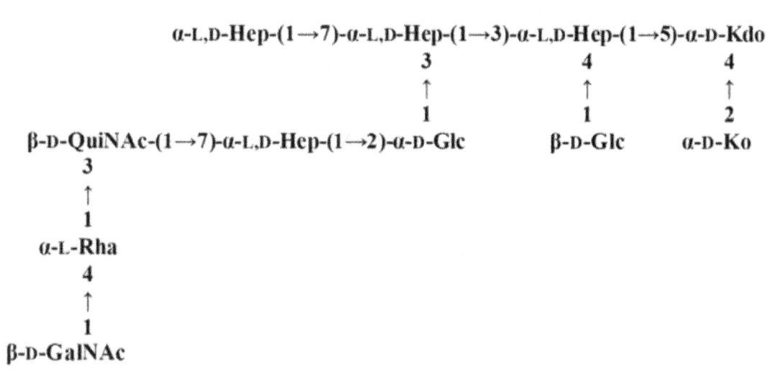

Figure 3.3 Structural diversity in the core oligosaccharide in different species of the same genus. *Top*: *Burkholderia multivorans* core oligosaccharide. *Bottom*: *Burkholderia cenocepacia* core oligosaccharide.

strong accumulation of negatively charged groups in the lipid A inner core region and has a number of novel features, including a galacturonyl phosphate attached at a 3-deoxy-D-*manno*-oct-2-ulosonic acid residue and a unique phosphoramide group.[49] Furthermore, in *Rhizobium*, *Acinetobacter* and *Agrobacterium* strains a 3-deoxy-D-*lyxo*-hept-2-ulosaric acid has been identified.[48,50]

The core oligosaccharide moiety together with the lipid A portion is the structural motive common to all LPSs; this suggests that these domains are indispensable for Gram-negative bacteria life; indeed, until a short while ago it was believed that the minimal requirement for bacterial survival was the tetrasaccharide formed by Kdo_2-lipid A. Thus mutants of Gram-negative bacteria without a minimal core oligosaccharide structure were considered not viable; on the contrary, mutants synthesizing only lipid A have been recently isolated.[44,45] However, in bacteria producing LOSs, as in many *Burkholderia* strains or in *Nesseria meningitidis*, *N. gonorrhoeae* and some chronic strains of *Pseudomonas aeruginosa*, the core moiety results in the most exposed portion of the LPS molecule and thus it is recognised by host adaptive immune system components and it accounts for serological specificity.[33,53] A further consequence of core moiety exposure to the environment is that both bacterial viability and resistance against antibiotic compounds are strictly related to its chemical structure. Some of the bacteria which possess a LOS and not a complete LPS are nasty microbes such as *N. meningitis* and *H. influenzae*, both of which are human pathogens responsible for meningitis. LOS produced by the first is the antigenic determinant of the macromolecule and contains residues such as *N*-acetylneuraminic acid (NeuAc) or L-fucose (L-Fuc) that is a reminder of human glycosphingolipids.[33,53] This mimicry of host cellular surface components is responsible for the increase of pathogen resistance against phagocytosis and host bactericidal activity, escaping attack from the host complement membrane complex.[33,54]

3.5 The O-Polysaccharide: The Antigenic Determinant of LPSs

The O-chain polysaccharide is the most variable portion of the LPS within bacteria belonging to the same genus. It has several roles, among which appears to be a protective one, acting as a defensive barrier and is indispensable in the case of *smooth*-type pathogenic bacteria present in tissues and body fluids where they persist only if they express the O-chain. It is also involved in the microbial adhesion to the host cells mediated by cell-surface molecules called adhesines. Being exposed on the cell surface, highly immunogenic and extremely variable, it is also considered an important virulence factor providing the major basis for serotyping methods for many Gram-negative bacteria.[55] Taking into account that the O-chain is the major antigen targeted by host adaptive responses, the structural elucidation of this LPS moiety is crucial since it contributes to a better understanding of the mechanisms of pathogenesis of infectious diseases, and it also provides pivotal information to develop novel vaccines and diagnostic reagents. In most Gram-negative bacteria the O-chain consists of up to 50 identical repeating oligosaccharide units of two to eight different glycosyl residues (heteroglycans) or, in some bacteria such as *L. pneumophila*,[56] of identical sugars (homoglycans). A single bacterium produces LPSs with O-chains characterised by a wide range of lengths as a result of incomplete synthesis of the polysaccharide chain;[15] this different degree of polymerisation is responsible for the ladder-like pattern, shown by SDS-PAGE,[57] typical of a *smooth* LPS.

The O-polysaccharide is considered the cell constituent with the highest structural variability, since it is composed of a large number of sugar residues (in both pyranose and furanose rings, anomeric and absolute configurations) which build up the repeating units, with different positions and stereochemistry of the glycosydic linkages and carrying non-carbohydrate substituents such as phosphate, amino acids, sulfate, acetyl or formamide groups, often present in a non-stoichiometric fashion.[58–61] As examples of peculiar monosaccharides that can characterise the LPS O-chain moiety are the C_{12} branched-chain sugar caryophyllose {3,6,10-trideoxy-4-*C*-[(*R*)-1-hydroxyethyl]-D-*erythro*-D-*gulo*-decose} and caryose (4,8-cyclo-3,9-dideoxy-L-*erythro*-D-*ido*-nonose), the only carbocyclic monosaccharide found in nature so far; both were isolated from the plant pathogen *Burkholderia caryophylli*, responsible for the wilting of carnations.[62,63]

A further example is represented by the O-chain from *Morganella morganii* LPS, a bacterium belonging to the normal human intestinal flora, which shows the presence of two rare sugars: a 5-*N*-acetimidoyl-7-*N*-acetyl derivative of 8-epilegionaminic acid and a branched ketouronamide termed shewanellose; this latter occurs in the furanose ring in some O-chain units and in the pyranose ring in others.[64] It is worthwhile to outline the unique LPS O-chain structure from the nitrogen-fixing soil bacterium *Bradyrhizobium* sp. BTAi1, characterised by a bicyclic monosaccharide, the bradyrhizose.[65] On the other hand, an example of particular substituents that can decorate the O-chain is the 3-hydroxy-2,3-dimethyl-5-oxoprolyl group found linked to the QuiNAc residue of the LPS O-chain moiety from the newly identified cystic fibrosis pathogen *Pandoraea pulmonicola*.[66] The function of these substituents is frequently unknown, although it can be speculated that bacteria can modify their LPS to mask themselves to the host immune system.[67,68] The classical example of this "stealth mechanism" involves the O-chain moiety from the human pathogen *Helicobacter pylori*, that express Lewis antigenic epitopes (mainly Le^x and Le^y) essential for bacterial survival in the human stomach acidic environment; this comes from the observation that *H. pylori* possessing *rough* LPSs is characterised by a strong reduced colonising ability.[67] On the other hand, it was observed that most bacteria can modify their O-polysaccharide structure in accordance with the host reaction in order to increase their virulence.[55] Finally, Bengoechea *et al.* demonstrated that the O-chain length and the proper distribution of the repeating units is important for the full virulence in pathogen bacteria and that the mere presence of an O-chain is not sufficient.[55,69]

3.6 Genetic and Biosynthesis of the LPS Molecule

The biosynthesis of the LPS molecule has been extensively investigated since it represents an attractive target in the development of new antibiotics and in the synthesis of molecules that can reduce or eliminate bacterial virulence and, moreover, that can work as an antibiotic adjuvant, potentiating antibiotic effects. The three structural domains of the LPS molecule are synthesized through different biosynthetic pathways that take place in the cytoplasm. Since endotoxins are distributed on the bacterial outer membrane, the cytoplasmic synthesis implies their subsequent export to the surface of bacteria cells. Herein, after assembling the lipid A and core oligosaccharide in the cytoplasm, the nascent LPS is translocated across the inner membrane. The O-chain is ligated to the core–lipid A moiety only at the periplasmic side of the membrane, thus concluding LPS synthesis.[15]

3.6.1 Kdo_2-Lipid A Biosynthesis

The first step of the biosynthetic pathway of the LPS molecule is the synthesis of Kdo_2-lipid A that takes place in the cytoplasm at first but ends up

Table 3.1 Enzymes involved in *E. coli* lipid A biosynthesis.

Enzyme	Function
LpxA	Fatty acylation of UDP-GlcNAc
LpxC	Deacetylation of UDP-3-*O*-(acyl)-GlcNAc
LpxD	Addition of a second R-3-hydroxymyristate chain to make UDP-2,3-diacyl-GlcN
LpxH	Cleavage of the pyrophosphate linkage of UDP-2,3-diacyl-GlcN to form lipid X
LpxB	Allowing condensation between UDP-2,3-diacyl-GlcN and lipid X
LpxK	Phosphorylation of the position 4′ of the disaccharide 1-phosphate to form lipid IV_A
KdtA	Addition of two Kdo residues to the position 6′ of lipid IV_A
LpxL	Addition of a secondary lauroyl fatty acid chain to the non-reducing glucosamine residue
LpxM	Addition of a secondary myristoyl fatty acid chain to the non-reducing glucosamine residue

on the cytoplasmic side of the inner membrane.[12,70,71] This biosynthetic pathway has been well characterised in *E. coli* and shown to be mediated by nine enzymes (Table 3.1), highly conserved among almost all Gram-negative bacteria. The initial reactions are catalysed by the soluble enzymes LpxA, LpxC and LpxD, which are responsible for the acylation of the small molecule UDP-GlcNAc, followed by the addition of two 3-OH acyl chains to the positions 2 and 3 of this UDP-GlcNAc, forming the UDP-diacyl-GlcN (Figure 3.4).[71]

In detail, *E. coli* LpxA catalyses the reversible transfer of a C14 hydroxyacyl chain to the 3-OH group of UDP-GlcNAc, thus forming an ester linkage. The capability of LpxA to incorporate only C14 fatty acid chains led to consideration of the active site of this enzyme as possessing a precise "hydrocarbon ruler" able to work with a β-hydroxymyristate [C14:0(3-OH)] at a rate two orders of magnitude faster than C12 or C16 chains,[72] as demonstrated by the structure of *E. coli* lipid A, thus explaining the occurrence of most of the lipid as possessing fatty acids with 14 carbon atoms. On the other hand, the "hydrocarbon ruler" of LpxA from *Pseudomonas aeruginosa* is set on acyl chains of 10 carbon atoms.[73] It was demonstrated that a single point mutation is required to convert the "hydrocarbon ruler" of *E. coli* to a C10 enzyme.[72]

LpxC is a Zn^{2+}-dependent enzyme that acts as a deacetylase of the UDP-3-*O*-acyl-GlcNAc to produce UDP-3-*O*-acylglucosamine.[73–77] Since LpxC is encoded from a single-copy gene present in several Gram-negative bacteria, displaying no sequence similarity to other deacetylases, it represents a promising target for development of broad-spectrum Gram-negative novel antibiotics aimed at interrupting lipid A biosynthesis.[77–80] LpxC is regulated by the metalloprotease FtsH which controls the cytoplasm levels of the deacetylase; this regulation is fundamental since mutation in FtsH degradation activity results in a lethal phenotype due to the high levels of LpxC in the cytoplasm.[76b]

The deacylation step is required since it allows the biosynthesis equilibrium to move towards the incorporation of the second acyl chain on the UDP-3-*O*-acylglucosamine by LpxD to form UDP-2,3-diacylglucosamine by

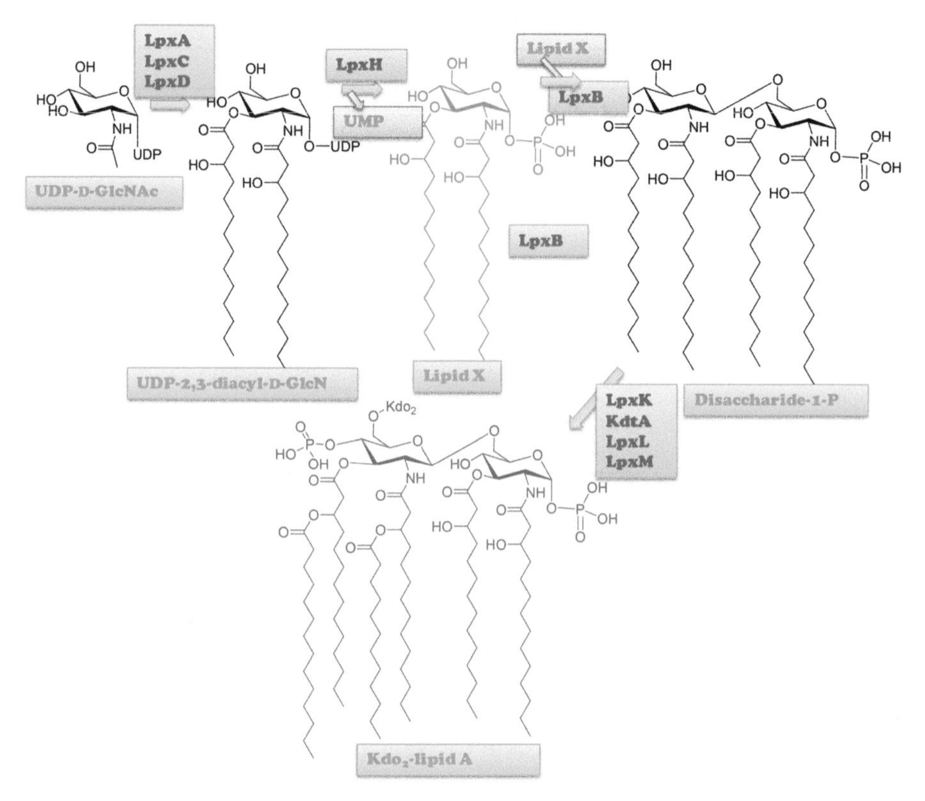

Figure 3.4 Schematised biosynthetic pathway of Kdo_2-lipid A in *E. coli*. The structure and names (*orange*) of the intermediates, as well as the names (*blue*) of the enzymes that catalyze each reaction, are highlighted.

using ACP thioesters as the obligate acyl donors.[81] As with LpxA, also the acyltransferase LpxD shows the presence of a "hydrocarbon ruler" specific for [C14:0(3-OH)].

The next step of lipid A biosynthesis consists in cleavage of the pyrophosphate bond of UDP-2,3-diacylglucosamine by the highly selective pyrophosphatase LpxH to form 2,3-diacylglucosamine-1-phosphate (lipid X) and UMP (Figure 3.4).[82] The disaccharide synthase LpxB is responsible for the subsequent condensation of one molecule of lipid X with one molecule of its precursor, the UDP-2,3-diacylglucosamine (Figure 3.4).[83,84] Both LpxB and LpxH are peripheral membrane proteins, in contrast to the subsequent-step enzymes LpxK, KdtA, LpxL and LpxM, which are integral proteins of the inner membrane.[71]

LpxK is a specific kinase that phosphorylates, in the presence of ATP, the 4′-position of the disaccharide-1-P to form the precursor lipid IV_A (Figure 3.4).[85,86] This latter is of great interest since it was observed that on mouse cells it acts as a potent agonist on the TLR4/MD-2 receptorial complex, whereas on human cells it plays as an antagonist (see below), underling a species-specific difference in the recognition of lipid IV_A by the receptor.[41] At this step, the synthesis of Kdo residues takes place and is considered part of the lipid A biosynthesis since its presence is required for the addition of the secondary fatty acids to the nascent lipid A moiety (see below). The transfer of the first sugar of the core is catalysed by KdtA that acts as a bifunctional enzyme, transferring two Kdo residues to the 6′-position of the non-reducing lipid IV_A glucosamine, by using the sugar nucleotide CMP-Kdo as the donor (Figure 3.4).[71,87] Previous studies showed that KdtA homologues can transfer from one to four Kdo residues to the nascent LPS molecule, but it is not possible to determine whether this Kdo transferase acts as mono-, bi- or even trifunctional enzyme by bioinformatic analyses. The capability to transfer more than two Kdo residues has only been observed to date in *Chlamydia* species, with *C. psitticae* capable of transferring up to four Kdo units.[77,88–90]

Subsequent steps relate to the addition of a secondary lauroyl residue and myristoyl residues by LpxL and LpxM enzymes, respectively, to the non-reducing glucosamine residue, producing the Kdo_2-lipid A (Figure 3.4).[91] As stated before, both *E. coli* transferases of secondary acyl chains require the presence of the Kdo disaccharide for their activity,[77] and in non-thermophilic bacteria they need the occurrence of a phosphate group linked to the Kdo; this latter phosphate group is transferred on lipid A by the enzyme KdkA.[92] In this context, it is worthwhile to underline that lipid A biosynthetic enzymes becoming active in certain conditions have been found. Indeed, *E. coli* presents one additional acyltransferase homologous to LpxL, known as LpxP, which is activated under cold shock conditions (12 °C), introducing a palmitoleate (C16:1) in place of a laurate (C12:0), thus incorporating an unsaturated acyl chain into the lipid A structure which increases membrane fluidity necessary at low temperatures.[93]

In parallel to the conserved biosynthetic pathway, other enzymes able to modulate the number of acyl chains of a given lipid A species have been described. In particular, LpxR and PagL are involved in the reduction in acyl chain numbers,[94,95] whereas the PagP enzyme is responsible for the increase in fatty acid numbers;[96,97] all three proteins are immersed in the bacterial outer membrane. Furthermore, in the periplasm or OM can occur several other modifications in the lipid A structure in order to confer an advantage to the bacteria in escaping the host innate immune system (see above). For example, the decoration of the phosphate groups with Ara*p*4N, by the enzyme ArnT, is typical of several species, among which are *Burkholderia*, *E. coli*, *Pseudomonas* and *Salmonella* species, and is involved in resistance to the most commonly used cationic antimicrobial peptides.[98] The transferase ArnT employs an undecaprenyl-linked Ara*p*4N as the donor substrate,[99] preferentially modifying the 4′-phosphate group of lipid A on the periplasmic side of the inner membrane.[100] Recently, it was demonstrated that in *Burkholderia* genus, ArnT is involved in the transfer of Ara*p*N residues both on the lipid A and on the core oligosaccharide moieties.[35] In other bacterial species, EptA transferase is responsible for the transfer of a phosphoethanolamine (PEtN) on the lipid A backbone; the incorporation of this zwitterionic residue may be advantageous at low pH when the 2-aminoethanol group (EtN) is protonated.[77,101]

3.6.2 Core Oligosaccharide Biosynthesis and Connection to Lipid A

Once the biosynthesis of the Kdo_2-lipid A is completed, the molecule acts as acceptor on which the core oligosaccharide is assembled; this process of assembly provides rapid and sequential glycosyl transfers, catalysed by specific membrane-associated glycosyltransferases, acting as a coordinated complex, which utilise nucleotide sugar precursors.[15] Genes involved in inner core synthesis as well as the outer core assembly are well studied in *Salmonella*, *E. coli* and *K. pneumonia* due to their location as a cluster on the chromosome. In particular, in *E. coli* and *Salmonella* these genes are grouped in three operons located in the *waa* region of the chromosome: gmhD, waaQ and kdtA.[102–104] The first operon contains four genes: gmhD, waaF and waaL are involved in biosynthesis and subsequent transfer of heptoses, whereas waaL is responsible for the subsequent ligation of the O-chain to the core-lipid A moieties.[110] The waaQ operon contains genes necessary for the biosynthesis of the outer core and for its decoration. Eventually, the operon kdtA presents the KdtA gene that is responsible for the addition of the two Kdo residues described above. Several studies have demonstrated that core oligosaccharide biosynthesis is regulated by the RfaH enzyme that is considered a positive regulator of core expression controlling the elongation of the initiated transcript;[105] however, the details of the mechanism still remain unclear. Very little is known also about the environmental-dependent core expression, although several pieces of evidence support the notion that high temperatures up-regulate some *E. coli* core genes.[106,107]

3.6.3 The O-Chain Biosynthesis and Assembly

Smooth-type LPS biosynthesis ends with the addition of the O-chain moiety to the nascent lipid A–core. Similarly to the core oligosaccharide, the O-polysaccharide is synthesised on the cytoplasmic surface of the inner membrane, but the ligation to the core occurs at the periplasmic face of the membrane; thus both lipid A–core and O-chain must be flipped to the periplasmic face by the proteins MsbA and Wzx, respectively, to complete the LPS biosynthesis. In both *E. coli* and *Salmonella* the enzymes involved in O-polysaccharide synthesis, assembly and export are encoded by genes composing the cluster *rfb*.[15] In detail, the peculiarity of such gene cluster is to encode for the glycosyltransferases required for O-chain assembly, the synthesis of the sugar nucleotide precursors and the proteins responsible for the translocation of the O-chain across the inner membrane.[108]

Unlike the core oligosaccharide, the repeating units of the O-chain are assembled on the undecaprenyl phosphate carrier (UndP) (Figure 3.5), embedded in the plasma membrane, by using sugar nucleotides as donors. The

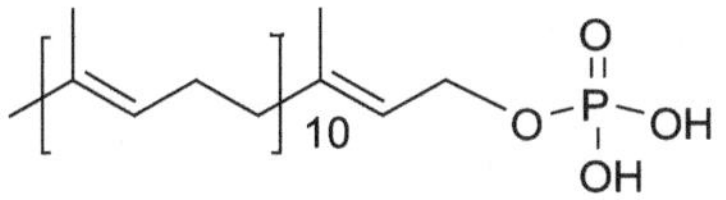

Figure 3.5 Undecaprenyl phosphate structure.

elongation of the polysaccharide chain can occur in three different pathways termed Wzy-dependent, ABC-transporter-dependent and synthase-dependent.[15]

Briefly, the Wzy-dependent process occurs in the case of heteropolymers or branched polysaccharides in which the single repeating unit is synthesised in the cytoplasm by sequential glycosyltransferases and then is transferred to the periplasm by the flippase Wzx, which it is supposed to recognise as the first sugar phosphate bound to the UndP.[109] Afterwards the O-chain is polymerised and ligated to the core-lipid A by the Wzy and Wzz proteins, respectively, producing the nascent O-polysaccharide.[15,108] The ABC-transporter pathway takes its name from the translocation of the chain, which is synthesised in the cytoplasm for homopolysaccharides, into the periplasm that requires the action of an ATP binding cassette transporter.[15]

Finally, the third pathway, the synthase-dependent one, is rarely found in O-polysaccharide biosynthesis; indeed, it was observed only for some species such as *S. enterica* serovar Borreze O:54[110] and typically regarded homopolymers. The mechanism provides the presence of synthases that are integral membrane proteins with a double function: to polymerise the chain elongation acting as sequential glycosyltransferases and to extrude the nascent chain across the membrane playing as an exit pore.[110,111] Thus, the sugars' addition to the growing end of the chain extrudes the polymer from the cytoplasm to the periplasm.

Independently from the pathway, once the UndP-linked O-chain is exported to the periplasm, the WaaL ligase links the O-chain to the preformed lipid A–core, thus concluding the LPS biosynthesis.

3.6.4 Export of the LPS to the Outer Leaflet of the OM

The nascent LPS, at this point, must be translocated from the periplasmic face of the inner membrane to the OM. It has been reported that eight proteins, constituting the "Lpt machinery", transport the LPS macromolecule across the periplasm, translocate it to the inner leaflet of the OM and then export it to the outer leaflet of the OM.[112–117]

This machinery is composed of the cytosolic protein LptB, the periplasmic protein LptA, the inner membrane proteins LptC, LptF and LptG, and the outer membrane proteins LptD and LptE.[112–117] Briefly, the ATP-binding cassette transporter LptBFG translocates nascent LPS to the inner leaflet of the outer membrane, functioning with LptC and LptA.[113,114] In particular, this latter appears the key protein in the Lpt machinery, representing the protein bridge between the inner and the outer membrane as well as the binding protein to the LPS.[118] This was demonstrated with several experiments in which it was observed that mutations in LptA result in accumulation of LPS in the periplasm.[118] Once the endotoxin is in the inner leaflet of the outer membrane, LptD and LptE export it to the outer leaflet.[118] It is worthwhile to underline that both the detailed mechanisms of LPS translocation across the periplasm and its insertion in the OM as well as the molecular requirements for LPS binding to LptA still remain unclear. In this scenario, a very interesting study demonstrated that the LptG protein, in *Burkholderia* genus, is responsible for the recognition of Ara*p*N-modified LPS, thus exporting to the OM only LPS possessing the Ara*p*N decoration.[35] Since previous publications have reported that the presence of Ara*p*N on *Burkholderia* LPS is essential for survival and viability of this bacterium, the above-mentioned study has demonstrated that this essentiality is related to the proper export mechanism and the assemble of the *Burkholderia* LPS to the bacterial outer membrane.[35]

3.7 LPS as Elicitor of Innate Immunity

In vertebrates, an early encounter with infective microorganisms triggers the intervention of the innate immune system, which generates a rapid inflammatory process in order to arrest the growth and proliferation of the invading microbe. Following on, the adaptive immune response takes place, leading to the elimination of the infectious agent and to the establishment of a memory of this infection in order to respond more quickly and with higher efficiency in the case of a possible new encounter with the same microbe. Despite the sophisticated adaptive immune response, the more crude innate immunity can count on diverse phagocytic cells, including monocytes and macrophages, that constitutively express on their surface pattern recognition receptors (PRRs) that recognise conserved microbial components known as pathogen-associated molecular patterns (PAMPs) and also recognise host-derived danger signals (DAMPs) released in response to tissue damage. The recognition of these highly conserved microbial molecules allows the host to unequivocally signal the presence of bacterial infection. In the case of Gram-negative bacteria, lipopolysaccharides, and in particular the lipid A moiety, are recognised as PAMPs by a family of PRRs termed Toll-like receptors (TLRs);[119,120] moreover, the core oligosaccharide or, in *smooth* bacterial colonies, the O-chain, are the antigenic determinants recognised by the adaptive immune system. Among all TLRs, TLR4 is the receptor designated to recognise the LPS macromolecule, but its detection requires several accessory molecules. Indeed, since LPS is present in oligomeric micelles, the first accessory protein, termed LBP (LPS binding protein), is responsible for the conversion to a monomeric LPS and to the delivery of the endotoxin to another protein, the glycosyl phosphatidylinositol-anchored CD14.[121] This latter can chaperone LPS from LBP to TLR4 at the cell surface. The extracellular domain of TLR4 is linked to a small glycoprotein, MD-2 (myeloid differentiation factor 2), which turns out to be

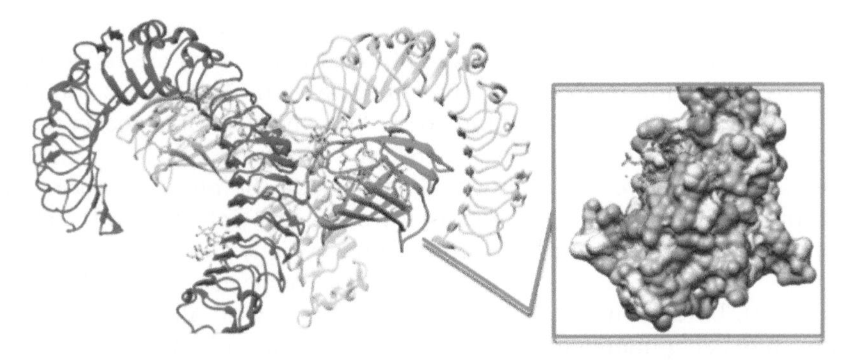

Figure 3.6 *Left*: 3D model of the dimer of *E. coli* LPS core in complex with TLR4/MD-2. *Right*: detail of the hydrophobic binding pocket of the MD-2 protein.

absolutely necessary for the proper functioning of the TLR4 receptor.[122] The binding of LPS to the TLR4/MD-2 receptorial complex (Figure 3.6) leads to activation of two different immune response pathways, both resulting in the amplification of the transduction signal with the consequent massive production of inflammatory proteins and thus eliciting the inflammatory process.[123]

As described above, the endotoxic properties of LPS principally reside in the lipid A moiety and are strongly influenced by its primary structure; thus each chemical variation that can occur in the general lipid A architecture can affect the interaction and the activation of the receptorial complex. A plethora of studies aimed at the elucidation of the LPS structure–activity relationship have highlighted that the hexa-acylated bis-phosphorylated lipid A with an asymmetric 4 : 2 fatty acid distribution, found in the majority of enterobacteria as the *E. coli* lipid A, is considered to have the highest immunostimulatory capacity in mammals cells. In contrast, the hypoacylated synthetic precursor of *E. coli* lipid A, the tetra-acylated lipid IV_A (Figure 3.2), showed weak agonistic effects for some species of mammals and is well known to possess a strong antagonistic activity on human cells.[41] These different biological effects are correlated to two interconnected structural parameters: the molecular shape of the lipid A and the tilt angle between the diglucosamine backbone and fatty acid chains that is the inclination of lipid A hydrophilic moiety with respect to the hydrophobic portion.[42] In detail, the molecular shape possessed by lipid A influences its ability to be recognised by host immune system receptors; indeed, it has been shown that at 37 °C, in aqueous solution and in physiological conditions, the most agonistic lipid A form has a truncated cone that drives to a hexagonal supra-structure, while lipid A's presenting an antagonistic activity assume a cylindrical shape leading to a lamellar structure.[42] Regarding the tilt angle, it has been found that the most active form has a conical structure with a tilt angle $>50°$, while the antagonist structures present smaller values of tilt angles:[39] species with a tilt angle $<25°$, such as lipid IV_A and the penta-acylated and symmetrically $(3+3)$ hexa-acylated lipid A's, act as antagonists; species with an angle between 25–50°, as monophosphorylated lipid A, have a low bioactivity.[38] These physical parameters are reflected in the different binding mode of the LPS molecule to the TLR4/MD-2 complex, since they regulate the lipid A affinity to the MD-2 hydrophobic pocket that is pivotal for the TLR4 activation. Briefly, the structure of *E. coli* lipid A provides that only five of its fatty acids are inside the binding task of the MD-2 protein while the sixth one is extruded and interacts with TLR4, leading to its activation and thus promoting the activation of the immune response.[121] On the other hand, antagonist lipid A species, such as lipid IV_A (Figure 3.2), binds completely to the MD-2 pocket with all four fatty acid chains, thus being unable to trigger any TLR4 activation.[122] In this scenario, a fundamental finding has been the discovery of monophosphoryl lipid A that is used as a vaccine adjuvant, since it has low toxicity and has been demonstrated to be safe and effective.[123]

3.8 LPSs in Plant Immunity

Elicitation of innate immunity in plants through recognition of LPSs is still an elusive and interesting issue. LPSs seem to have diverse roles in the bacterial pathogenesis of plants. Several experiments showed that LPS-deficient mutants present an increased sensitivity to antibiotics and a reduction in bacterial adherence to plant surfaces.[124] LPSs can act as other established elicitors, including flagellin and periplasmic oligosaccharides, being recognised by plants in order to elicit or potentiate plant defence-related responses.[125] This capability of LPSs has been observed in both a local and systemic fashion, although differences among plants have been found. It was previously assessed that LPSs are able to retard or block what is known as "hypersensitive response" (HR). HR is a programmed cell death that is induced by non-pathogenic bacteria and is considered as a plant host resistance since is associated with a rapid decline of the number of viable microbes present in plant tissues, followed by cell and tissue necrosis. Dow *et al.*[126] demonstrated that inoculation of leaves with heat-killed bacteria followed by inoculation with living bacteria prevents the HR effects in a localised fashion. For instance, it was recently demonstrated that pretreatment with lipid A from *Xanthomonas campestris* pv. *campestris* (Xcc) of *A. thaliana* leaves prevents the HR induced by *P. syringae* pv. *tomato.*[49] Moreover, it was shown also that core oligosaccharides from this bacterium induce plant defence but through a mechanism that is independent from that of lipid A.[49] Interestingly, lipid A from Xcc did not protect from HR in tobacco cells, but rather it was the core oligosaccharide part that was responsible.[127,128] Using synthetic O-antigen polysaccharides (oligorhamnans) it has been shown that the O-chain of LPSs is recognised by *Arabidopsis*, and that this recognition elicits a specific gene transcription response associated with defence.[129] Several studies using modified LPSs demonstrated that variation in LPS structure strongly influence its activity on plants. As example the de-phosporylation of lipid A from Xcc results in an

LPS unable to prevent HR in *Arabidopsis* leaves, suggesting that the negative charged groups are fundamental in the process of plant defence response induction.[49,127] A single pathogen-associated compound can consequently originate multiple signals, indicating the existence of multiple receptors for several general elicitors. Further on, LPS is able to elicit systemic resistance in plants through two different responses: the induced systemic resistance (ISR) and the systemic acquired resistance (SAR). This latter consists in the activation of defence-related genes upon necrosis caused by a localising pathogen, whereas the former response is only activated after pathogen challenge.[127] The ISR has been observed in carnation and radish by *P. fluorescens*, whereas treatment of *Arabidopsis* with *P. aeruginosa* LPS was associated with SAR.[127]

The comprehension of the molecular mechanisms underlying the effects that LPSs have on plants has enormous importance in the control of bacterial plant diseases as well as in the study of the beneficial bacteria–plants and bacteria–fungi interactions, which in some cases is demonstrated to occur as a consequence of recognition of LPSs.[65,130] Therefore, one of the major goals of research on LPSs is aimed to identify the yet unknown LPS receptors involved in recognition and successive signal transduction in plant cells.

References

1. R. Pfeiffer, *Z. Hyg.*, 1892, **11**, 393–412.
2. E. Th. Rietschel and O. Westphal, in *Endotoxin in Health and Disease*, ed. H. Brade, S. M. Opal, S. N. Vogel and D. C. Morrisson, Marcel Dekker, New York, 1999, pp. 1–30.
3. O. Westphal and K. Jann, *Methods Carbohydr. Chem.*, 1965, **5**, 83–91.
4. K. Tanamoto, U. Zähringer, G. R. Mckenzie, C. Galanos, E. T. Rietschel, O. Lüderitz, S. Kusumoto and T. Shiba, *Infect. Immun.*, 1984, **44**, 421–426.
5. O. Lüderitz, M. A. Freudenberg, C. Calanos, V. Lehmann, E. Th. Rietschel and D. H. Shaw, in *Current Topics in Membranes and Transport*, Academic Press Inc., New York, 1982, vol. 17, pp. 79–51.
6. O. Westphal, O. Lüderitz, C. Calanos, H. Mayer and E.Th. Rietschel, in *Advances in Immunopharmacology*, ed. L. Chedid, J. W. Hadden and F. Spreafico, Pergamon Press, Oxford, 1986, pp. 13–34.
7. S. C. Wilkinson, in *Surface carbohydrates of the procaryotic cell*, ed. I. W. Sutherland, Academic Press Inc., New York, 1977, pp. 97–105.
8. C. Alexander and E. T. Rietschel, *J. Endoxin Res.*, 2001, **7**, 167–202.
9. E. Schneck, R. G. Oliveira, F. Rehfeldt, B. Demé, K. Brandenburg, U. Seydel and M. Tanaka, *Phys. Rev. E: Stat., Nonlinear, Soft Matter Phys.*, 2009, **80**, 041929.
10. M. Vaara, W. Z. Plachy and H. Nikaido, *Biochim. Biophys. Acta*, 1990, **1024**(1), 152–158.
11. R. Hiruma, A. Yamaguchi and T. Sawai, *FEBS Lett.*, 1984, **170**(2), 268–272.

12. L. S. Cardoso, M. I. Araujo, A. M. Góes, L. G. Pacífico, R. R. Oliveira and S. C. Oliveira, *Microb. Cell Fact.*, 2007, **6**, 1.
13. A. Silipo, C. De Castro, R. Lanzetta, M. Parrilli and M. Molinaro, in *Prokaryotic cell wall compounds structure and biochemistry*, ed. H. Konig, C. Herald and A. Varma, Springer-Verlag, Berlin, Germany, 2010, 133–154.
14. C. R. Raetz, *Annu. Rev. Biochem.*, 1990, **59**, 129–170.
15. C. R. Raetz and C. Whitfield, *Annu. Rev. Biochem.*, 2002, **71**, 635–700.
16. O. Holst, A. Molinaro, in *Microbial Glicobiol.*, ed. A. Moran, O. Holst, P. J. Brennan and M. von Itzstein, Elsevier, London, 2009, pp. 29–56.
17. O. Lüderitz, C. Galanos, H. J. Risse, E. Ruschmann, S. Schlecht, G. Schmidt, H. Schulte-Holthausen, R. Wheat, O. Westphal and J. Schlosshardt, *Ann. N. Y. Acad. Sci.*, 1966, **133**, 347–349.
18. U. Zähringer, B. Lindner and E. T. Rietschel, in *Endotoxin in Health and Disease*, ed. H. Brade, S. M. Opal, S. N. Vogel and D. C. Morrison, Marcel Dekker, New York, Basel, 1999, 93–114.
19. C. G. Hellerqvist, B. Lindberg, K. Samuelsson and A. A. Lindberg, *Acta Chem. Scand.*, 1971, **25**, 955–961.
20. C. G. Hellerqvist, B. Lindberg, S. Svensson, T. Holme and A. A. Lindberg, *Carbohydr. Res.*, 1969, **9**, 237–241.
21. S. B. Svenson, J. Lönngren, N. Carlin and A. A. Lindberg, *J. Virol.*, 1979, **32**, 583–592.
22. J. Szafranek, J. Kumirska, M. Czerwicka, D. Kunikowska, H. Dziadziuszko and R. Glosnicka, *FEMS Immunol. Med. Microbiol.*, 2006, **48**, 223–236.
23. Z. Kaczynski, J. Gajdus, H. Dziadziuszko and P. Stepnowski, *J. Pharm. Biomed. Anal.*, 2009, **50**, 679–682.
24. C. G. Hellerqvist, O. Larm, B. Lindberg, T. Holme and A. A. Lindberg, *Acta Chem. Scand.*, 1969, **23**, 2217–2222.
25. J. L. Di Fabio, J.-R. Brisson and M. B. Perry, *Biochem. Cell Biol.*, 1989, **67**, 278–280.
26. B. Lindberg, K. Leontein, U. Lindquist, S. B. Svenson, G. Wrangsell, A. Dell and M. Rogers, *Carbohydr. Res.*, 1988, **174**, 313–322.
27. D. Schwudke, M. Linscheid, E. Strauch, B. Appel, U. Zähringer, H. Moll, M. Müller, L. Brecker, S. Gronow and B. Lindner, *J. Biol. Chem.*, 2003, **278**, 27502–27512.
28. T. Ieranò, A. Silipo, E. L. Nazarenko, R. P. Gorshkova, E. P. Ivanova, D. Garozzo, L. Sturiale, R. Lanzetta, M. Parrilli and A. Molinaro, *Glycobiology*, 2010, **20**, 586–593.
29. A. Sonesson, E. Jantzen, K. Bryn, L. Larsson and J. Eng, *Arch. Microbiol.*, 1989, **153**(1), 72–78.
30. I. A. Kaltashov, V. Doroshenko, R. J. Cotter, K. Takayama and N. Qureshi, *Anal. Chem.*, 1997, **69**(13), 2317–2322.
31. J. H. Krauss, U. Seydel, J. Weckesser and H. Mayer, *Eur. J. Biochem.*, 1989, **180**(3), 519–526.

32. I. Komaniecka, A. Choma, B. Lindner and O. Holst, *Chemistry*, 2010, **16**(9), 2922–2929.
33. J. P. van Putten, *EMBO J.*, 1993, **12**, 4043–4051.
34. C. R. Raetz, C. M. Reynolds, M. S. Trent and R. E. Bishop, *Annu. Rev. Biochem.*, 2007, **76**, 295–329.
35. M. A. Hamad, F. Di Lorenzo, A. Molinaro and M. A. Valvano, *Mol. Microbiol.*, 2012, **85**(5), 962–974.
36. K. Brandenburg, H. Mayer, M. H. Koch, J. Weckesser, E. T. Rietschel and U. Seydel, *Eur. J. Biochem.*, 1993, **218**, 555–563.
37. E. T. Rietschel, T. Kirikae, F. U. Schade, U. Mamat, G. Schmidt, H. Loppnow, A. J. Ulmer, U. Zähringer, U. Seydel, F. Di Padova, M. Schreier and H. Brade, *FASEB J.*, 1994, **8**, 217–225.
38. U. Seydel, M. Oikawa, K. Fukase, S. Kusumoto and K. Brandenburg, *Eur. J. Biochem.*, 2000, **267**, 3032–3039.
39. S. Fukuoka, K. Brandenburg, M. Müller, B. Lindner, M. H. Koch and U. Seydel, *Biochim. Biophys. Acta*, 2001, **1510**, 185–197.
40. M. Oikawa, T. Shintaku, N. Fukuda, H. Sekljic, Y. Fukase, H. Yoshizaki, K. Fukase and S. Kusumoto, *Org. Biomol. Chem.*, 2004, **2**, 3557–3565.
41. D. T. Golenbock, R. Y. Hampton, N. Qureshi, K. Takayama and C. R. Raetz, *J. Biol. Chem.*, 1991, **266**, 19490–19498.
42. M. G. Netea, M. van Deuren, B. J. Kullberg, J. M. Cavaillon and J. W. van der Meer, *Trends Immunol.*, 2002, **23**, 135–139.
43. O. Holst, in *Endotoxin in Health and Disease*, ed. H. Brade, D. C. Morrsion, S. Opal and S. Vogel, Marcel Dekker, New York, 1999, pp. 115–154.
44. T. C. Meredith, P. Aggarwal, U. Mamat, B. Lindner and R. W. Woodard, *ACS Chem. Biol.*, 2006, **1**, 33–42.
45. U. Mamat, T. C. Meredith, P. Aggarwal, A. Kühl, P. Kirchhoff, B. Lindner, A. Hanuszkiewicz, J. Sun, O. Holst and R. W. Woodard, *Mol. Microbiol.*, 2008, **67**, 633–648.
46. E. V. Vinogradov, K. Bock, B. O. Petersen, O. Holst and H. Brade, *Eur. J. Biochem.*, 1997, **243**, 122–127.
47. E. V. Vinogradov, A. Korenevsky and T. J. Beveridge, *Carbohydr. Res.*, 2003, **338**, 1991–1997.
48. A. De Soyza, A. Silipo, R. Lanzetta, J. R. Govan and A. Molinaro, *Innate Immun.*, 2008, **14**, 127–144.
49. A. Silipo, A. Molinaro, L. Sturiale, J. M. Dow, G. Erbs, R. Lanzetta, M. A. Newman and M. Parrilli, *J. Biol. Chem.*, 2005, **280**, 33660–33668.
50. C. De Castro, A. Molinaro, R. Lanzetta, A. Silipo and M. Parrilli, *Carbohydr. Res.*, 2008, **343**(12), 1924–1933.
51. O. Holst, *FEMS Microbiol. Lett.*, 2007, **271**, 3–11.
52. E. Vinogradov and K. Bock, *Carbohydr. Res.*, 1999, **319**, 92–101.
53. C. M. Tsai, E. Jankowska-Spephens, R. M. Mizamur and J. F. Cipollo, *J. Biol. Chem.*, 2009, **284**, 4616–4625.

54. A. Preston, R. E. Mamdrell, B. W. Gibson and M. A. Apicella, *Crit. Rev. Microbiol.*, 1996, **22**, 139–180.
55. L. Wang, Q. Wang and P. R. Reeves, in *Endotoxins: Structure, Function and Recognition*, ed. X. Wang and P. J. Quinn, Springer, London, New York, 2010, pp. 123–152.
56. U. Zähringer, Y. A. Knirel, B. Lindner, J. H. Helbig, A. Sonesson, R. Marre and E. T. Rietschel, *Prog. Clin. Biol. Res.*, 1995, **392**, 113–139.
57. R. Kittelberger and F. Hilbink, *J. Biochem. Biophys. Methods*, 1993, **26**(1), 81–86.
58. M. Adinolfi, M. M. Corsaro, C. De Castro, A. Evidente, R. Lanzetta, A. Molinaro, P. Lavermicocca and M. Parrilli, *Carbohydr. Res.*, 1996, **284**, 119–133.
59. P. E. Jansson, H. Brade, D. C. Morrison, S. Opal and S. Vogel, in *Endotoxin in Health and Disease*, New York, 1999, pp. 155–178.
60. C. Whitfield, *Annu. Rev. Biochem.*, 2006, **75**, 39–68.
61. I. Lerouge and J. Vanderleyden, *FEMS Microbiol. Rev.*, 2001, **26**, 17–47.
62. A. Adinolfi, M. M. Corsaro, C. De Castro, A. Evidente, R. Lanzetta, L. Mangoni and M. Parrilli, *Carbohydr. Res.*, 1995, **274**, 223–232.
63. A. Adinolfi, M. M. Corsaro, C. De Castro, A. Evidente, R. Lanzetta, A. Molinaro and M. Parrilli, *Carbohydr. Res.*, 1996, **284**, 111–118.
64. A. S. Shashkov, V. I. Torgov, E. L. Nazarenko, V. A. Zubkov, N. M. Gorshkova, R. P. Gorshkova and G. Widmalm, *Carbohydr. Res.*, 2002, **337**, 1119–1127.
65. A. Silipo, M. R. Leone, G. Erbs, R. Lanzetta, M. Parrilli, W. S. Chang, M. A. Newman and A. Molinaro, *Angew. Chem., Int. Ed.*, 2011, **50**(52), 12610–12612.
66. F. Di Lorenzo, A. Silipo, A. Costello, L. Sturiale, D. Garozzo, M. Callaghan, R. Lanzetta, M. Parrilli, S. McClean and A. Molinaro, *Eur. J. Org. Chem.*, 2012, **11**, 2243–2249.
67. Y. Amano and T. Kanda, *Trends Glycosci. Glycotechnol.*, 2002, **14**, 105–114.
68. Y. A. Knirel, in *Bacterial lipopolysaccharides. Structure, chemical synthesis, biogenesis and interaction with host cells*, ed. Y. A. Knirel and M. A. Valvano, Springer Wien, New York, 2011, pp. 41–118.
69. J. A. Bengoechea, H. Najdenski and M. Skurnik, *Mol. Microbiol.*, 2004, **52**, 451–469.
70. W. T. Doerrler, *Mol. Microbiol.*, 2006, **60**, 542–552.
71. X. Wang and P. J. Quinn, *Prog. Lipid Res.*, 2010, **49**, 97–107.
72. T. J. Wyckoff, S. Lin, R. J. Cotter, G. D. Dotson and C. R. Raetz, *J. Biol. Chem.*, 1998, **273**, 32369–32372.
73. G. D. Dotson, I. A. Kaltashov, R. J. Cotter and C. R. H. Raetz, *J. Bacteriol.*, 1998, **180**, 330–337.
74. M. S. Anderson, H. G. Bull, S. M. Galloway, T. M. Kelly, S. Mohan, K. Radika and C. R. Raetz, *J. Biol. Chem.*, 1993, **268**, 19858–19865.
75. M. S. Anderson, A. D. Robertson, I. Macher and C. R. Raetz, *Biochemistry*, 1988, **27**, 1908–1917.

76. (a) K. Young, L. L. Silver, D. Bramhill, P. Cameron, S. S. Eveland, C. R. Raetz, S. A. Hyland and M. S. Anderson, *J. Biol. Chem.*, 1995, **270**, 30384–30391; (b) T. Ogura, K. Inoue, T. Tatsuta, T. Suzaki, K. Karata, K. Young, L. H. Su, C. A. Fierke, J. E. Jackman, C. R. Raetz, J. Coleman, T. Tomoyasu and H. Matsuzawa, *Mol. Microbiol.*, 1999, **31**, 833–844.
77. C. M. Stead, A. C. Pride and M. S. Trent, in *Bacterial lipopolysaccharides Structure, chemical synthesis, biogenesis and interaction with host cells*, ed. Y. A. Knirel and M. A. Valvano, Springer Wien, New York, 2011, pp. 163–194.
78. H. R. Onishi, B. A. Pelak, L. S. Gerckens, L. L. Silver, F. M. Kahan, M. H. Chen, A. A. Patchett, S. M. Galloway, S. A. Hyland, M. S. Anderson and C. R. Raetz, *Science*, 1996, **274**, 980–982.
79. J. E. Jackman, C. A. Fierke, L. N. Tumey, M. Pirrung, T. Uchiyama, S. H. Tahir, O. Hindsgaul and C. R. Raetz, *J. Biol. Chem.*, 2000, **275**, 11002–11009.
80. J. M. Clements, F. Coignard, I. Johnson, S. Chandler, S. Palan, A. Waller, J. Wijkmans and M. G. Hunter, *Antimicrob. Agents Chemother.*, 2002, **46**, 1793–1799.
81. T. M. Kelly, S. A. Stachula, C. R. H. Raetz and M. S. Anderson, *J. Biol. Chem.*, 1993, **268**, 19866–19874.
82. K. J. Babinski and C. R. H. Raetz, *FASEB J.*, 1998, **12**, A1288.
83. B. L. Ray, G. Painter and C. R. Raetz, *J. Biol. Chem.*, 1984, **259**, 4852–4859.
84. K. Radika and C. R. Raetz, *J. Biol. Chem.*, 1988, **263**, 14859–14867.
85. T. A. Garrett, N. L. Que and C. R. Raetz, *J. Biol. Chem.*, 1998, **273**, 12457–12465.
86. T. A. Garrett, J. L. Kadrmas and C. R. Raetz, *J. Biol. Chem.*, 1997, **272**, 21855–21864.
87. K. A. Brozek, K. Hosaka, A. D. Robertson and C. R. Raetz, *J. Biol. Chem.*, 1989, **264**, 6956–6966.
88. C. J. Belunis, T. Clementz, S. M. Carty and C. R. Raetz, *J. Biol. Chem.*, 1995, **270**, 27646–27652.
89. W. Brabetz, B. Lindner and H. Brade, *Eur. J. Biochem.*, 2000, **267**, 5458–5465.
90. S. Rund, B. Lindner, H. Brade and O. Holst, *Eur. J. Biochem.*, 2000, **267**, 5717–5726.
91. K. A. Brozek and C. R. H. Raetz, *J. Biol.Chem.*, 1990, **265**, 15410–15417.
92. K. A. White, S. Lin, R. J. Cotter and C. R. Raetz, *J. Biol. Chem.*, 1999, **274**, 31391–31400.
93. S. M. Carty, K. R. Sreekumar and C. R. Raetz, *J. Biol. Chem.*, 1999, **274**, 9677–9685.
94. C. M. Reynolds, A. A. Ribeiro, S. C. McGrath, R. J. Cotter, C. R. Raetz and M. S. Trent, *J. Biol. Chem.*, 2006, **281**, 21974–21987.
95. M. S. Trent, W. Pabich, C. R. Raetz and S. I. Miller, *J. Biol. Chem.*, 2001, **276**, 9083–9092.
96. L. Guo, K. B. Lim, C. M. Poduje, M. Daniel, J. S. Gunn, M. Hackett and S. I. Miller, *Cell*, 1998, **95**, 189–198.

97. R. E. Bishop, H. S. Gibbons, T. Guina, M. S. Trent, S. I. Miller and C. R. Raetz, *EMBO J.*, 2000, **19**, 5071–5080.
98. J. S. Gunn, K. B. Lim, J. Krueger, K. Kim, L. Guo, M. Hackett and S. I. Miller, *Mol. Microbiol.*, 1998, **27**, 1171–1182.
99. M. S. Trent, A. A. Ribeiro, W. T. Doerrler, S. Lin, R. J. Cotter and C. R. Raetz, *J. Biol. Chem.*, 2001, **276**, 43132–43144.
100. W. T. Doerrler, H. S. Gibbons and C. R. Raetz, *J. Biol. Chem.*, 2004, **279**, 45102–45109.
101. H. Lee, F. F. Hsu, J. Turk and E. A. Groisman, *J. Bacteriol.*, 2004, **186**, 4124–4133.
102. D. E. Heinrichs, J. A. Yethon and C. Whitfield, *Mol. Microbiol.*, 1998, **30**, 221–232.
103. M. Regué, N. Climent, N. Abitiu, N. Coderch, S. Merino, L. Izquierdo, M. Altarriba and J. M. Tomás, *J. Bacteriol.*, 2001, **183**, 3564–3573.
104. C. Roncero and M. J. Casadaban, *J. Bacteriol.*, 1992, **174**, 3250–3260.
105. A. A. Lindberg and C. G. Hallerqvist, *J. Gen. Microbiol.*, 1980, **116**, 25–32.
106. M. Karow, S. Raina, C. Georgopoulos and O. Fayet, *Res. Microbiol.*, 1991, **142**, 289–294.
107. U. Mamat, M. Skurnik and J. A. Bengoechea, in *Bacterial lipopolysaccharides Structure, chemical synthesis, biogenesis and interaction with host cells*, ed. Y. A. Knirel and M. A. Valvano, Springer Wien, New York, 2011, pp. 237–273.
108. X. Wang and P. J. Quinn, in *Endotoxins: Structure, Function and Recognition*, ed. X. Wang and P. J. Quinn, Springer, London, New York, 2010, pp. 3–25.
109. C. L. Marolda, J. Vicarioli and M. A. Valvano, *Microbiology*, 2004, **150**, 4095–4105.
110. W. J. Keenleyside and C. Whitfield, *J. Biol. Chem.*, 1996, **271**, 28581–28592.
111. C. Whitfield, P. A. Amor and R. Köplin, *Mol. Microbiol.*, 1997, **23**, 629–638.
112. W. T. Doerrler, *Mol. Microbiol.*, 2006, **60**, 542–552.
113. M. P. Bos, B. Tefsen, J. Geurtsen and J. Tommassen, *Proc. Natl. Acad. Sci. U. S. A.*, 2004, **101**, 9417–9422.
114. T. Wu, A. C. McCandlish, L. S. Gronenberg, S. S. Chng, T. J. Silhavy and D. Kahne, *Proc. Natl. Acad. Sci. U. S. A.*, 2006, **103**, 11754–11759.
115. P. Sperandeo, F. K. Lau, A. Carpentieri, C. De Castro, A. Molinaro, G. Dehò, T. J. Silhavy and A. Polissi, *J. Bacteriol.*, 2008, **190**, 4460–4469.
116. N. Ruiz, L. S. Gronenberg, D. Kahne and T. J. Silhavy, *Proc. Natl. Acad. Sci. U. S. A.*, 2008, **105**, 5537–5542.
117. S. S. Chng, N. Ruiz, G. Chimalakonda, T. J. Silhavy and D. Kahne, *Proc. Natl. Acad. Sci. U. S. A.*, 2010, **107**, 5363–5368.
118. P. Sperandeo, G. Dehò, A. Polissi, in *Bacterial lipopolysaccharides. Structure, chemical synthesis, biogenesis and interaction with host cells*, ed. Y. A. Knirel and M. A. Valvano, Springer Wien, New York, 2011, pp. 262–282.

119. R. Medzhitov and C. Janeway Jr., *Immunol. Rev.*, 2000, **173**, 89–97.
120. S. Akira and K. Takeda, *Nat. Rev. Immunol.*, 2004, **4**, 499–511.
121. B. S. Park, D. H. Song, H. M. Kim, B. S. Choi, H. Lee and J. O. Lee, *Nature*, 2009, **458**, 1191–1195.
122. J. Meng, J. R. Drolet, B. G. Monks and D. T. Golenbock, *J. Biol. Chem.*, 2010, **285**, 27935–27943.
123. V. Mata-Haro, C. Cekic, M. Martin, P. M. Chilton, C. R. Casella and T. C. Mitchell, *Science*, 2007, **316**(5831), 1628–1632.
124. M. A. Newman, J. M. Dow, A. Molinaro and M. Parrilli, *J. Endotoxin Res.*, 2007, **13**, 69–84.
125. U. Conrath, O. Thulke, V. Katz, S. Schwindling and A. Kohler, *Eur. J. Plant Pathol.*, 2001, **107**, 113–119.
126. J. M. Dow, M. A. Newman and E. von Roepenack, *Annu. Rev. Phytopathol.*, 2000, **38**, 241–261.
127. G. Erbs, A. Molinaro, J. M. Dow, M. A. Newman, in *Endotoxins: Structure, Function and Recognition*, ed. X. Wang and P. J. Quinn, Springer, London, New York, 2010, pp. 387–403.
128. S. G. Braun, A. Meyer, O. Holst, A. Pülher and K. Niehaus, *Mol. Plant-Microbe Interact.*, 2005, **18**, 674–681.
129. E. Bedini, C. De Castro, G. Erbs, L. Mangoni, J. M. Dow, M. A. Newman, M. Parrilli and C. Unverzagt, *J. Am. Chem. Soc.*, 2005, **127**(8), 2414–2416.
130. M. R. Leone, G. Lackner, A. Silipo, R. Lanzetta, A. Molinaro and C. Hertweck, *Angew. Chem., Int. Ed.*, 2010, **49**(41), 7476–7480.

CHAPTER 4

Molecular Basis of Mycobacterium tuberculosis *Recognition by the C-Type Lectin DC-SIGN: from the Modulation of Innate Immune Response to the Design of Innovative Anti-inflammatory Drugs*

EMILYNE BLATTES,[a,b,†] ALAIN VERCELLONE,[a,b] SANDRO SILVA-GOMES,[a,b] JACQUES PRANDI[a,b] AND JÉRÔME NIGOU*[a,b]

[a] CNRS, Institut de Pharmacologie et de Biologie Structurale, 205 route de Narbonne, F-31077 Toulouse, France; [b] Université de Toulouse, UPS, Institut de Pharmacologie et de Biologie Structurale, F-31077 Toulouse, France

*Email: jerome.nigou@ipbs.fr

†Present address: Department of Chemistry and Biochemistry, University of Bern, Freiestrasse 3, 3012 Bern, Switzerland.

RSC Drug Discovery Series No. 43

Carbohydrates in Drug Design and Discovery

Edited by Jesús Jiménez-Barbero, F. Javier Cañada and Sonsoles Martín-Santamaría

Published by the Royal Society of Chemistry, www.rsc.org

4.1 DC-SIGN C-Type Lectin Receptor

The dendritic cell-specific intracellular adhesion molecules (ICAM)-3 grabbing non-integrin (DC-SIGN, CD209) is a C-type lectin receptor (CLR) restricted to potent antigen-presenting cells (APCs) and involved in multiple immune functions. It is a type II membrane protein able to recognize glycosylated antigens in a Ca^{2+}-dependent manner and cause their internalization.

4.1.1 Cell-type Distribution and Expression

DC-SIGN is highly expressed in immature dendritic cells (DCs) in peripheral tissues and at a lower level in mature DCs localized in dermis, lymphoid and interstitial tissues.[1,2] Originally, DC-SIGN was described as a phenotypic marker of these cells. However, it is not expressed either at follicular and plasmacytoid DC membranes or on Langerhans cells.[3] Extensive studies highlighted the expression of the receptor on several macrophage populations, like the lung alveolar macrophages,[4] the tumour-associated M2-polarized macrophages,[5] the $CD14^{+}$ subpopulation in intestinal lamina propria[6] and in dermis,[7] and also on antigen-presenting cells from the placenta[8] and on microglia in healthy human brain.[9] The basal and tissue-specific expression of DC-SIGN involves the PU.1 transcription factor whose expression is restricted to myeloid and B-lymphoid cells.[10] Therefore DC-SIGN is no longer a DC-specific receptor but rather a myeloid cell phenotypic marker with a high antigen-presenting capacity. DC-SIGN has one human homologue, the liver/lymph node-specific intercellular adhesion molecule-3-grabbing integrin (L-SIGN) with 77% amino acid identity. It shares the same properties and is abundantly expressed on endothelial cells of liver and lymph nodes but not on APCs.

In vitro DC-SIGN is found on monocyte-derived DCs and its expression is dependent on a Th2-derived cytokine, interleukin (IL)-4.[11] Early surface expression of the receptor appears after 24 h incubation of peripheral blood monocytes with the interleukin and the granulocyte-macrophage colony-stimulating factor (GM-CSF). IL-4 also induces *de novo* DC-SIGN expression on the THP-1 monocytic cell line pre-treated with differentiation-inducing agents such as phorbol esters (PMA) and on activated monocyte-derived macrophages.[12] By contrast, DC-SIGN is negatively regulated in Th1-mediated inflammatory responses by transforming growth factor (TGF)-β, interferon (INF)-α and -γ, tumour necrosis factor (TNF)-α, the inflammatory endotoxin lipopolysaccharide (LPS) and the anti-inflammatory agent dexamethasone.[11]

4.1.2 Structure and Mechanisms of Carbohydrate Recognition

DC-SIGN receptor is a member of the asialoglycoprotein receptor (ASGR) subgroup II of the CLR superfamily in which are also found dectin-1 and

MGL (macrophage galactose lectin). These proteins present a single C-type lectin domain (CTLD) and a low molecular weight of 45 kDa, compared with the mannose receptor (MR), another C-type lectin involved in immune functions and microorganism recognition, which harbours eight CTLDs and a molecular weight of 180 kDa. However, like the MR, DC-SIGN recognizes high-mannose as well as Lewis-type structures with fucose units.

The C-terminal extracellular domain of DC-SIGN contains a single CTLD supported by $7\frac{1}{2}$ repeats of neck domains (Figure 4.1). The crystal structure of the CTLD bound to a pentasaccharide containing three mannopyranoside units was solved by Feinberg *et al.* in 2001[13] and found to share the typical overall structure of long-form C-type lectins.[14] Like MBP-A (for review, see Zelensky and Gready[15]), the DC-SIGN CTLD structure folds on itself in a double-loop stabilized by two highly conserved disulfide bridges located at the bases of the loops. The whole peptide sequence of the domain fits in the first loop, whereas the second loop, called the long loop region (LLR), is intramolecular and lies within the domain. This latter is crucial for sugar binding and is involved in Ca^{2+}-dependent carbohydrate recognition. The Ca^{2+} ion (in site 2) forms coordination bonds with the carbonyl groups of amino acids highly conserved in the "EPN...E...WND" sequence motif. The sugar unit binding to the Ca^{2+} ion and its positioning in the site are identical to that of all C-type lectins. The equatorial 3- and 4-OHs of a mannopyranoside unit each form coordination bonds with the Ca^{2+}. In the case of a fucofuranoside unit, the equatorial 3-OH and the axial 4-OH interact with the Ca^{2+}, inducing a 90° rotation of the sugar in the binding site compared with the mannose unit.[16] Then the complex is stabilized by hydrogen bonds between the side chains of amino acids carrying the Ca^{2+} ion and the hydroxyls. However, unlike the MR and MBP-A, DC-SIGN sugar binding is unusual because it involves an internal, rather than a terminal, sugar. It binds to a broad set of fucosylated oligosaccharides, including the Lewis^{a} and Lewis^{x} blood epitopes and to high-mannose-type N-glycans (see Section 1.3.1).[16] The interaction of the overall oligosaccharide in the primary binding site is stabilized with a secondary hydrogen-bond network and van der Waals contacts.

The neck region contains seven complete and one incomplete repeats of a 23 amino acid sequence and projects DC-SIGN CTLD ~350 Å above the cell membrane.[17] Half of the domain folds in an α-helical conformation and bears a strongly conserved heptad motif of hydrophobic residues spaced at regular intervals. Several analytical studies converge towards the same conclusions that the neck domains form extended stalks, allowing the self-assembly of four α-helical regions stabilized by lateral hydrophobic interactions.[18–20] Hence, the neck region induces the tetramerization of DC-SIGN (Figure 4.1). The clustering of four CTLDs on the extracellular side allows a multivalent interaction with various mannose- and fucose-containing glycoconjugates and therefore provides high-affinity binding. Indeed, the CTLDs of each monomer are independent and dispose of a certain flexibility for conformational changes and repositioning upon ligand binding.[21]

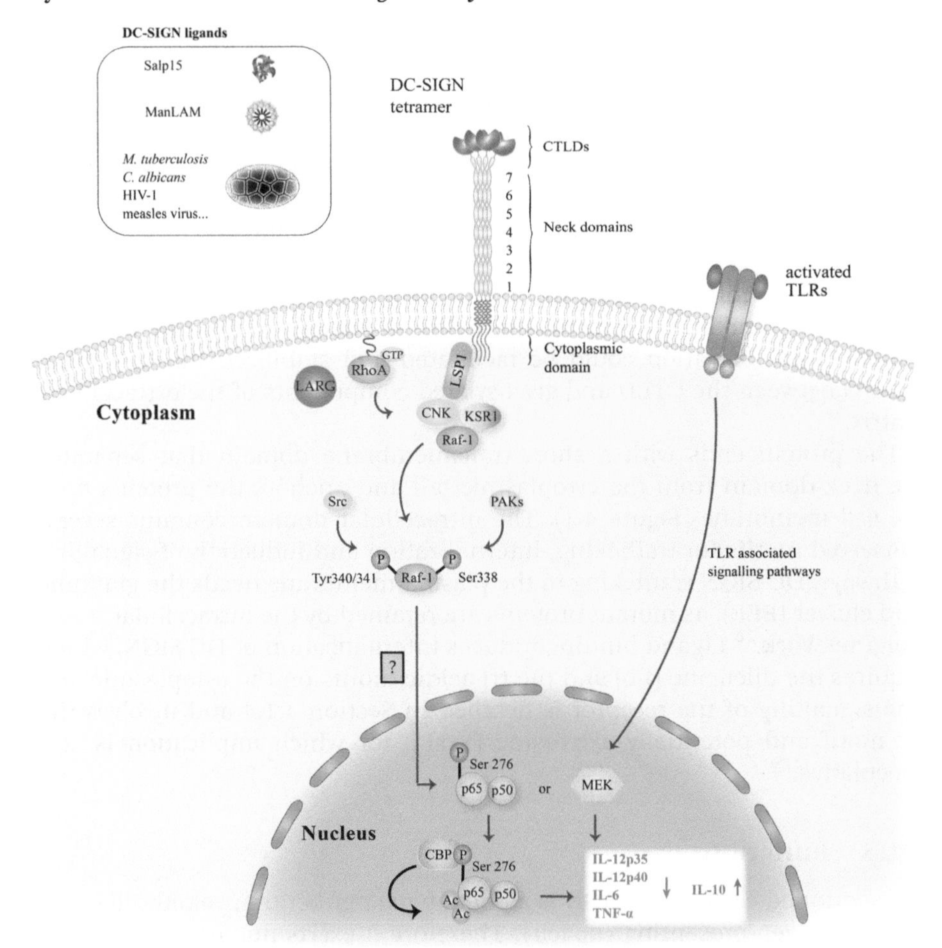

Figure 4.1 DC-SIGN receptor and associated intracellular signalling pathways. DC-SIGN is a CLR with a C-terminal extracellular domain containing a single CTLD supported by $7\frac{1}{2}$ repeats of neck domains. The neck region induces the tetramerization of the receptor, allowing a multivalent interaction with various mannose- and fucose-containing glycoconjugates and therefore providing high-affinity binding. The intracellular domain contains several conserved motifs for trafficking, internalization and induction of signalling pathways. In DCs, DC-SIGN is constitutively associated with the protein adaptor LSP1, which recruits the triplex KSR1-CNK-Raf-1. Upon ligand binding, the upstream effectors LARG and RhoA activate Raf-1 while TLR activation induces NF-κB translocation into the nucleus. Signalling *via* DC-SIGN leads to the acetylation of the p65 subunit of NF-κB, prolonging and increasing the transcription rate of anti-inflammatory IL-10 cytokine. In the case of the Salp15 protein, Raf-1 activates MEK instead to induce anti-inflammatory signals. The present scheme depicts DC-SIGN signalling pathways initially described to occur in response to *M. tuberculosis* natural ligand ManLAM, Salp15 and pathogens. The dichotomy of DC-SIGN signalling in response to fucosylated *vs.* mannosylated ligands is not represented here.

In addition, the DC-SIGN tetramer has a higher level of organization and clusters in nanodomains randomly distributed on the plasma membrane of live immature DCs or cell lines ectopically expressing DC-SIGN.[22–25] Such clustering of DC-SIGN is thought to improve binding avidity by multiplying simultaneous interactions. However, very recently, Liu *et al.* studied the number of DC-SIGN molecules per nanodomains thanks to a total internal reflection fluorescence microscopy (TIRFM) imaging approach on immature DCs and NIH 3T3 transfected cells.[26] They found an average occupancy of 4–8 proteins, corresponding to one or two tetramers per nanodomains sufficient to bind and efficiently internalize the dengue virus. The DC-SIGN nanodomain formation could be facilitated and stabilized through interactions between the CTLD and glycosylated components of the extracellular matrix.[27]

The protein ends with a short transmembrane domain that separates the neck domain from the cytoplasmic tail and anchors the proteins onto the cell membrane (Figure 4.1). The intracellular domain contains several conserved motifs for trafficking, internalization and induction of signalling pathways. DC-SIGN trafficking to the plasma membrane needs the glutamic acid cluster (EEE), as mutant proteins are retained by the intracellular trans-Golgi network.[28] Ligand binding induces internalization of DC-SIGN, which requires the dileucine (LL) and the tri-acidic motifs on the cytoplasmic tail. The signalling of the receptor is detailed in Section 4.1.4 and involves the LL motif and potentially a tyrosine (YxxL), for which implication is still speculative.[29]

4.1.3 Functions

As mentioned earlier, DC-SIGN is a phenotypic marker of myeloid cells with a high antigen-presenting capacity. Therefore it exerts numerous processes of continuous immune body surveillance and its functions are well studied in DCs, where it is involved in the antigen capture from pathogens and intercellular adhesion of these cells with endothelial cells (migration process) or T cells and neutrophils (adaptive immune response). Many of the carbohydrates recognized by DC-SIGN are found either in pathogens or in self-glycoproteins. That said, DC-SIGN has obviously two sets of well-distinguishable ligands: pathogenic and self or endogenous ligands.

4.1.3.1 Physiological Role

The list of endogenous ligands features all Lewis-type sequences and comprises the proteins intercellular adhesion molecule (ICAM)-2[2] and ICAM-3,[1,30] the carcinoembriogenic antigen (CEA)-related cell adhesion molecule (CEACAM1) and Mac-1 on neutrophils.[31] DC-SIGN was first described through its interaction with ICAM-3 on the surface of resting T cells[1] and this highlighted its major role in the intercellular adhesion of DCs. ICAM-3 is an N-glycosylated membrane protein expressed on all leukocytes, yet only the peripheral granulocytes bear the Lewis^x oligosaccharide which is recognized by DC-SIGN.[30] The complex ICAM-3/DC-SIGN mediates a transient contact between T cells and DCs, allowing T cells receptor recognition of the peptide antigen presented on the major histocompatibility complex (MHC). DC-SIGN also recognizes ICAM-2 through the Lewis^y oligosaccharide and thus mediates the tethering, rolling and adhesion of DCs to vascular and lymphoid endothelium for tissue migration.[2,32] This event is of major importance in the trafficking from blood to the peripheral tissues of DC precursors, as well as recruitment to inflammatory sites. Finally, DC-SIGN interacts with Mac-1 and CEACAM1 on activated neutrophils, translating a close contact between these cells and DCs. Once activated, neutrophils release cytokines such as TNF-α and induce the maturation of DCs and the stimulation of T cell proliferation to Th1 polarization.

However, DC-SIGN also recognizes endogenous ligands in non-immunological tissues like the MHC-encoded butyrophilin,[33] the bile salt-stimulated lipase (BSSL) from human milk,[34] the CEA proteins expressed on the surface of colorectal cancer cells[35] and, finally, several seminal plasma glycoproteins (clusterin, galectin-3 binding glycoprotein, prostatic acid phosphatase, *etc.*).[36] Very recently, a constituent of the central nervous system, the myelin oligodendrocyte glycoprotein (MOG), was described as a ligand of DC-SIGN.[9] This protein acts as an autoantigen in the neuroinflammatory disease multiple sclerosis. The glycosylation status and the MOG/DC-SIGN interaction could be involved in immune homeostasis control in the healthy human brain. A glycan array study showed that *in vitro* DC-SIGN can interact with all kinds of Lewis-type antigens (Lewis^a, Lewis^b, Lewis^x, Lewis^y, sulfo-Lewis^a and pseudo-Lewis^y),[16] but it seems it prefers the Lewis^x oligosaccharide on endogenous ligands as all the above-mentioned except ICAM-2 carry it.

4.1.3.2 Role in the Immune System

DC-SIGN is highly expressed in immature dendritic cells (see Section 4.1.1), which are a central element for the protection against pathogens. Immature DCs are localized in the peripheral tissues, where they serve as sentinels that monitor for pathogens and drive the development of a specific immune response. Upon recognition and capture of a pathogen, DCs undergo maturation and migrate to lymphoid organs to initiate immunity.[37] This event is one of the most crucial steps in the induction of protective immunity. DCs use pattern-recognition receptors (PRRs) to recognize conserved molecules of microbes, referred to as pathogen-associated molecular patterns (PAMPs).[38] As described in Sections 4.1.2 and 4.1.3.1, DC-SIGN recognizes fucosylated glycans and high-mannose structures on endogenous glycoproteins. However, these structures are widely distributed in nature and are present in several pathogens. Consequently, DC-SIGN plays a role of PRR on immature DCs.

DC-SIGN was initially identified in a screening for the binding of the HIV envelope glycoprotein gp120,[39] and since then it has been shown to function as a PRR for several other viruses, including Ebola virus, hepatitis C virus, cytomegalovirus and Dengue virus through recognition of N-linked glycans on envelope proteins (reviewed in Van Breedam *et al.*[40]). DC-SIGN also recognizes bacteria such as mycobacteria (see Section 4.3), *Helicobacter pylori*[41] and *Klebsiella pneumoniae*[42] through Lewis^{x}-containing LPS and capsular polysaccharide, respectively, and parasites such as *Leishmania* and *Schistosoma mansoni* through fucose-containing glycan epitopes on soluble egg antigen.[43]

Engagement of DC-SIGN results in rapid receptor-mediated endocytosis and targeting to late endosomes/lysosomes, leading to antigen processing and presentation to $CD4^{+}$ T cells.[3] The outcome of DC-SIGN triggering depends on the pathogen involved. It has been reported to lead to the inhibition[41] or induction[44] of T helper (Th) 1 cell development, induction of a Th2 response[45] or induction of regulatory T cells.[46] However, most of the pathogens recognized by DC-SIGN cause chronic infections and thereby dictate a Th1 *versus* Th2 balance to support their persistence.[47] The convergence of DC-SIGN and other PRR pathways such as Toll-like receptors (TLRs) allows the adaptation of the immune response to different stimuli (see Section 4.1.4). However, the potential of DC-SIGN to induce an anti-inflammatory response might be hijacked by pathogens to reduce the magnitude of the immune response and escape immune surveillance. A striking example is the virus HIV-1, which has been shown to target DC-SIGN to promote its survival. Immature DCs at the infection site capture HIV-1 through the high-affinity interaction of DC-SIGN with the envelope glycoprotein gp120,[48] and migrate to lymphoid tissues. The DC-SIGN bound HIV-1 is protected from degradation and retains competence to infect target cells.[49] The virus can then be transmitted to $CD4^{+}$ T cells, the primary target cell of HIV-1,[48] through the formation of DC/T cell junctions.[50] Notably, DC-SIGN enhances the infection of T cells, which are not infected at low virus titres without the assistance of DC-SIGN.[48]

4.1.4 Signalling Pathway Associated

DC-SIGN intracellular signalling pathways were thoroughly investigated over recent years with pathogen ligands. However, it seems that the triggering of DC-SIGN alone does not cause DC maturation or cytokine production. Instead, it shapes the immune response induced by other PRRs, such as TLRs, and its activation could manipulate the innate immune response to other purposes (see Figure 4.1). As mentioned earlier, DC-SIGN displays two motifs (LL and YxxL) on the cytoplasmic tail that are important for signalling. These motifs are able to constitutively recruit the lymphocyte-specific adaptor protein LSP1, which associates with the complex KSR1-CNK-Raf-1 (RAF proto-oncogene serine/threonine-protein kinase).[51] Upon ligand interaction, upstream effectors, the Rho guanine nucleotide-exchange factor

(LARG) and the small GTPase RhoA, are recruited and activate Raf-1. Phosphorylation of Raf-1 on both key residues Ser338 and Tyr340/341 by p21-activated kinases (PAK) and Src kinases, respectively, is essential to enhance its kinase activity. Raf-1 activation results in the phosphorylation of the p65 subunit of the nuclear transcription factor (NF)-κB at Ser276, which leads to the acetylation of p65 by two histone acetyltransferases.[52] NF-κB activity is then prolonged and increases the transcription rate at the IL-10 anti-inflammatory cytokine promoter. However, Raf-1 signalling alone does not induce cytokine expression. The p65 subunit translocates as a dimer with p50 to the nucleus only after DC activation by other PRRs such as TLRs. Thus, NF-κB subunits p65–p50 initially dedicated to the transcription of the pro-inflammatory IL-12p35, IL-12p40, IL-6 and TNF-α cytokine-coding genes are reoriented on anti-inflammatory promoter targets, resulting in the decrease of these cytokines to the benefit of IL-10.[52] Another DC-SIGN ligand, the immunomodulatory protein Salp15, induces a different signalling pathway through Raf-1 that does not lead to p65 acetylation but instead to the mitogen-activated protein kinase kinase (MEK) activation.[53] Likewise, Salp15 regulates TLR-induced production of pro-inflammatory cytokines by DCs and DC-induced T cell activation.

It has been proposed that DC-SIGN can discriminate among mannosylated and fucosylated ligands and modulate the TLR signalling into a pro- or anti-inflammatory response accordingly.[51,54] However, that statement remains contradictory with the set of data obtained using natural mannosylated ligands from pathogens like the mannose-capped lipoarabinomannan (ManLAM)[55–57] from *Mycobacterium tuberculosis*, intact pathogens[52] such as *Mycobacterium leprae*, *Candida albicans*, measles virus and human immunodeficiency virus (HIV)-1 or synthetic mannosylated ligands (mannodendrimers)[56] engaging DC-SIGN and leading to anti-inflammatory immune responses with low IL-12, IL-6 and increased IL-10 production. Nonetheless, Lewis^{x}-expressing pathogens such as *Helicobacter pylori* and *Schistosoma mansoni* were shown to use a different signalling route where LSP1 is not paired with the KSR1-CNK-Raf-1 complex.[58] Subsequently, LSP1 activation by MK2-mediated phosphorylation allows the recruitment of the kinase IKKε and the deubiquitinase CYLD, leading to the nuclear accumulation of the transcription factor Bcl3. This results in an increase of IL-10 and a down-regulation of IL-12p35, IL-12p40, IL-6 and IL-23 cytokine production.

4.1.5 Murine DC-SIGN Homologues

Unlike humans, mice have eight DC-SIGN homologues clustered within the same genomic region. Seven genes, *Signr1–5* and *Signr7–8*, code for seven proteins whereas the DC-SIGN-related protein 6 (SIGNR6) is considered a pseudogene. Based on amino acid sequence and expression patterns, it remains difficult to assign human DC-SIGN properties to one murine homologue. In 2006, Powlesland *et al.* described the structural features and carbohydrate binding properties of DC-SIGN homologues.[59] Among the seven proteins, SIGNR2 and -4 were rapidly excluded from the shortlist. Indeed, SIGNR2 is a soluble protein containing only two copies of the neck domain and preferring *N*-acetylglucosamine containing ligands over mannose and fucose units. Its role is unclear as it lacks accessory domains that could initiate cell signalling pathways. SIGNR4 is a small membrane protein with a short cytoplasmic tail unable to bind mannose residues. As a matter of fact, SIGNR4 is devoid of the key residues that chelate the Ca^{2+} ion in the sugar binding domain. Yet the *Signr4* gene is overexpressed in mice lungs after allergen exposure, hinting at a role in the inflammation process.[60,61] SIGNR5 was initially proposed as the mouse orthologue of human DC-SIGN. However, SIGNR5 interacts with few mannosylated ligands and does not recognize the DC-SIGN mannosylated ligands of *Mycobacterium tuberculosis* (see Section 4.3)[62] or fucose units. It is a short membrane-anchored protein that lacks endocytic activity, contrasting with DC-SIGN. Little is known about SIGNR7 and SIGNR8 apart from their preferential binding with high affinity to 6′-sulfo-sialyl Lewis^x and fucosylated oligosaccharides, respectively. Both proteins present a wide range tissue expression in mice. In the end, SIGNR1 and SIGNR3 are the mice proteins the most likely similar to DC-SIGN and have several comparable properties. SIGNR1 possesses four repeats in the neck region, allowing it self-association, whereas SIGNR3 is a short membrane-anchored protein with only one neck domain unable to form oligomers in solution. Despite diverse structural features, both receptors bind fucose and mannose units with a similar efficiency to DC-SIGN,[59,63] recognize mycobacterial cell envelope mannosylated compounds[62] and show endocytic activity. They are expressed on several cell subpopulations of DCs and macrophages in the dermis, lymph nodes and spleen. In addition, SIGNR3 is expressed on monocytes from blood and bone marrow,[64] and SIGNR1 on lymphoid endothelium, DCs in the lymph node medulla and peritoneal macrophages and immature DC-like cells.[65–67] The involvement in pathogen recognition and specificity of each homologue is still an open question, although the area covered by scientific communications investigating their immunological roles in mice has been growing recently. SIGNR1 recognizes mannans from *Candida albicans* and *Saccharomyces cerevisiae*,[68] the lipopolysaccharide of Gram-negative bacteria[69] and inactivated influenza virus.[67] In the latest case, SIGNR1 is essential for the capture of lymph-borne influenza virus and for humoral immune response. SIGNR3 is expressed in mice lungs during *M. tuberculosis* infection,[62] favours parasite resilience during *Leishmania infantum* pathogenesis[70] and interacts with intestinal fungi in commensal microbiota, thereby increasing colitis inflammation in SIGNR3$^{-/-}$ mice.[71]

4.2 *Mycobacterium tuberculosis* Recognition by DC-SIGN

Mycobacterium tuberculosis, the causative agent of human tuberculosis, is one of the most successful pathogens worldwide. It has evolved extensive mechanisms to evade eradication by the immune system. Most particularly, it has the ability to parasitize and manipulate phagocytic cells of its human host, macrophages and DCs.[72] It is now recognized that the phagocyte CLRs, MR and DC-SIGN play key roles in this process.[72,73] Non-opsonized *M. tuberculosis* bacilli generally use complement receptor (CR)-3 and MR to bind to and enter macrophages. However, surprisingly, DC-SIGN is the major receptor of *M. tuberculosis* on the surface of DCs, although CR3 and MR are also expressed by these cells.[74]

4.2.1 Molecular Basis and Mechanisms

Deciphering the molecular bases of *M. tuberculosis* binding to DC-SIGN revealed an underestimated complexity. In addition to be the *M. tuberculosis* major internalization receptor on the DC surface,[74] DC-SIGN selectively binds to the mycobacterial species of the *M. tuberculosis* complex.[75] Indeed, only the closest relatives of *M. tuberculosis* are recognized by DC-SIGN, whereas other mycobacterial species, either pathogenic or non-pathogenic, are not.

Among carbohydrates, DC-SIGN has a high affinity for mannose-rich oligosaccharides (see Section 4.1.2). The envelope of mycobacteria, and most particularly its cell surface, is exceptionally rich in mannoconjugates, including glycolipids, lipoglycans and glycoproteins.[76–79] Most of these mannoconjugates, which bear α-(1$\rightarrow$2)-oligomannosides, have been found to be able to interact with DC-SIGN as purified molecules.[74,75,80,81] They include the mannose-capped lipoarabinomannan (ManLAM), lipomannan (LM), phosphatidylinositol hexamannosides (PIM_6) and mannoproteins.[77,79,82] α-Glucan, a major polysaccharide of the cell surface, was also described as binding DC-SIGN.[83] To investigate the role played by these putative ligands on the binding of live *M. tuberculosis* bacilli to DC-SIGN, knock-out mutants have been tested. However, mutants deficient for either ManLAM,[84] PIM_6,[85] mannoproteins[79] or both ManLAM and PIM_6[85] bind DC-SIGN as efficiently as the wild-type strains. Even if a mutant deficient for α-glucan production has still to be evaluated, all the data converge to indicate that DC-SIGN ligands are most probably redundant at the *M. tuberculosis* cell surface. To test this hypothesis, multiple knock-out mutants are currently being investigated. Since DC-SIGN is thought to modulate the immune response to *M. tuberculosis*,[72,77,81] possibly favouring its survival inside the infected host, it is tentative to speculate that the pathogen has developed a redundant arsenal of ligands to ensure recognition by the receptor.

However, another level of complexity is at play because none of these DC-SIGN ligands are specific to the *M. tuberculosis* complex species. Indeed, all the mycobacterial species produce mannoconjugates that bind DC-SIGN very well as purified molecules.[75] Accordingly, we found that a recombinant soluble tetrameric form of DC-SIGN (extracellular domain; sDC-SIGN) recognizes all the mycobacterial species tested (unpublished). We thus examined the difference of selectivity between sDC-SIGN and cell

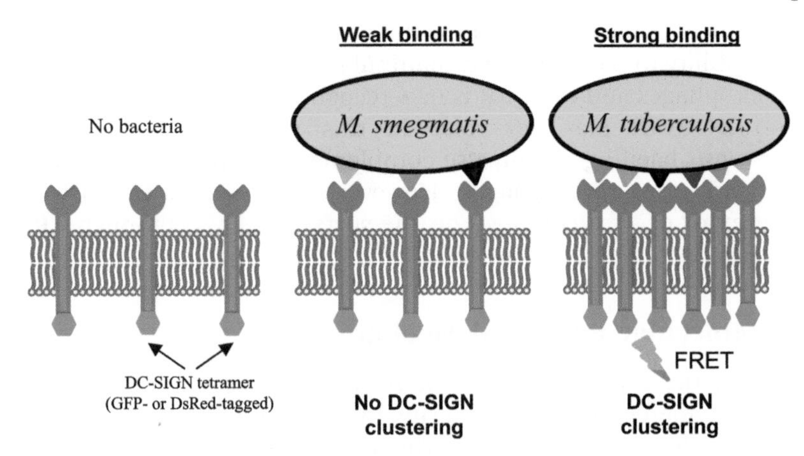

Figure 4.2 The selective binding of *M. tuberculosis* complex species to DC-SIGN relies on the unique ability of these species to induce DC-SIGN clustering at the host cell surface. An *M. tuberculosis* complex species, but not *M. smegmatis*, a fast growing mycobacterial species, induces a FRET signal in HEK cells expressing DC-SIGN tagged with either GFP or DsRed.

surface-expressed, membrane-anchored, DC-SIGN (mDC-SIGN). It has been reported that mDC-SIGN clusters in nanodomains and such clustering is thought to improve the binding avidity by multiplying simultaneous interactions (see Section 4.1.2).[22] Mimicking these nanodomains *in vitro* by a high-density coating of sDC-SIGN on microplates restored the selective recognition by the receptor of the *M. tuberculosis* complex species *versus* other mycobacterial species (unpublished). Interestingly, this correlated with the selective ability of *M. tuberculosis* complex species to induce DC-SIGN clustering in living cells, as demonstrated by Förster resonance energy transfer (FRET) experiments (unpublished).

Altogether, the selective binding of *M. tuberculosis* complex species to DC-SIGN seems to rely on the unique ability of these species to induce DC-SIGN clustering at the host cell surface (Figure 4.2). However, the underlying molecular bases are still unclear. *M. tuberculosis* complex species expose redundant DC-SIGN ligands at their cell envelope surface, but none of these ligands is specific for these species. Whether these ligands show a specific distribution, organization or mobility at the surface of the *M. tuberculosis* complex species envelope still needs to be investigated.

4.2.2 Impact on Host Immune Response

Macrophages rather than DCs are the main cell target for *M. tuberculosis*. Despite representing less than 1% of the total cells in the lung,[86] DCs capture and internalize *M. tuberculosis* and drive the development of acquired cellular immune response.[87] As mentioned above, DC-SIGN is the major receptor for the entry of mycobacteria into DCs.[74] Upon DC-SIGN-mediated internalization into DCs, mycobacteria persist in an immature vacuole, which lacks phagosome acidification, as observed in macrophages. However, in DCs only, the vacuole appears to be disconnected from the endocytic and exocytic pathways, a possible consequence of DC-SIGN-mediated entrance.[72,88] As discussed above in Sections 4.1.2.3 and 4.1.4, DC-SIGN signalling can alter TLR-mediated activation of DCs *in vitro*. One can speculate that mycobacteria may target this mechanism to interfere with DC function. Indeed, *M. tuberculosis* ManLAM binding to DC-SIGN prevents LPS-induced DC maturation, increases the LPS-induced secretion of the anti-inflammatory cytokine IL-10[81] and decreases the LPS-induced secretion of the pro-inflammatory cytokines IL-12[55] and TNF-α.[56,57] Macrophages infected with *M. tuberculosis* can release ManLAM,[89,90] which could be interpreted as an attempt by the bacilli to interfere with the activation of bystander DCs.[55,91,92] The block of DC maturation and increased production of IL-10 might contribute to the immune evasion of *M. tuberculosis* as immature and IL-10-treated DCs are less efficient at stimulating a specific $CD4^+$ cell response and can induce antigen-specific tolerance.[91,93]

In addition to being expressed by immature DCs, DC-SIGN was also shown to be induced in alveolar macrophages of patients with tuberculosis, and, notably, DC-SIGN-expressing macrophages constitute a preferential target for the bacillus.[4] Thus, DC-SIGN may contribute not only to the development of an acquired immunity to tuberculosis mediated by DCs, but also influence the entry and colonization of the bacilli in the host alveolar macrophages.

In the case of HIV-1, binding to DC-SIGN is considered to be clearly beneficial for the pathogen. However, the situation for mycobacteria is still unclear.[94,95] Although the interaction of *M. tuberculosis* with DC-SIGN might have an immunosuppressive outcome, which initially was viewed as a mechanism to promote pathogen survival, this suppression might be beneficial for the host. Indeed, while an inflammatory response is necessary to control *M. tuberculosis* infection, excessive inflammation is deleterious for the host and therefore needs to be strictly regulated. This question was addressed by using the mouse model of tuberculosis. Of the DC-SIGN functional homologues in mice, SIGNR1 and SIGNR3 (see Section 4.1.5), only SIGNR3 appears to have a role in the resistance to tuberculosis.[62,96] SIGNR3-deficient mice are more susceptible to infection with *M. tuberculosis*.[62] However, these mice did not die earlier than wild-type mice and their adaptive immune response was intact, suggesting that SIGNR3 function may be restricted to the early immune response, when it collaborates with TLR2 to induce inflammatory cytokines.[62] Thus, the mouse homologue of DC-SIGN appears to have a protecting role during *M. tuberculosis* infection. This view was corroborated by another study which reported that transgenic mice expressing DC-SIGN under the control of the CD11c promoter showed prolonged survival during a challenge with *M. tuberculosis*.[97] The beneficial effect of DC-SIGN was related to decreased tissue damage and not to a direct antibacterial mechanism, as the authors did not find differences in bacterial proliferation. These results suggest that instead of favouring immune evasion of the pathogen, DC-SIGN interaction with mycobacteria may be beneficial for the host by limiting *M. tuberculosis* induced pathology.

The contradiction of the results found in mice with the observations obtained with isolated cells, which suggested that DC-SIGN engagement would have an anti-inflammatory effect, may be explained by the use of LPS in these experiments.[98] Indeed, *M. tuberculosis* is a potent agonist of TLR2, both *in vitro*[99] and *in vivo*,[100] but not of TLR4, and, as such, the experiments with LPS may not reflect the signalling events occurring during an infection with *M. tuberculosis*.

Epidemiological studies have shown that polymorphisms in *CD209*, the gene coding for DC-SIGN, may be associated with susceptibility to tuberculosis. However, different studies have given conflicting results on the benefits of increasing or decreasing DC-SIGN expression. One study on a South African cohort suggested that a single nucleotide polymorphism (SNP) in the promoter region that leads to increased DC-SIGN expression is protective against tuberculosis.[101] Identical results were found in a cohort of Iranian individuals.[102] In contrast, studies in individuals from sub-Saharan Africa[103] and Brazil[104] showed that the variant allele of the same SNP associated with lower DC-SIGN expression is protective against tuberculosis. Others studies on Colombian,[105] Tunisian[106] and Chinese[107] cohorts did not find a correlation of this SNP with tuberculosis. These results can reflect differences in the origin and genetic background of the populations included in these reports and further genetic studies may help to clarify the impact of DC-SIGN expression in human tuberculosis.

4.3 Mimicking a Pathogen Strategy to Design Innovative Anti-inflammatory Molecules

We have previously shown that *M. tuberculosis* exposes a surface lipoglycan at its cell envelope,[78] namely ManLAM, which inhibits the production of pro-inflammatory cytokines IL-12 and TNF-α by LPS-stimulated human DCs,[55,57] *via* binding to DC-SIGN[80,81] (see Section 4.2.2; Figure 4.3). The strategy used by *M. tuberculosis* to undermine the host inflammatory response prompted us to design synthetic molecules that mimic the bioactive supramolecular structure of ManLAM with the objective of developing anti-inflammatory molecules (Figure 4.3).[56]

4.3.1 Molecular Basis of Mannosylated Lipoarabinomannan Anti-inflammatory Activity

ManLAM is a heterogeneous complex amphipathic macromolecule with an average molecular weight of 17 kDa that is composed of three domains: a mannosyl-phosphatidyl-*myo*-inositol (MPI) anchor, a heteropolysaccharidic core composed of D-mannan and D-arabinan, and mannose caps consisting of mono-, α-(1→2)-di- and α-(1→2)-trimannosides.[82] The immunomodulatory

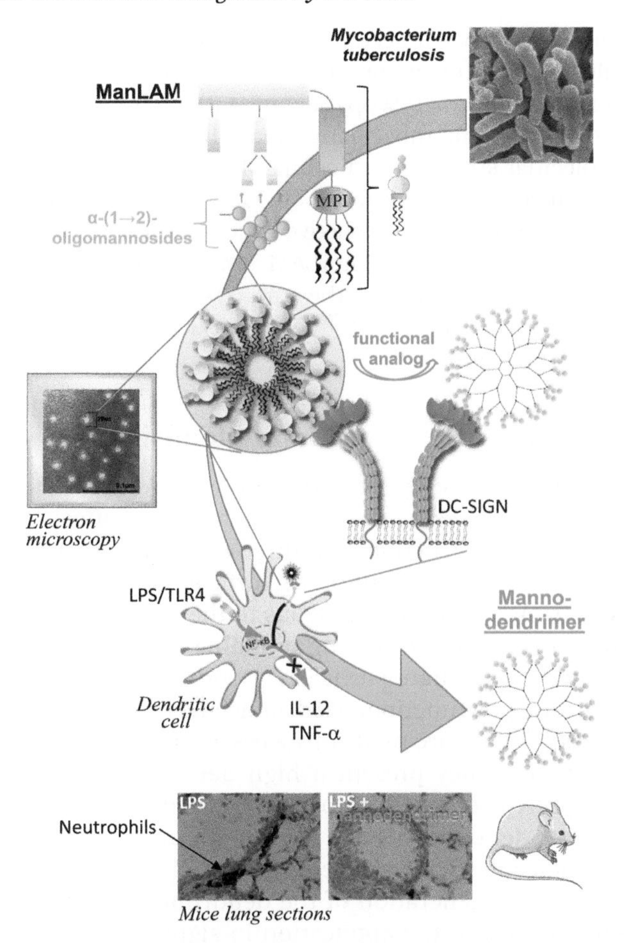

Figure 4.3 Mimicking a pathogen strategy to design innovative anti-inflammatory molecules. *M. tuberculosis* exposes a surface lipoglycan at its cell envelope, namely ManLAM, which inhibits the production of pro-inflammatory cytokines IL-12 and TNF-α by LPS-stimulated human DC, *via* binding to DC-SIGN. We designed synthetic molecules, namely mannodendrimers, that mimic the bioactive supramolecular structure of ManLAM with the objective of developing anti-inflammatory molecules. The third generation mannodendrimer capped with α-(1→2)-trimannopyranosides (3T) is a powerful anti-inflammatory compound that reduces lung accumulation of neutrophils in a mouse model of acute lung inflammation. Mannodendrimer 3T therefore represents an innovative fully synthetic compound for the treatment of lung inflammatory diseases.

properties of ManLAM rely on the presence of both the mannose caps and the fatty acids on the MPI anchor.[55,57] If the importance of the mannose caps for the interaction with DC-SIGN is obvious, the precise role of the fatty acids was puzzling until the demonstration that these fatty acids are responsible for the formation of a supramolecular structure of ManLAM. Indeed, MPI anchor fatty acids induce a supramolecular organization of ManLAM in aqueous solution, resulting in the formation of a 30 nm spherical structure, as observed by transmission electron microscopy (Figure 4.3), composed of approximately 450 molecules with the mannose caps exposed at the surface.[108] This multivalent supramolecular structure is presumably the bioactive form of ManLAM, as the critical micellar concentration (CMC) measured for its formation perfectly fits with the minimum bioactive concentration of ManLAM.[55,57,108] It allows multipoint attachment of ManLAM, *via* mannose caps, to multimeric DC-SIGN receptors[13,19] expressed at the surface of DCs, thereby ensuring high affinity binding to the receptor[13,19] and induction of anti-inflammatory activity[55,57,77,108] (see Figure 4.3).

4.3.2 Chemical Synthesis of Mannodendrimers

Mimicking the supramolecular structure of ManLAM in order to obtain a synthetic molecule harbouring the anti-inflammatory effects of the natural compound required construction of a three-dimensional spherical scaffold of a suitable size, and then to graft on its surface synthetic oligomannosides (mono-, di- and trimannosides) analogous to the mannose caps of ManLAM for interaction with DC-SIGN. Among available scaffolds were found polymers, dendrimers and nanoparticles. Dendrimers are monodisperse, hyperbranched and polyfunctionalized molecules obtained by an iterative chemical process.[109] They are made of successive identical "layers", carrying branching points and they present a high density of end groups at their surface, each layer corresponding to a generation. Compared to polymers and nanoparticles, dendrimers have two main crucial advantages: they are monodisperse molecules with a precisely defined structure, and their size, which is related to the generation of the dendrimer, can be easily adapted, within certain margins, to the application in sight. Furthermore, they offer some flexibility during their synthesis, allowing the incorporation of various markers on the dendrimer skeleton (a fluorescent dye, for example) and making available bifunctional compounds.[110] Among known and available dendrimers, we chose the poly(phosphorhydrazone) scaffold with a central N_3P_3 core for its known biocompatibility, its easy preparation and the possibility to introduce various chemical groups at the end of the branches.[111] Their size is also known with good precision: a fifth generation poly(phosphorhydrazone) dendrimer has been shown by DOSY NMR experiments to have a diameter around 8 nm.[112]

The oligomannosides were assembled using a linear sequence of glycosylation/deprotection,[113] with either thiomannoside or trichloroacetimidate groups as active groups on the anomeric position of the mannosidic glycosyl donor (compounds **2** and **3**, Scheme 4.1). As α-(1$\rightarrow$2) linkages are present in the caps of ManLAM, a participating group was introduced on the 2-position of the mannosyl donors **2** and **3** to secure the stereochemical outcome of the glycosylation reactions as α, while positions 3, 4 and 6 were protected by benzyl groups during all the synthesis. The mannosyl donors **2** and **3** were

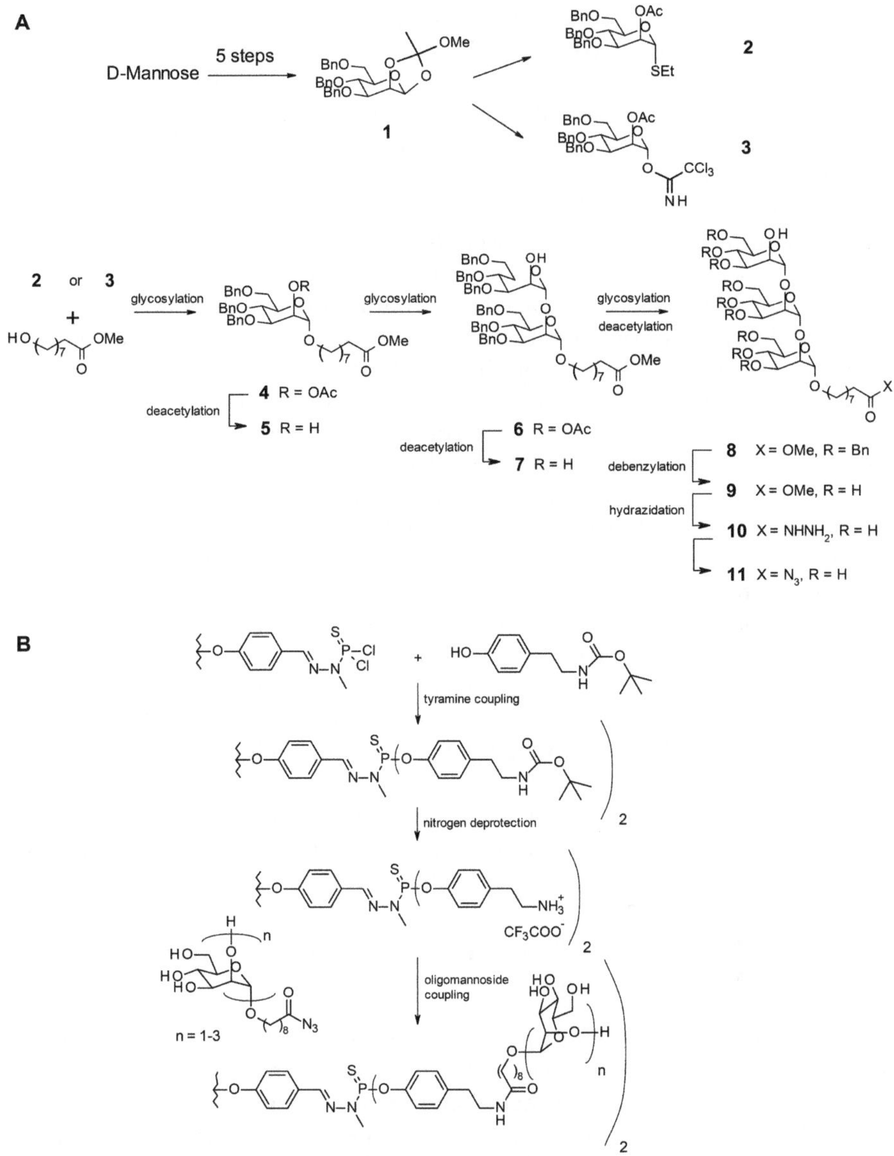

Scheme 4.1 Synthesis of mannodendrimers. (A) Elaboration of the oligomannosides. (B) Coupling of the oligomannosides to the poly(phosphorhydrazone) dendrimers [only one terminal $P(=S)Cl_2$ group of the dendrimer is represented].

obtained in one or two steps from orthoester **1**, available in five steps from D-mannose using optimized described procedures.[114]

Finally, to allow some flexibility of the glycosidic ligands at the dendrimer surface for their efficient interaction with clustered DC-SIGN, a nine-carbon-long linker was introduced between the end groups of the dendrimer and the oligomannosides. Once the desired oligomannosides, the mono- (**4**), di- (**6**) and trimannosides (**8**), had been elaborated, they were fully deprotected in two steps before coupling to the dendrimer.

Our first mannosylated dendrimer was obtained by coupling a hydrazide (**10**) with a dendrimer carrying aldehyde groups at its surface. Although the chemical efficiency of this coupling was very high (total substitution of the surface groups of the dendrimer in one step, using only a small excess of mannoside moiety), the products were found totally insoluble in water after purification. As water solubility was mandatory for biological evaluation of the mannodendrimers, we replaced this coupling reaction by the formation of a more classical amide bond between an amino-derived dendrimer and an acyl azide **11**, easily obtained from hydrazide **10** (Scheme 4.1). This coupling reaction is well documented for the preparation of modified proteins.[115,116] Despite its lower efficiency, satisfactory yields of mannodendrimers were obtained after three coupling cycles. All products were purified by size-exclusion chromatography and were characterized by ^{1}H, ^{13}C and ^{31}P NMR. Using mono-, di- and trimannosidic building blocks and dendrimers of various generations (from generation 1 to 4), we obtained a family of mannodendrimers, which differed by their size, density and the number of oligomannosides at their surface.[56] Mannodendrimers were named for simplification purposes according to their generation (1 to 4) and to the structure of the caps: mono- (M), di- (D) or trimannosides (T). The dendrimer of the third generation substituted by trimannosides was called **3T** and its partial structure is depicted in Figure 4.4A. Homogeneity and polydispersity of the mannodendrimers could be assessed using analytical size-exclusion chromatography with multi-angle light-scattering detection. Figure 4.4B shows the chromatograph of **3T** using this technique. This dendrimer was found to be pure and its measured polydispersity to be 1.17 ± 0.33.

4.3.3 *In Vitro* and *In Vivo* Anti-inflammatory Activity of Mannodendrimers

The synthesis of mannodendrimers was oriented towards the most active molecules as identified by a series of bio-assays with increasing stringency:[56] ability to (i) bind a recombinant soluble tetrameric form of DC-SIGN (sDC-SIGN); (ii) bind the full-length membrane-expressed DC-SIGN receptor (mDC-SIGN) using stably transfected HEK cells; (iii) bind DC-SIGN on the surface of human monocyte-derived DCs; and (iv) inhibit pro-inflammatory cytokine production by LPS-stimulated DCs (Figure 4.3). The dendrimers of the third generation and beyond that were capped with mono-, di- or tri-mannoside caps were the best ligands for sDC-SIGN. However, the mannose cap length became a critical parameter for binding to mDC-SIGN. Interestingly, the third generation mannodendrimers carrying α-(1→2)-di- (**3D**) or -trimannoside (**3T**) units (with a total of 48 caps) bound to mDC-SIGN as efficiently as the ManLAM bioactive particle. The higher avidity of di- and trimannoside-capped dendrimers *versus* single mannose-capped is in agreement with the preference of DC-SIGN for oligo- rather than mono-mannosides (see Section 4.1.2). The dendrimer generation has an impact on both their size and valency. The high functionalization of third generation dendrimers with 48 mannosylated caps (Figure 4.4A) seems to provide optimum size and multipoint attachment to the receptor, since avidity was not improved by a fourth generation scaffold. However, the avidity of manno-dendrimers for DC-SIGN was not completely sufficient for prediction of their capacity to inhibit production of pro-inflammatory cytokines. Indeed, signalling *via* DC-SIGN required a dendrimer scaffold size of the third generation substituted by trimannosides (**3T**). Sub-micromolar concentrations

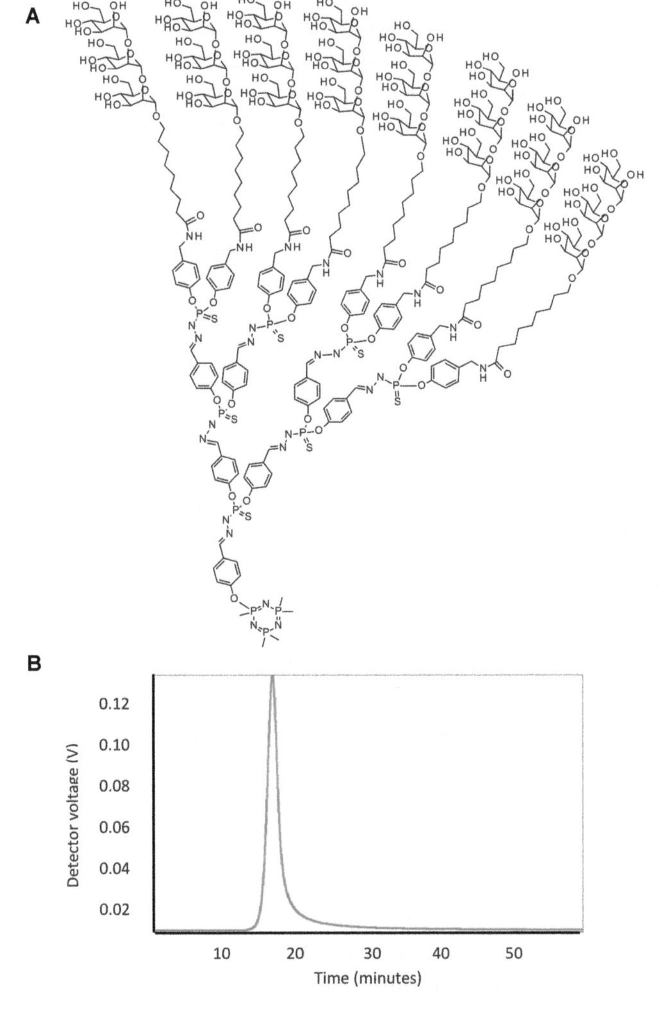

Figure 4.4 (A) Partial structure of mannodendrimer **3T**; only one out of the six branches present on the N_3P_3 core is represented. (B) Size-exclusion analytical chromatograph of **3T** with light-scattering detection.

of **3T** inhibited TLR4-mediated production of pro-inflammatory cytokines (TNF-α, IL-6, IL-8) and stimulated production of the anti-inflammatory cytokine IL-10 by LPS-stimulated DCs in a DC-SIGN-dependent manner.

Our most potent mannodendrimer, **3T**, was thus tested for its potency *in vivo* using a validated mouse model of acute lung inflammation, which involved exposure of mice to aerosolized LPS. TLR4 plays a critical role in the pathogenesis of inflammation and is a key receptor involved in the LPS-induced neutrophil accumulation into lungs that underlies pulmonary failure and leading to sepsis-related death.[117] Neutrophil sequestration into the lungs and associated inflammatory damage is believed to contribute to the pathogenesis of diverse lung diseases, including acute lung injury and the acute respiratory distress syndrome, chronic obstructive lung disease, and cystic fibrosis.[118] Oral delivery of **3T** mannodendrimer, as ManLAM, was found to prevent lung inflammation by reducing neutrophil recruitment (Figure 4.3) and TNF-α release in mice exposed to aerosolized LPS. This anti-inflammatory effect was dependent on the murine DC-SIGN homologue SIGNR1.

Mannodendrimer **3T** is thus a novel type of fully synthetic powerful anti-inflammatory molecule. Its mode of action is original and different from that of corticosteroids, which are classically used to fight inflammation, but are non-specific, immunosuppressive and can be deleterious during prolonged or high-dose therapy.[119] Mannodendrimer **3T** should now be tested in different pathologically models to determine the broader applicability of its therapeutic use.

References

1. T. B. Geijtenbeek, R. Torensma, S. J. van Vliet, G. C. van Duijnhoven, G. J. Adema, Y. van Kooyk and C. G. Figdor, *Cell*, 2000, **100**, 575–585.
2. T. B. Geijtenbeek, D. J. Krooshoop, D. A. Bleijs, S. J. van Vliet, G. C. van Duijnhoven, V. Grabovsky, R. Alon, C. G. Figdor and Y. van Kooyk, *Nat. Immunol.*, 2000, **1**, 353–357.
3. A. Engering, T. B. Geijtenbeek, S. J. van Vliet, M. Wijers, E. van Liempt, N. Demaurex, A. Lanzavecchia, J. Fransen, C. G. Figdor, V. Piguet and Y. van Kooyk, *J. Immunol.*, 2002, **168**, 2118–2126.
4. L. Tailleux, N. Pham-Thi, A. Bergeron-Lafaurie, J. L. Herrmann, P. Charles, O. Schwartz, P. Scheinmann, P. H. Lagrange, J. de Blic, A. Tazi, B. Gicquel and O. Neyrolles, *PLoS Med.*, 2005, **2**, e381.
5. A. Dominguez-Soto, E. Sierra-Filardi, A. Puig-Kroger, B. Perez-Maceda, F. Gomez-Aguado, M. T. Corcuera, P. Sanchez-Mateos and A. L. Corbi, *J. Immunol.*, 2011, **186**, 2192–2200.
6. N. Kamada, T. Hisamatsu, H. Honda, T. Kobayashi, H. Chinen, M. T. Kitazume, T. Takayama, S. Okamoto, K. Koganei, A. Sugita, T. Kanai and T. Hibi, *J. Immunol.*, 2009, **183**, 1724–1731.
7. M. T. Ochoa, A. Loncaric, S. R. Krutzik, T. C. Becker and R. L. Modlin, *J. Invest. Dermatol.*, 2008, **128**, 2225–2231.

8. E. J. Soilleux, L. S. Morris, B. Lee, S. Pohlmann, J. Trowsdale, R. W. Doms and N. Coleman, *J. Pathol.*, 2001, **195**, 586–592.
9. J. J. Garcia-Vallejo, J. M. Ilarregui, H. Kalay, S. Chamorro, N. Koning, W. W. Unger, M. Ambrosini, V. Montserrat, R. J. Fernandes, S. C. Bruijns, J. R. van Weering, N. J. Paauw, T. O'Toole, J. van Horssen, P. van der Valk, K. Nazmi, J. G. Bolscher, J. Bajramovic, C. D. Dijkstra, B. A. t Hart and Y. van Kooyk, *J. Exp. Med.*, 2014, **211**, 1465–1483.
10. A. Dominguez-Soto, A. Puig-Kroger, M. A. Vega and A. L. Corbi, *J. Biol. Chem.*, 2005, **280**, 33123–33131.
11. M. Relloso, A. Puig-Kroger, O. M. Pello, J. L. Rodriguez-Fernandez, G. de la Rosa, N. Longo, J. Navarro, M. A. Munoz-Fernandez, P. Sanchez-Mateos and A. L. Corbi, *J. Immunol.*, 2002, **168**, 2634–2643.
12. A. Puig-Kroger, D. Serrano-Gomez, E. Caparros, A. Dominguez-Soto, M. Relloso, M. Colmenares, L. Martinez-Munoz, N. Longo, N. Sanchez-Sanchez, M. Rincon, L. Rivas, P. Sanchez-Mateos, E. Fernandez-Ruiz and A. L. Corbi, *J. Biol. Chem.*, 2004, **279**, 25680–25688.
13. H. Feinberg, D. A. Mitchell, K. Drickamer and W. I. Weis, *Science*, 2001, **294**, 2163–2166.
14. K. Drickamer, *Curr. Opin. Struct. Biol.*, 1999, **9**, 585–590.
15. A. N. Zelensky and J. E. Gready, *FEBS J.*, 2005, **272**, 6179–6217.
16. Y. Guo, H. Feinberg, E. Conroy, D. A. Mitchell, R. Alvarez, O. Blixt, M. E. Taylor, W. I. Weis and K. Drickamer, *Nat. Struct. Mol. Biol.*, 2004, **11**, 591–598.
17. Q. D. Yu, A. P. Oldring, A. S. Powlesland, C. K. Tso, C. Yang, K. Drickamer and M. E. Taylor, *J. Mol. Biol.*, 2009, **387**, 1075–1080.
18. H. Feinberg, Y. Guo, D. A. Mitchell, K. Drickamer and W. I. Weis, *J. Biol. Chem.*, 2005, **280**, 1327–1335.
19. D. A. Mitchell, A. J. Fadden and K. Drickamer, *J. Biol. Chem.*, 2001, **276**, 28939–28945.
20. G. A. Snyder, M. Colonna and P. D. Sun, *J. Mol. Biol.*, 2005, **347**, 979–989.
21. S. Menon, K. Rosenberg, S. A. Graham, E. M. Ward, M. E. Taylor, K. Drickamer and D. E. Leckband, *Proc. Natl. Acad. Sci. U. S. A.*, 2009, **106**, 11524–11529.
22. A. Cambi, F. de Lange, N. M. van Maarseveen, M. Nijhuis, B. Joosten, E. M. van Dijk, B. I. de Bakker, J. A. Fransen, P. H. Bovee-Geurts, F. N. van Leeuwen, N. F. Van Hulst and C. G. Figdor, *J. Cell Biol.*, 2004, **164**, 145–155.
23. M. Koopman, A. Cambi, B. I. de Bakker, B. Joosten, C. G. Figdor, N. F. van Hulst and M. F. Garcia-Parajo, *FEBS Lett.*, 2004, **573**, 6–10.
24. B. I. de Bakker, F. de Lange, A. Cambi, J. P. Korterik, E. M. van Dijk, N. F. van Hulst, C. G. Figdor and M. F. Garcia-Parajo, *ChemPhysChem*, 2007, **8**, 1473–1480.
25. M. S. Itano, A. K. Neumann, P. Liu, F. Zhang, E. Gratton, W. J. Parak, N. L. Thompson and K. Jacobson, *Biophys. J.*, 2011, **100**, 2662–2670.
26. P. Liu, X. Wang, M. S. Itano, A. K. Neumann, A. M. de Silva, K. Jacobson and N. L. Thompson, *Traffic*, 2014, **15**, 179–196.

27. P. Liu, X. Wang, M. S. Itano, A. K. Neumann, K. Jacobson and N. L. Thompson, *Traffic*, 2012, **13**, 715–726.
28. A. K. Azad, J. B. Torrelles and L. S. Schlesinger, *J. Leukocyte Biol.*, 2008, **84**, 1594–1603.
29. A. Hodges, K. Sharrocks, M. Edelmann, D. Baban, A. Moris, O. Schwartz, H. Drakesmith, K. Davies, B. Kessler, A. McMichael and A. Simmons, *Nat. Immunol.*, 2007, **8**, 569–577.
30. V. Bogoevska, P. Nollau, L. Lucka, D. Grunow, B. Klampe, L. M. Uotila, A. Samsen, C. G. Gahmberg and C. Wagener, *Glycobiology*, 2007, **17**, 324–333.
31. K. P. van Gisbergen, I. S. Ludwig, T. B. Geijtenbeek and Y. van Kooyk, *FEBS Lett.*, 2005, **579**, 6159–6168.
32. J. J. Garcia-Vallejo, E. van Liempt, P. da Costa Martins, C. Beckers, B. van het Hof, S. I. Gringhuis, J. J. Zwaginga, W. van Dijk, T. B. Geijtenbeek, Y. van Kooyk and I. van Die, *Mol. Immunol.*, 2008, **45**, 2359–2369.
33. G. Malcherek, L. Mayr, P. Roda-Navarro, D. Rhodes, N. Miller and J. Trowsdale, *J. Immunol.*, 2007, **179**, 3804–3811.
34. M. A. Naarding, A. M. Dirac, I. S. Ludwig, D. Speijer, S. Lindquist, E. L. Vestman, M. J. Stax, T. B. Geijtenbeek, G. Pollakis, O. Hernell and W. A. Paxton, *Antimicrob. Agents Chemother.*, 2006, **50**, 3367–3374.
35. K. P. van Gisbergen, C. A. Aarnoudse, G. A. Meijer, T. B. Geijtenbeek and Y. van Kooyk, *Cancer Res.*, 2005, **65**, 5935–5944.
36. G. F. Clark, P. Grassi, P. C. Pang, M. Panico, D. Lafrenz, E. Z. Drobnis, M. R. Baldwin, H. R. Morris, S. M. Haslam, S. Schedin-Weiss, W. Sun and A. Dell, *Mol. Cell. Proteomics*, 2012, **11**, M111 008730.
37. R. M. Steinman and H. Hemmi, *Curr. Top. Microbiol. Immunol.*, 2006, **311**, 17–58.
38. S. Akira, S. Uematsu and O. Takeuchi, *Cell*, 2006, **124**, 783–801.
39. B. M. Curtis, S. Scharnowske and A. J. Watson, *Proc. Natl. Acad. Sci. U. S. A.*, 1992, **89**, 8356–8360.
40. W. Van Breedam, S. Pohlmann, H. W. Favoreel, R. J. de Groot and H. J. Nauwynck, *FEMS Microbiol. Rev.*, 2014, **38**, 598–632.
41. M. P. Bergman, A. Engering, H. H. Smits, S. J. van Vliet, A. A. van Bodegraven, H. P. Wirth, M. L. Kapsenberg, C. M. Vandenbroucke-Grauls, Y. van Kooyk and B. J. Appelmelk, *J. Exp. Med.*, 2004, **200**, 979–990.
42. B. Evrard, D. Balestrino, A. Dosgilbert, J. L. Bouya-Gachancard, N. Charbonnel, C. Forestier and A. Tridon, *Infect. Immun.*, 2010, **78**, 210–219.
43. C. M. van Stijn, S. Meyer, M. van den Broek, S. C. Bruijns, Y. van Kooyk, R. Geyer and I. van Die, *Mol. Immunol.*, 2010, **47**, 1544–1552.
44. L. Steeghs, S. J. van Vliet, H. Uronen-Hansson, A. van Mourik, A. Engering, M. Sanchez-Hernandez, N. Klein, R. Callard, J. P. van Putten, P. van der Ley, Y. van Kooyk and J. G. van de Winkel, *Cell. Microbiol.*, 2006, **8**, 316–325.

45. E. van Liempt, S. J. van Vliet, A. Engering, J. J. Garcia Vallejo, C. M. Bank, M. Sanchez-Hernandez, Y. van Kooyk and I. van Die, *Mol. Immunol.*, 2007, **44**, 2605–2615.
46. H. H. Smits, A. Engering, D. van der Kleij, E. C. de Jong, K. Schipper, T. M. van Capel, B. A. Zaat, M. Yazdanbakhsh, E. A. Wierenga, Y. van Kooyk and M. L. Kapsenberg, *J. Allergy Clin. Immunol.*, 2005, **115**, 1260–1267.
47. Y. van Kooyk and T. B. H. Geijtenbeek, *Nat. Rev. Immunol.*, 2003, **3**, 697–709.
48. T. B. Geijtenbeek, D. S. Kwon, R. Torensma, S. J. van Vliet, G. C. van Duijnhoven, J. Middel, I. L. Cornelissen, H. S. Nottet, V. N. KewalRamani, D. R. Littman, C. G. Figdor and Y. van Kooyk, *Cell*, 2000, **100**, 587–597.
49. D. S. Kwon, G. Gregorio, N. Bitton, W. A. Hendrickson and D. R. Littman, *Immunity*, 2002, **16**, 135–144.
50. D. McDonald, L. Wu, S. M. Bohks, V. N. KewalRamani, D. Unutmaz and T. J. Hope, *Science*, 2003, **300**, 1295–1297.
51. S. I. Gringhuis, J. den Dunnen, M. Litjens, M. van der Vlist and T. B. Geijtenbeek, *Nat. Immunol.*, 2009, **10**, 1081–1088.
52. S. I. Gringhuis, J. den Dunnen, M. Litjens, B. van Het Hof, Y. van Kooyk and T. B. Geijtenbeek, *Immunity*, 2007, **26**, 605–616.
53. J. W. Hovius, M. A. de Jong, J. den Dunnen, M. Litjens, E. Fikrig, T. van der Poll, S. I. Gringhuis and T. B. Geijtenbeek, *PLoS Pathog.*, 2008, **4**, e31.
54. J. J. Garcia-Vallejo and Y. van Kooyk, *Trends Immunol.*, 2013, **34**, 482–486.
55. J. Nigou, C. Zelle-Rieser, M. Gilleron, M. Thurnher and G. Puzo, *J. Immunol.*, 2001, **166**, 7477–7485.
56. E. Blattes, A. Vercellone, H. Eutamene, C. O. Turrin, V. Theodorou, J. P. Majoral, A. M. Caminade, J. Prandi, J. Nigou and G. Puzo, *Proc. Natl. Acad. Sci. U. S. A.*, 2013, **110**, 8795–8800.
57. J. Nigou, M. Gilleron, M. Rojas, L. F. Garcia, M. Thurnher and G. Puzo, *Microbes Infect.*, 2002, **4**, 945–953.
58. S. I. Gringhuis, T. M. Kaptein, B. A. Wevers, A. W. Mesman and T. B. Geijtenbeek, *Nat. Commun.*, 2014, **5**, 3898.
59. A. S. Powlesland, E. M. Ward, S. K. Sadhu, Y. Guo, M. E. Taylor and K. Drickamer, *J. Biol. Chem.*, 2006, **281**, 20440–20449.
60. E. Di Valentin, C. Crahay, N. Garbacki, B. Hennuy, M. Gueders, A. Noel, J. M. Foidart, J. Grooten, A. Colige, J. Piette and D. Cataldo, *Am. J. Physiol.: Lung Cell. Mol. Physiol.*, 2009, **296**, L185–197.
61. K. Fredriksson, A. Mishra, J. K. Lam, E. M. Mushaben, R. A. Cuento, K. S. Meyer, X. Yao, K. J. Keeran, G. Z. Nugent, X. Qu, Z. X. Yu, Y. Yang, N. Raghavachari, P. K. Dagur, J. P. McCoy and S. J. Levine, *J. Immunol.*, 2014, **192**, 4497–4509.
62. A. Tanne, B. Ma, F. Boudou, L. Tailleux, H. Botella, E. Badell, F. Levillain, M. E. Taylor, K. Drickamer, J. Nigou, K. M. Dobos, G. Puzo, D. Vestweber, M. K. Wild, M. Marcinko, P. Sobieszczuk, L. Stewart, D. Lebus, B. Gicquel and O. Neyrolles, *J. Exp. Med.*, 2009, **206**, 2205–2220.
63. Y. Kawauchi, Y. Kuroda and N. Kojima, *Int. Immunopharmacol.*, 2014, **19**, 27–36.

64. K. Nagaoka, K. Takahara, K. Minamino, T. Takeda, Y. Yoshida and K. Inaba, *J. Leukocyte Biol.*, 2010, **88**, 913–924.
65. Y. Zhou, H. Kawasaki, S. C. Hsu, R. T. Lee, X. Yao, B. Plunkett, J. Fu, K. Yang, Y. C. Lee and S. K. Huang, *Nat. Med.*, 2010, **16**, 1128–1133.
66. Y. Kawauchi, M. Igarashi and N. Kojima, *Cell. Immunol.*, 2014, **287**, 121–128.
67. S. F. Gonzalez, V. Lukacs-Kornek, M. P. Kuligowski, L. A. Pitcher, S. E. Degn, Y. A. Kim, M. J. Cloninger, L. Martinez-Pomares, S. Gordon, S. J. Turley and M. C. Carroll, *Nat. Immunol.*, 2010, **11**, 427–434.
68. K. Takahara, T. Arita, S. Tokieda, N. Shibata, Y. Okawa, H. Tateno, J. Hirabayashi and K. Inaba, *Infect. Immun.*, 2012, **80**, 1699–1706.
69. K. Nagaoka, K. Takahara, K. Tanaka, H. Yoshida, R. M. Steinman, S. Saitoh, S. Akashi-Takamura, K. Miyake, Y. S. Kang, C. G. Park and K. Inaba, *Int. Immunol.*, 2005, **17**, 827–836.
70. L. Lefevre, G. Lugo-Villarino, E. Meunier, A. Valentin, D. Olagnier, H. Authier, C. Duval, C. Dardenne, J. Bernad, J. L. Lemesre, J. Auwerx, O. Neyrolles, B. Pipy and A. Coste, *Immunity*, 2013, **38**, 1038–1049.
71. M. Eriksson, T. Johannssen, D. von Smolinski, A. D. Gruber, P. H. Seeberger and B. Lepenies, *Front. Immunol.*, 2013, **4**, 196.
72. L. Tailleux, N. Maeda, J. Nigou, B. Gicquel and O. Neyrolles, *Trends Microbiol.*, 2003, **11**, 259–263.
73. J. D. Ernst, *Infect. Immun.*, 1998, **66**, 1277–1281.
74. L. Tailleux, O. Schwartz, J. L. Herrmann, E. Pivert, M. Jackson, A. Amara, L. Legres, D. Dreher, L. P. Nicod, J. C. Gluckman, P. H. Lagrange, B. Gicquel and O. Neyrolles, *J. Exp. Med.*, 2003, **197**, 121–127.
75. S. Pitarque, J. L. Herrmann, J. L. Duteyrat, M. Jackson, G. R. Stewart, F. Lecointe, B. Payre, O. Schwartz, D. B. Young, G. Marchal, P. H. Lagrange, G. Puzo, B. Gicquel, J. Nigou and O. Neyrolles, *Biochem. J.*, 2005, **392**, 615–624.
76. P. J. Brennan and H. Nikaido, *Annu. Rev. Biochem.*, 1995, **64**, 29–63.
77. M. Gilleron, M. Jackson, J. Nigou and G. Puzo, in *The Mycobacterial Cell Envelope*, ed. M. Daffe and J. Reyrat, ASM Press, Washington DC, 2008, pp. 75–105.
78. S. Pitarque, G. Larrouy-Maumus, B. Payre, M. Jackson, G. Puzo and J. Nigou, *Tuberculosis (Edinb)*, 2008, **88**, 560–565.
79. C. F. Liu, L. Tonini, W. Malaga, M. Beau, A. Stella, D. Bouyssie, M. C. Jackson, J. Nigou, G. Puzo, C. Guilhot, O. Burlet-Schiltz and M. Riviere, *Proc. Natl. Acad. Sci. U. S. A.*, 2013, **110**, 6560–6565.
80. N. Maeda, J. Nigou, J. L. Herrmann, M. Jackson, A. Amara, P. H. Lagrange, G. Puzo, B. Gicquel and O. Neyrolles, *J. Biol. Chem.*, 2003, **278**, 5513–5516.
81. T. B. Geijtenbeek, S. J. Van Vliet, E. A. Koppel, M. Sanchez-Hernandez, C. M. Vandenbroucke-Grauls, B. Appelmelk and Y. Van Kooyk, *J. Exp. Med.*, 2003, **197**, 7–17.
82. J. Nigou, M. Gilleron and G. Puzo, *Biochimie*, 2003, **85**, 153–166.

83. J. Geurtsen, S. Chedammi, J. Mesters, M. Cot, N. N. Driessen, T. Sambou, R. Kakutani, R. Ummels, J. Maaskant, H. Takata, O. Baba, T. Terashima, N. Bovin, C. M. Vandenbroucke-Grauls, J. Nigou, G. Puzo, A. Lemassu, M. Daffe and B. J. Appelmelk, *J. Immunol.*, 2009, **183**, 5221–5231.
84. B. J. Appelmelk, J. den Dunnen, N. N. Driessen, R. Ummels, M. Pak, J. Nigou, G. Larrouy-Maumus, S. S. Gurcha, F. Movahedzadeh, J. Geurtsen, E. J. Brown, M. M. Eysink Smeets, G. S. Besra, P. T. Willemsen, T. L. Lowary, Y. van Kooyk, J. J. Maaskant, N. G. Stoker, P. van der Ley, G. Puzo, C. M. Vandenbroucke-Grauls, C. W. Wieland, T. van der Poll, T. B. Geijtenbeek, A. M. van der Sar and W. Bitter, *Cell. Microbiol.*, 2008, **10**, 930–944.
85. N. N. Driessen, R. Ummels, J. J. Maaskant, S. S. Gurcha, G. S. Besra, G. D. Ainge, D. S. Larsen, G. F. Painter, C. M. Vandenbroucke-Grauls, J. Geurtsen and B. J. Appelmelk, *Infect. Immun.*, 2009, **77**, 4538–4547.
86. L. Cochand, P. Isler, F. Songeon and L. P. Nicod, *Am. J. Respir. Cell Mol. Biol.*, 1999, **21**, 547–554.
87. A. M. Cooper, *Annu. Rev. Immunol.*, 2009, **27**, 393–422.
88. L. Tailleux, O. Neyrolles, S. Honore-Bouakline, E. Perret, F. Sanchez, J. P. Abastado, P. H. Lagrange, J. C. Gluckman, M. Rosenzwajg and J. L. Herrmann, *J. Immunol.*, 2003, **170**, 1939–1948.
89. W. L. Beatty, E. R. Rhoades, H. J. Ullrich, D. Chatterjee, J. E. Heuser and D. G. Russell, *Traffic*, 2000, **1**, 235–247.
90. U. E. Schaible, F. Winau, P. A. Sieling, K. Fischer, H. L. Collins, K. Hagens, R. L. Modlin, V. Brinkmann and S. H. Kaufmann, *Nat. Med.*, 2003, **9**, 1039–1046.
91. Y. van Kooyk, B. Appelmelk and T. B. Geijtenbeek, *Trends Mol. Med.*, 2003, **9**, 153–159.
92. N. Dulphy, J. L. Herrmann, J. Nigou, D. Rea, N. Boissel, G. Puzo, D. Charron, P. H. Lagrange and A. Toubert, *Cell. Microbiol.*, 2007, **9**, 1412–1425.
93. H. Jonuleit, E. Schmitt, G. Schuler, J. Knop and A. H. Enk, *J. Exp. Med.*, 2000, **192**, 1213–1222.
94. J. den Dunnen, S. I. Gringhuis and T. B. Geijtenbeek, *Cancer Immunol. Immunother.*, 2009, **58**, 1149–1157.
95. S. Ehlers, *Eur. J. Cell Biol.*, 2010, **89**, 95–101.
96. C. W. Wieland, E. A. Koppel, J. den Dunnen, S. Florquin, A. N. McKenzie, Y. van Kooyk, T. van der Poll and T. B. Geijtenbeek, *Microbes Infect.*, 2007, **9**, 134–141.
97. M. Schaefer, N. Reiling, C. Fessler, J. Stephani, I. Taniuchi, F. Hatam, A. O. Yildirim, H. Fehrenbach, K. Walter, J. Ruland, H. Wagner, S. Ehlers and T. Sparwasser, *J. Immunol.*, 2008, **180**, 6836–6845.
98. A. Tanne and O. Neyrolles, *Virulence*, 2010, **1**, 285–290.
99. S. Jang, S. Uematsu, S. Akira and P. Salgame, *J. Immunol.*, 2004, **173**, 3392–3397.
100. N. Reiling, C. Holscher, A. Fehrenbach, S. Kroger, C. J. Kirschning, S. Goyert and S. Ehlers, *J. Immunol.*, 2002, **169**, 3480–3484.

101. L. B. Barreiro, O. Neyrolles, C. L. Babb, L. Tailleux, H. Quach, K. McElreavey, P. D. Helden, E. G. Hoal, B. Gicquel and L. Quintana-Murci, *PLoS Med.*, 2006, **3**, e20.
102. M. Naderi, M. Hashemi, M. Taheri, H. Pesarakli, E. Eskandari-Nasab and G. Bahari, *J. Microbiol., Immunol. Infect.*, 2014, **47**, 171–175.
103. F. O. Vannberg, S. J. Chapman, C. C. Khor, K. Tosh, S. Floyd, D. Jackson-Sillah, A. Crampin, L. Sichali, B. Bah, P. Gustafson, P. Aaby, K. P. McAdam, O. Bah-Sow, C. Lienhardt, G. Sirugo, P. Fine and A. V. Hill, *PLoS One*, 2008, **3**, e1388.
104. R. C. da Silva, L. Segat, H. L. da Cruz, H. C. Schindler, L. M. Montenegro, S. Crovella and R. L. Guimaraes, *Mol. Biol. Rep.*, 2014, **41**, 5449–5457.
105. L. M. Gomez, J. M. Anaya, E. Sierra-Filardi, J. Cadena, A. Corbi and J. Martin, *Hum. Immunol.*, 2006, **67**, 808–811.
106. M. Ben-Ali, L. B. Barreiro, A. Chabbou, R. Haltiti, E. Braham, O. Neyrolles, K. Dellagi, B. Gicquel, L. Quintana-Murci and M. R. Barbouche, *Hum. Immunol.*, 2007, **68**, 908–912.
107. R. Zheng, Y. Zhou, L. Qin, R. Jin, J. Wang, J. Lu, W. Wang, S. Tang and Z. Hu, *Hum. Immunol.*, 2011, **72**, 183–186.
108. M. Riviere, A. Moisand, A. Lopez and G. Puzo, *J. Mol. Biol.*, 2004, **344**, 907–918.
109. A. M. Caminade, C. O. Turrin, R. Laurent, A. Ouali and B. Delavaut-Nicot, *Dendrimers: towards catalytic, material and biomedical uses*, John Wiley & Sons, Chichester, 2011.
110. M. Sowinska and Z. Urbanczyk-Lipkowska, *New J. Chem.*, 2014, **38**, 2168–2203.
111. A.-M. Caminade, R. Laurent, M. Zablocka and J.-P. Majoral, *Molecules*, 2012, **17**, 13605–13621.
112. J. Leclaire, Y. Coppel, A.-M. Caminade and J.-P. Majoral, *J. Am. Chem. Soc.*, 2004, **126**, 2304–2305.
113. T. Peters, *Liebigs Ann. Chem.*, 1991, 135–141.
114. N. E. Franks and R. Montgomery, *Carbohydrate Res.*, 1968, **6**, 286–298.
115. J. K. Inman, B. Merchant, L. Claflin and S. E. Tacey, *Immunochemistry*, 1973, **10**, 165–174.
116. R. U. Lemieux, D. R. Bundle and D. A. Baker, *J. Am. Chem. Soc.*, 1975, **97**, 4076–4083.
117. G. Andonegui, C. S. Bonder, F. Green, S. C. Mullaly, L. Zbytnuik, E. Raharjo and P. Kubes, *J. Clin. Invest.*, 2003, **111**, 1011–1020.
118. R. L. Zemans, S. P. Colgan and G. P. Downey, *Am. J. Respir. Cell Mol. Biol.*, 2009, **40**, 519–535.
119. P. J. Barnes, *Br. J. Pharmacol.*, 2011, **163**, 29–43.

CHAPTER 5

Glyconanotechnology and Disease: Gold Nanoparticles Coated with Glycosides as Multivalent Systems for Potential Applications in Diagnostics and Therapy

MARCO MARRADI,*[a,b,†] FABRIZIO CHIODO[a,‡] AND ISABEL GARCÍA[a,b]

[a] Laboratory of GlycoNanotechnology, Biofunctional Nanomaterials Unit, CIC biomaGUNE, Paseo Miramón 182, 20009 San Sebastián, Spain;
[b] Biomedical Research Networking Center in Bioengineering, Biomaterials and Nanomedicine (CIBER-BBN), Paseo Miramón 182, 20009 San Sebastián, Spain
*Email: marcomarradi76@gmail.com

†Present address: Biomaterials Unit, Materials Division, IK4-CIDETEC, Paseo Miramón 196, 20009 Donostia-San Sebastián, Spain.
‡Present address: Department of Parasitology, Leiden University Medical Center, 2333 ZA Leiden, Netherlands.

RSC Drug Discovery Series No. 43
Carbohydrates in Drug Design and Discovery
Edited by Jesús Jiménez-Barbero, F. Javier Cañada and Sonsoles Martín-Santamaría

Published by the Royal Society of Chemistry, www.rsc.org

5.1 Introduction

5.1.1 Nanotechnology and Carbohydrates

Engineering materials at nanoscale (nanotechnology) is a rapidly expanding field that allows the development of novel systems for different applications.[1] One of the emerging areas is the generation of hybrid (inorganic/organic) nanomaterials for biomedical applications.[2,3] Among these systems, nanoparticles consisting of an inorganic core functionalised with biomolecules are of great interest because they combine the optical/electronic[4] and/or magnetic[5,6] characteristics of nano-sized nuclei with the properties of the (bio)organic shell.[7–9] Gold nanoparticles coated with biomolecules are probably the best known type of hybrid nanomaterials.[10–14] The ease of gold functionalisation by thiol chemistry[15] converts gold nanostructures into ideal scaffolds for the conjugation of biomolecules with targeting and/or therapeutic properties. Furthermore, gold nanoparticles are relatively inert and little toxic and can be potentially used not only in biosensing[16] but also in cellular imaging,[17] drug delivery[14] and therapy, as well as *in vivo* diagnostics.[11–13,18] Last, but not least, the control of gold nanoparticles morphology[19] and size allows the modulation of nanoparticle–light interactions (nanoplasmonics), with potential applications in photothermal therapy.

While the conjugation of peptides and nucleotides to gold nanoparticles has been broadly exploited for biomedical purposes, carbohydrates have received less attention in spite of their recognition properties and signalling functions in cells.[20] Both inside the cells and at their surface, carbohydrates are usually covalently conjugated to other biomolecules such as proteins and lipids (glycoproteins, proteoglycans and glycosphingolipids). These naturally occurring glycoconjugates play key roles in several biological events that involve the interaction between cells and their external environment, including physiological and pathological processes.[21] Owing to their structural diversity, the variability of the sugar units sequence and linkage points, the anomeric configuration, the substitution at different positions and with different chemical groups, and the interconversion of conformers, carbohydrates can store and transmit biological information through a complex "glyco-code" which is much bigger than that provided by amino acid and nucleotide sequences.[22] Furthermore, glycoconjugates usually appear as clusters in order to develop their biological function. This is because carbohydrate-based interactions, which comprise carbohydrate–protein[23] and carbohydrate–carbohydrate[24] interactions, are generally weak and need "multivalency" to be strengthened.[25] The renewed interest in the therapeutic applications of glycans[26] has promoted a symposium between nanotechnology and glycoscience. Glyconanotechnology was originally defined as "an integrated approach that included both chemical models mimicking glycosphingolipid patches at the cell membrane and new surface analytical techniques to evaluate their interactions"[27] and the application of this technology to different fields has been pursued in the last decade.[28]

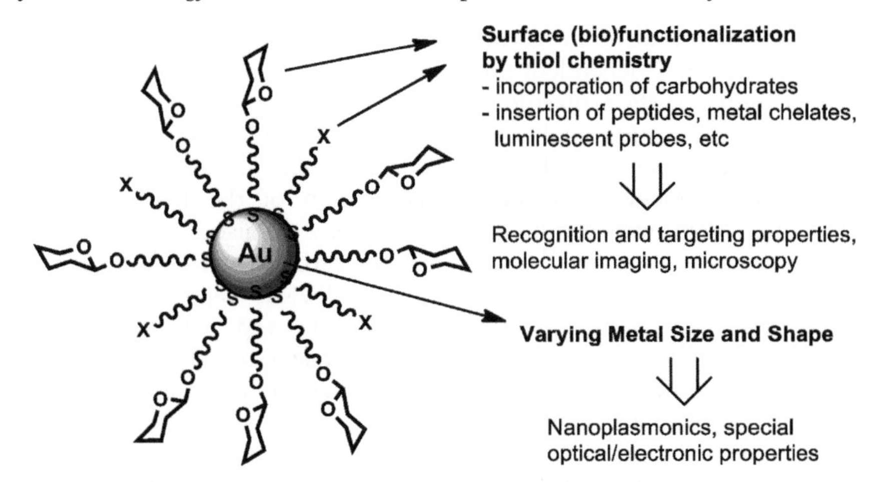

Figure 5.1 Gold glyconanoparticles (GNPs): hybrid nanomaterials that combine the properties of the gold nano-sized nucleus with the properties of the biomolecule-functionalised shell.

Gold nanoparticles coated with defined glycosides (gold glyconanoparticles, GNP)[29] combine the multivalent presentation of carbohydrates (glycoclusters) with the special chemical and physical properties of the nanosized metallic core (Figure 5.1). The opportunity of tailoring different types of carbohydrates and other molecules (such as luminescent probes, peptides and metal chelates) onto the same gold nanoparticles makes these multivalent glyco-scaffolds suitable for carrying out studies on carbohydrate-mediated interactions and for translational science.[30] This chapter surveys the recent advances in the applications of gold nanoparticles functionalised with carbohydrates by thiol chemistry. We will focus on GNPs as antipathogenic agents and their potential use in diagnostics and therapy. Before going through these sections, a brief overview on the origin of GNPs is presented in order to understand why gold glyconanotechnology was designed and developed.

5.1.2 The Seminal Idea of Gold Glyconanotechnology

The seminal work with GNPs was presented to the scientific community in 2001 by Penadés' group.[31] Inspired by the Brust–Schiffrin method for the functionalisation of gold nanoparticles with alkanethiolates, water-dispersible 2 nm gold nanoparticles coated with β-glycosides of lactose [Lac, β-D-Gal*p*-(1→4)-β-D-Glc*p*-(1→)] or Lewis X trisaccharide antigen [Le^{X}, α-L-Fuc*p*-(1→3)-β-D-Gal*p*-(1→4)-β-D-Glc*p*NAc-(1→)] were prepared *in situ* by reducing an Au(III) salt with an excess of sodium borohydride in the presence of thiol-ending neoglycoconjugates. GNPs were used as synthetic and multivalent systems to mimic the globular carbohydrate display at the cell surface and worked as a chemical model to corroborate Hakomori's[32] hypothesis that carbohydrate–carbohydrate interactions in water are multivalent, specific, Ca^{2+}-dependent and reversible. To that aim, different techniques including surface plasmon resonance (SPR), transmission electron microscopy (TEM), isothermal titration calorimetry (ITC) and atomic force microscopy (AFM) were employed.[27] To prove evidence of carbohydrate–carbohydrate interactions in the self-association of the marine sponge *Microciona prolifera* proposed by Burger,[33] GNPs were also used as a model system by Kamerling's group.[34] In this case, the carbohydrates involved in the interactions are a sulfated disaccharide [β-D-Glc*p*NAc-3*S*-(1→3)-L-Fuc] and a pyruvylated trisaccharide [β-D-Gal*p*-4,6(*R*)-Pyr-(1→4)-β-D-Glc*p*NAc-(1→3)-L-Fuc], which are present in the extracellular proteoglycans of the cells of *M. prolifera* and determine species-specific self-recognition and adhesion. The same group, in collaboration with Jiménez-Barbero and colleagues, performed NMR experiments using diffusion-ordered spectroscopy (DOSY) and transferred nuclear Overhauser enhancement spectroscopy (TR-NOESY).[35] They provided evidence of Ca^{2+}-mediated carbohydrate–carbohydrate interactions by monitoring the changes in the diffusion of the free pyruvylated trisaccharide in solution after the addition of the corresponding GNPs and calcium salts, as well as the changes in the sign of the sugar NOE peaks. Also, Russell *et al.* made use of gold nanoparticles as synthetic scaffolds for multivalent presentation of β-lactosides to study calcium-mediated carbohydrate–carbohydrate self-interactions in water by UV and TEM.[36]

In recent years, the number of publications on the use of glycoside-coated gold nanoparticles has increased, together with new synthetic methodologies which make use of gold–sulfur affinity to attach the carbohydrate-based ligands to the gold surface of NPs of different sizes.[37] Particular care has been dedicated to the gold–biomolecule linkage in order to allow optimal presentation of carbohydrates.[38] Most of the works deal with the intervention of GNPs in protein–carbohydrate interactions.

Since the beginning, gold glyconanotechnology based on thiol chemistry has anticipated the opportunity to generate a variety of multivalent GNPs presenting different carbohydrates at the gold surface, together with other (bio)molecules, in order to study the role of carbohydrate presentation and density on recognition events.[39] In this article, we will focus on the main results which have been published in the last five years on gold glyconanoparticles, highlighting the contributions to solve biological problems. Other reviews comprehensively resume the precedent works on GNPs, including those related to glyconanoparticles whose metal core is not gold as well as the different synthetic protocols which are used for their preparation.[37,40–43]

5.2 Gold Glyconanoparticles as Anti-pathogenic Agents

Multivalent protein–carbohydrate interactions[44] are involved in many biological processes. High-affinity carbohydrate-based ligands are necessary to study, inhibit and control the processes governed by carbohydrate interactions with proteins, mainly lectins. Lectins are glycan binding proteins (GBP)[45] with relatively well-defined carbohydrate-recognition domains and usually recognise specific terminal moieties of glycan chains by fitting them into binding pockets.[46] The low affinity of lectin–carbohydrate interactions (millimolar range for monosaccharides) is compensated by multivalency, which can come from the architecture of the lectin, containing more than one carbohydrate binding site, and/or by the presentation of carbohydrate ligands as clusters at the cell surface.[25,47] This phenomenon is known as the cluster–glycoside effect.[23,48,49] In general, a combination of several binding events is involved in the resulting "multivalent" interaction.[25,50,51] The first example of GNPs to study carbohydrate–protein interactions was reported by Kataoka's group.[52] Gold nanoparticles coated with α-lactosyl-ω-mercapto-poly(ethylene glycol) were employed in aggregation tests with *Ricinus communis* agglutinin (RCA_{120}), a bivalent GBP which specifically recognises terminal β-D-galactopyranosides.[53] The selective, concentration-dependent and reversible aggregation was followed by changes in the UV/Vis absorption spectrum. A visible colour change from pink (dispersion) to purple (aggregation) due to the shift in the surface plasmon absorption band of the metal clusters allowed monitoring the phenomenon "at a glance" (see Section 5.3.1.1).

As multivalency is critical in protein–carbohydrate interactions, strategies to amplify the number of glycans at the gold surface have been proposed. For example, Roy constructed 2 nm gold nanoparticles from poly(amidoamine) generation G0 dendrimers scaffolded on cystamine and further coupled *p*-isothiocyanatophenyl α-D-mannopyranoside to the amine functional groups.[54] Also critical is the simplification of the synthetic preparation, especially when challenging oligosaccharides need to be exposed at the gold surface. In this sense, Yan and Ramström[55] have reported a straightforward procedure which consists in photochemically induced coupling of unmodified saccharides to perfluorophenyl azide (PFPA)-functionalised gold nanoparticles.

Several pathogens make use of multivalent protein–carbohydrate interactions to facilitate infection processes.[56] For this reason, the design of multivalent glycoconjugates that mimic the carbohydrate clusters at the cell surface is important in order to generate synthetic systems with anti-adhesive and targeting properties, *e.g.* able to interfere with the recognition between host cells and pathogens.[57,58] Synthetic glycosides have been tailored on gold nanoparticles to present carbohydrate ligands in a multivalent fashion and enhance the affinity of the monovalent ligand towards its protein receptor. In general, two main strategies can be envisaged to interfere with host–pathogen interactions in which carbohydrates are involved (Figure 5.2). The first one is to construct multivalent systems which mimic the presentation of glycans at the pathogens' surface. The other is to develop multivalent glycoconjugates that can avidly target proteins which are used by pathogens to invade and infect host cells.

In this part of the chapter we will present some examples of GNP-based *in vitro* studies which are promising but still far from *in vivo* applications.

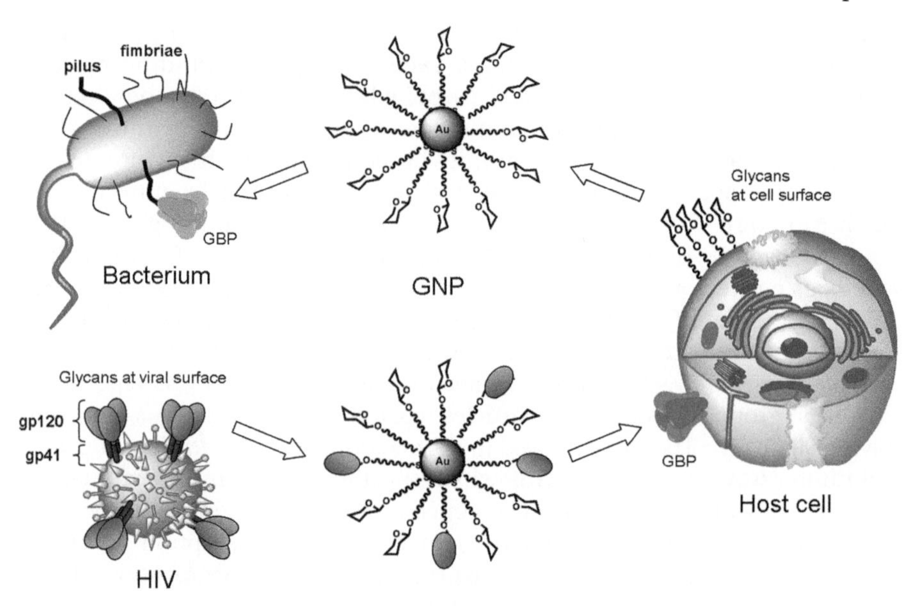

Figure 5.2 Main strategies to use multivalent gold glyconanoparticles (GNPs) as anti-pathogenic agents. *Top*: GNPs that mimic glycans at the host cell surface can target glycan binding proteins (GBPs) which are used by pathogens (a bacterium in this case) to adhere to host cells. *Bottom*: GNPs that mimic glycans at the pathogens' surface (in this case, carbohydrates at HIV envelope glycoprotein gp120) can target GBPs of the host to create an anti-adhesion barrier.

5.2.1 GNPs as Pathogen Mimetics

Mimicking glycans at the pathogens' surface is a viable strategy for the design of carbohydrate-based systems that target host GBPs. This strategy is based on the assumption that a multivalent presentation of glycans by synthetic glycoconjugates should afford efficient competitors in order to avoid pathogen adhesion and inhibit infection. A drawback of this strategy is that the target lectins have natural functions in host signalling which can be permanently unbalanced or blocked. In the case of gold nanoparticles, concerns may come also from metal accumulation in vital organs, such as liver and lungs, which discourages a systemic use and suggests that their use should be limited to sporadic administration. In this sense, the use of biodegradable nanoparticles which have been already approved by competent institutions is preferable. However, topical use remains a feasible option to address the early steps of infection with metallic nanoparticles. Most of the examples of mimicking the pathogens' surface are related to viral glycoproteins and some examples will be shown in this section. However, the presentation of fragments of bacterial capsular polysaccharides or tumour-associated carbohydrate antigens on gold NPs has also been developed, mainly as a strategy for inducing immune response, and will be presented in Section 5.4.1. In this latter context, as well as in diagnostics, the translation of gold nanoparticles to the clinic seems to be achievable in the near future.

5.2.1.1 *Mimicking Glycans at the Viral Surface: the Case of HIV*

The human immunodeficiency virus (HIV) is still one of the big medical problems and the highly active antiretroviral therapy (HAART) is nowadays the only admitted policy against the HIV pandemic.[59] Owing to viral resistance, adverse effects and high costs, HAART is not viable in poor countries where acquired immunodeficiency syndrome (AIDS) originated by HIV is still a cause of death. In the strategies to fight against HIV, carbohydrates play a key role due to the fact that the viral envelope is highly glycosylated. Indeed, mimicking the carbohydrate presentation of the viral envelope with multivalent glycoconjugates can afford a way towards anti-HIV carbohydrate vaccines or potential anti-adhesion systems which may block the early interaction of HIV with host cells.

Regarding this latter concept, a key protein is the mannose-binding lectin DC-SIGN (dendritic cell-specific ICAM 3-grabbing non-integrin). DC-SIGN is a cellular receptor that mediates the interaction between the high mannose-type glycans of the HIV envelope glycoprotein gp120 and DCs.[60] HIV–DC interaction is one of the early steps involved in viral infection and targeting DC-SIGN is a strategy to intervene in the uptake of pathogens by dendritic cells, as well as in the signalling processes mediated by this receptor. In order to mimic the cluster presentation of high mannose-type glycans on the HIV-1 envelope, 2 nm gold nanoparticles functionalised with high mannose-type glycans (*manno*-GNPs, Figure 5.3), which are present on gp120, were prepared.[61] Biosensor-based experiments demonstrated that *manno*-GNPs are able to inhibit the DC-SIGN/gp120 binding in the nanomolar range, while the corresponding monovalent oligomannosides required millimolar concentrations. Selected *manno*-GNPs were further used in a relevant infection model where DC-SIGN is involved and were able to inhibit the DC-SIGN-mediated HIV *trans*-infection of human activated peripheral blood mononuclear cells at nanomolar concentrations in an experimental setting, which mimics the natural route of virus transmission from DCs to T lymphocytes.[62] In this direction, also glycodendritic compounds have shown good inhibition against HIV *trans*-infection of T cells.[63] The search for multivalent glycoconjugates of different types is very important because it gives the opportunity to have in hand a broad arsenal of different scaffolds that present multiple copies of a carbohydrate ligand and can be chosen against pathogens as a function of the concrete application.

In Section 5.4.1.3 we will see how *manno*-GNPs can be also used in route to carbohydrate vaccines against HIV, following the examples of other groups which used different scaffolds to multivalently present high mannose-type oligosaccharides.

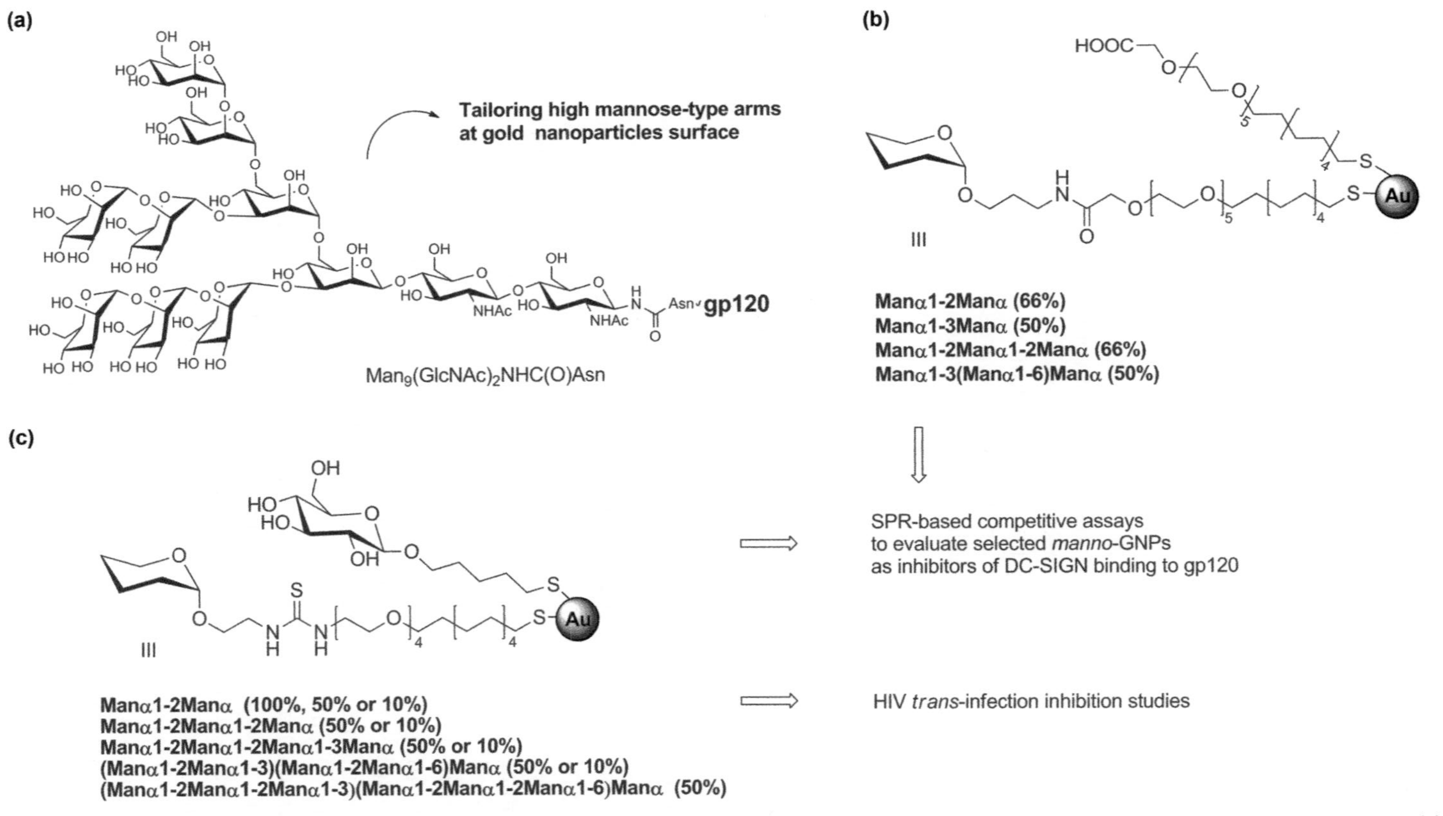

Figure 5.3 GNPs that mimic high mannose-type glycans at the HIV envelope glycoprotein gp120 and target the GBP DC-SIGN. (a) Undecasaccharide present in the HIV envelope gp120; (b) *manno*-GNPs bearing different percentages of a carboxylic linker and of dimannosides attached to a thiol-ending spacer by amide coupling; (c) *manno*-GNPs bearing different amounts of a neoglucoconjugate and high mannose-type oligosaccharides (from a di- to a heptamannoside) attached to a thiol-ending spacer by thiourea coupling (see Martinez-Avila *et al.*[61]).

5.2.2 GNPs to Target Pathogen Proteins

To target proteins at a pathogen's surface or secreted by bacteria in order to proliferate and invade the host is a strategy to prevent, limit or block pathological states. The resistance of bacterial and viral strains to antibiotics is one of the main reasons why it is necessary to search for alternative systems which can impede pathogen infections. In the following part of this article, we will describe the main examples of GNPs (see Table 5.1) that have high affinity for the GBP of pathogens and thus are promising anti-adhesive systems or microbicides. We will limit the discussion to those papers that have appeared in the last five years. Other interesting examples can be found in previous reviews.[37,41,64,65]

5.2.2.1 *Targeting Bacterial Proteins*

Adhesion of bacteria to cells can be promoted through organelles (such as fimbriae and pili) that are present at the bacterial surface. This is the case for type 1 fimbriae of *Escherichia coli*, which mediates α-D-mannoside-specific adhesion through adhesin FimH and for this reason synthetic mannosylated ligands have been designed to develop potential anti-adhesive compounds.[66] The specific binding of gold nanoparticles coated with multiple copies of a thiol-ending pentyl α-mannoside to FimH was first observed using TEM by Lin *et al.*,[67] after incubation of the nanoparticles with an *E. coli* strain (ORN178) which expresses wild-type type 1 pili. The avidity of this kind of nanoparticles has been widely exploited for *E. coli* detection by means of colorimetric assays (see Section 5.3.1.1).

Among lung pathogens, *Pseudomonas aeruginosa* (PA) is a bacterium whose virulence and host invasion is mediated by the tetrameric lectins PA-IL (LecA) and PA-IIL (LecB) that have affinity for galactose and fucose, respectively. LecA is a cytotoxic lectin that binds galactose with the participation of a Ca^{2+} ion[68] and seems to promote the formation of bacterial micro-colonies and biofilms.[69] Several multivalent glycoconjugates have been constructed for targeting LecA in order to inhibit its binding to natural galactosylated surfaces.[58] Among these systems, 2 nm gold nanoparticles with different β-D-galactoside loadings have been prepared and evaluated for their binding to LecA.[70] Hemagglutination inhibition assay, surface plasmon resonance-based biosensors and isothermal titration calorimetry indicated that high-density galactose presentation on the gold nanoplatform allows exceptionally high avidity towards this lectin, affording a 3000-fold increase in binding activity compared with a monovalent reference carbohydrate.

Burkholderia cenocepacia is another opportunistic bacterium responsible for nosocomial infections.[71] It is often implicated in co-infection with *Pseudomonas aeruginosa* and makes use of a member of the PA-IIL lectin family called lectin A, BC2L-A, to promote infections. BC2L-A is homodimeric, calcium-dependent and specific for mannose (oligo)saccharides.[72] Gold nanoparticles were used as a scaffold to multimerise mannoside

Table 5.1 Some examples of GNPs that target pathogen proteins.

Glycosyl moiety	Neoglycoconjugates at GNP surface	Target protein/ pathogen	Observations	Ref.
High mannose-type (oligo)saccharides		DC-SIGN/HIV	Au-NPs coated with di- and trimannosides showed up to 20 000-fold increased activity with respect to the corresponding monomeric saccharide as inhibitors of DC-SIGN binding to sensor chip-immobilized gp120 in SPR experiments	61
β-D-Galactoside		LecA/ *Pseudomonas aeruginosa*	Au-NPs coated with galactose conjugates showed up to 3000-fold increased activity with respect to the corresponding monomer as binders of LecA	70
α-D-Mannoside		BC2L-A/ *Burkholderia cenocepacia*	Au-NPs coated with multiple copies of mannose conjugates are good binders of BC2L-A	73
Globotriose		Shiga-like toxins/ *Escherichia coli*	Au-NPs of different sized and coated with globotriose showed high avidity towards Stx1 and Stx2	75, 76

ligands at different presentation densities.[73] A mannose presentation density greater than 25%, or an inter-ligand distance less than 2.4 nm, was required to induce a significant cluster glycoside effect on recognition by lectin BC2L-A, as observed by SPR and ITC assays.

Another interesting example of targeting bacterial proteins is the case of toxins. Among bacterial toxins, Shiga and Shiga-like (verotoxin) ones are AB5-type systems that consist of an enzymatically active A-subunit and five symmetric B-subunits. Shiga toxin is produced by the bacterium *Shigella dysenteriae* and Shiga-like toxins (SLT-1 and SLT-2 or Stx1 and Stx2) are generated by some strains of *Escherichia coli*, such as serotypes O157:H7 and O104:H4. These toxins specifically recognise the glycolipid globotriaosylceramide (Gb3 or P^k), *e.g.* the blood group antigen which contains the globotriose trisaccharide α-D-Gal*p*-(1→4)-β-D-Gal*p*-(1→4)-β-D-Glc-(1→). Inspired by Bundle's work,[74] GNPs coated with a globotriose derivative were constructed and showed high avidity towards the pentameric B-subunits of Shiga-like toxins as a function of gold core size (4, 13 and 20 nm) and linker nature (short and aliphatic or long and hydrophilic).[75] The 20 nm core-sized GNPs coated with the long linker interacted more efficiently with the binding sites, probably due to the flattening of the gold surface. Similar GNPs protected the Vero monkey kidney cell line from Stx-mediated toxicity in a dose-dependent manner.[76] Inhibition of Stx1 and the more clinically relevant Stx2 was achieved. Natural variants of Stx2 (Stx2c and Stx2d) having minimal amino acid variation in the receptor binding site of the B-subunit or changes in the A-subunit were not neutralised by either the Stx1- or Stx2-specific gold glyconanoparticles.

These examples demonstrate that GNPs can be used in basic glycobiology studies with a promising perspective towards biomedical applications.

5.2.2.2 Targeting Other Proteins

Polyanionic compounds are known to inhibit HIV entry by binding the viral envelope glycoprotein gp120 and have been proposed as microbicide candidates.[77] Gold nanoparticles coated with multiple copies of sulfated ligands and a β-glucoside [5-(thio)pentyl β-D-glucopyranoside] were able to bind gp120 in SPR-based assays and to inhibit the *in vitro* HIV infection of T lymphocytes in the nanomolar range.[78] The percentage of sulfate-ended ligands was modulated with glucose, which also served to confer better stability in biological media and guarantee high cell viability. As polyanionic compounds display inhibitory effects not only against HIV but also on a variety of enveloped viruses, such as the herpes simplex virus (HSV) and the vesicular stomatitis virus (VSV), other examples of polysulfonated[79] and polysulfated[80] gold nanoparticles have been reported. In particular, the work of Haag and collaborators[80] showed that the inhibition of binding and infection of VSV strongly depends on the particle size of polysulfated gold colloids (Figure 5.4).

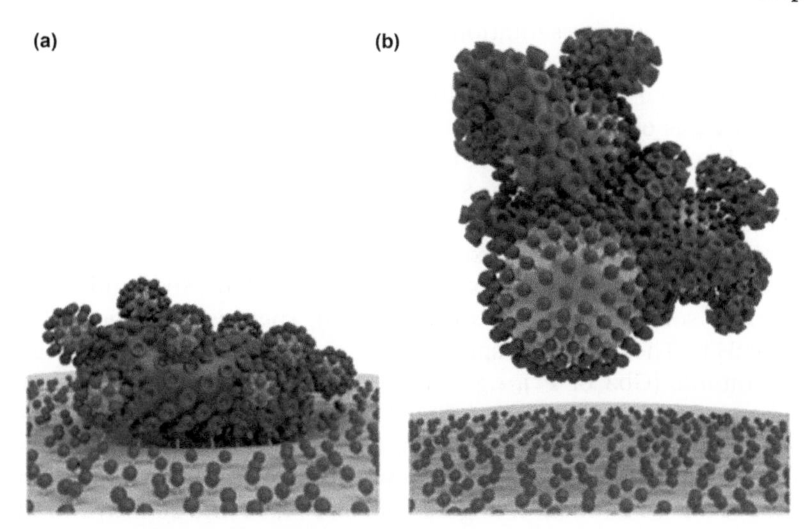

Figure 5.4 Size-dependent VSV inhibition by polysulfated gold nanoparticles: (a) 19 nm gold NPs do not efficiently inhibit virus–cell binding; (b) 110 nm gold NPs induce the formation of virus–inhibitor clusters, inhibiting the virus–cell binding.[80]

Another example that is worth mentioning is related to the GBPs called selectins,[81,82] which bind Lewis blood group-type glycans and are over-expressed in inflammation states and metastasis. Gold nanoparticles (6 nm) functionalised with different sulfated monocyclic aminopyrans and aminofurans (Figure 5.5) were studied as novel L- and P-selectin binders.[83] The idea was to simplify the structural motifs with respect, for example, to heparin (a sulfated glycosaminoglycan that binds with high affinity both L- and P-selectin) in order to develop easily accessible selectin ligands. Competitive SPR-based assays showed that the multivalent presentation of sulfated aminopyrans and aminofurans on the gold nanoparticles (NP-sulf) produced strong binding and moderate selectivities for either P- or L-selectin. In particular, a biotinylated conjugate of the selectin ligand sialyl Lewis X {SiaLe^X, α-Neu*p*5Ac-(2→3)-β-D-Gal*p*-(1→4)-[α-L-Fuc*p*-(1→3)]-β-D-Glc*p*NAc-(1→)} was immobilised on the sensor chip and the binding of selectin-coated NPs (15 nm) was referred to as 100%. Pre-incubation of these selectin-coated NPs with increasing concentrations of the sulfated gold NPs (NP-sulf, the inhibitors in these experiments) decreased the binding signal. In all cases, the half-maximal inhibitory concentration (IC_{50}, *i.e.* the molar concentration of the inhibitor needed to reduce the binding signal to 50% of the initial value) was in the sub-nanomolar range.

Usually the carbohydrate density, the nanoparticle core size and the linker nature are of key importance in the recognition of GNPs by GBPs, as recently highlighted.[38] These characteristics are in fact related to the carbohydrate presentation at the gold surface and critical for a broad range of proteins, including enzymes such as glycosidases. Gold nanoparticles coated with

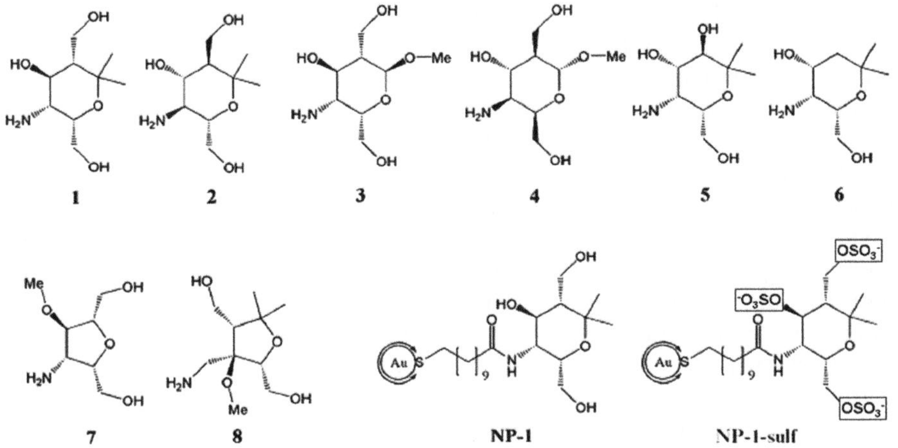

Figure 5.5 Aminopyrans (1–6) and aminofurans (7 and 8) were coupled to gold nanoparticles coated with *N*-hydroxysuccinimido-11-mercaptoundecanoate. The resulting nanoparticles, such as NP-1 depicted here, were then sulfated with $SO_3 \cdot$ DMF complex to obtain NP-1-sulf-type nanoparticles.[83]

different lactose densities (100%, 30% and 5%) were practically not susceptible to *E. coli* β-galactosidase-mediated hydrolysis, in comparison with the free lactose neoglycoconjugate.[84] Jensen's group demonstrated that maltotriose oxyamine-GNPs were more stable than the corresponding maltotriose oxime-GNPs towards glucoamylase from the fungus *Aspergillus niger* (1,4-α-D-glucan glucohydrolase), probably due to their different packaging of neoglycoconjugates at the gold surface.[85] The enzymatic stability of glycans on synthetic scaffolds is an important issue and needs be considered for applications of GNPs in living systems.

In the next section we will survey the applications of GNPs as probes for detecting pathogens or carbohydrate-mediated phenomena which can be related to disease states. We anticipate that the same principle of targeting pathogens' lectins can be used to develop sensitive detection methods, but it is not the only way to obtain diagnostic tools based on GNPs.

5.3 Gold Glyconanoparticles for Detection

5.3.1 Biosensing and Biolabelling with GNPs

The multimerisation of carbohydrates on gold nanoparticles presents the advantage of using the properties associated with the metal core. For example, we have already discussed how the exploitation of GNPs for demonstrating carbohydrate–carbohydrate interactions made use of different techniques such as surface plasmon resonance (SPR) and transmission electron microscopy (TEM).[27,34] These and other techniques have been widely employed for basic studies on carbohydrate–protein interactions with GNPs.[37] Recently, GNPs have been used in "proof of concept" experiments with the quartz crystal microbalance (QCM),[86,87] surface enhanced Raman scattering (SERS)[88] and screen-printed carbon electrodes for impedimetric biosensing.[89] A scanometric approach to detect carbohydrates at the cell surface has been also developed by combining the recognition properties of lectins and the further binding with GNPs.[90] In this work, human gastric cancer cells (BGC-823) were incubated with the mannose-binding lectin concanavalin A (ConA). The lectin supernatants, collected after the incubation with cells, were treated with Man-GNPs to quantify the inhibition of aggregation with respect to the GNP/lectin control by gold-catalyzed silver enhancement and spot tests. This label-free biosensor approach may be generalised and help in reading out the amount of cell-surface carbohydrate groups by choosing other lectin/GNP partners.

Now, we will focus on the recent examples concerning the recognition of pathogen-associated GBPs or the detection of enzymatic activity by *in vitro* tests which make use of gold nanoparticles in colorimetric assays. We will also present new approaches to use GNPs in ELISA-type assays and glyco-blotting.

5.3.1.1 Colorimetric-based Assays

The special characteristics of the UV/Vis absorption spectra of colloidal metallic elements[91] can be used for monitoring the specific aggregation of gold NPs by spectrophotometry.[92] Usually a colour change by 10/16 nm for gold nanoparticles from a pinkish-red colour (in dispersion) to a purple colour (after aggregation) is observed at a glance. This phenomenon is due to the decrease of the inter-particle distance. One of the most notable examples has been reported by Mirkin with DNA-functionalised gold nanoparticles, where colour shifts are triggered by aggregative processes induced by selective hybridisation of two complementary DNA strands on different gold NPs.[93] We have seen how lactose-GNPs generated a colour change associated with a shift in the surface plasmon absorption band of the metal cluster after incubation with protein RCA_{120}.[52] Many other examples based on aggregation of gold nanoparticles coated with carbohydrates have appeared in the literature,[37] such as the use of lactose-GNPs for the colorimetric detection of cholera toxin.[94] More recently, GNP-based linear discriminant analysis (LDA) was used as a statistical method for identifying unknown samples of lectins.[95] The strategy is based on the creation of a pattern recognition database of selected lectins by using GNPs that have strong, weak or no affinity towards them, depending on the carbohydrate coating. UV/Vis spectroscopy was used to measure the wavelength of maximum absorption (λ_{max}) of each GNP in solution before and after treating with lectins. In this way, shifts ($\Delta\lambda_{max}$, nm) in the localised surface plasmon resonance (LSPR) could be recorded and used for generating "training matrices". The ability of a training matrix to identify unknown lectins was assessed by blind tests using lectin samples blindly prepared. After treating each sample with the GNPs, the $\Delta\lambda_{max}$ were recorded under the same conditions of the selected training matrix. Probability calculations were then used to successfully classify the lectins in these blind tests.

In order to perform highly sensitive and accurate colorimetric bioassays, a uniform size of the GNPs is the key requisite to maximise the initial intensity of the surface plasmon absorption band. Then, it is important to adjust the density of the "active" carbohydrate ligands in order to obtain the maximum effect in the minimum time.

GNPs have been also explored as sensitive tools for the detection and discrimination of bacterial lectins in order to develop new tools for the identification of pathogens. In a recent study,[96] α-Man- and β-Gal-coated gold nanoparticles (50 nm gold core diameter) were able to discriminate between different strains of *E. coli* on the basis of their different expression levels of the FimH adhesion molecule that has a specific affinity for mannose residues but not for galactose. Notably, when poly(ethylene glycol) linkers were employed to attach the carbohydrates to the gold surface, improved stability was found. However, the convenient visual read-out commonly associated with the nano-gold red–blue colour shift was prevented, probably due to steric shielding effects of the PEG layer. Thus, the authors monitored the significant change in absorption at 700 nm with UV/Vis spectroscopy as a specific output. Man-GNPs generated an increase of adsorption at 700 nm in the presence of K-12 bacteria (expressing FimH), but not with TOP10 ones that do not express FimH. Too highly concentrated samples of bacteria led to a signal reduction, suggesting that a calibration for the detection/discrimination of bacterial phenotypes has to be accurately carried out to find a suitable working range when using this methodology.

This example indicates that the pinkish-red colour of well-dispersed gold nanoparticles with a suitable gold core (usually >10 nm) not always turns to purple (bathochromic shift) as a consequence of aggregation, *e.g.* of a change in the inter-particle distance induced by binding. Another way to follow this phenomenon, instead of monitoring the increase of adsorption at 700 nm, is to measure the decrease of the gold surface plasmon band at around 520 nm. Lin and collaborators have developed a novel GNP-based strategy for highly sensitive detection of the activity of α-2,8-polysialyltransferase (PST).[97] They constructed gold nanoparticles (13 nm gold diameter) decorated with disialyl lactoside [α-Neu*p*5Ac-(2→8)-α-Neu*p*5Ac-(2→3)-β-D-Gal*p*-(1→4)-β-D-Glc*p*-(1→)] conjugated to an amphiphilic thiol-ending linker (Figure 5.6). This neoglycoconjugate was used as a primer for a PST which catalyses the successive addition of sialyl moieties from cytidine monophosphate-sialic acid (CMP-Sia) to the non-reducing end of a growing polysialic acid (PSA) chain on the NP surface. An inactive PSA-specific trimeric phage tailspike endosialidase (EndoNF-DM), which has the function to induce aggregation of the formed PSA-GNPs, was used as a "cross-linker" for facilitating colorimetric detection. However, since PST catalyses the formation of PSA-GNPs with several sialic acid residues and EndoNF-DM binds a minimum of 5–8 sialic acid moieties, the inter-particle distance of the "cross-linked" PSA-GNPs was not short enough to induce strong coupling between the adjacent nanoparticles. Nevertheless, a decrease in absorbance at 525 nm resulted from the precipitation of PSA-GNP

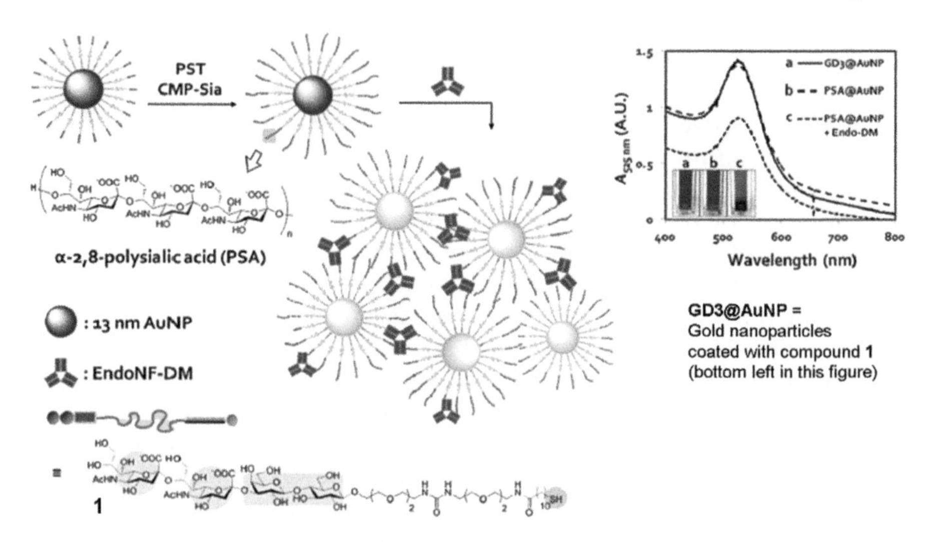

Figure 5.6 *Left*: Biosensing strategy for α-2,8-polysialyltransferase (PST) activity. The growth of a polysialic acid (PSA) chain on the 13 nm gold NP surface is obtained by PST-mediated addition of sialyl moieties from cytidine monophosphate-sialic acid (CMP-Sia). An inactive and PSA-specific trimeric phage tailspike endosialidase (EndoNF-DM) binds to the PSA chain at the gold surface and induces precipitation of gold nanoparticles, *e.g.* a decrease in absorbance at 525 nm. *Right*: UV/Vis spectra of (a) disialyl lactoside-GNPs (GD3@AuNP), (b) PSA-GNPs (PSA@AuNP) and (c) PSA-GNPs after addition of EndoNF-DM.[97]

aggregates, which was facilitated by centrifugation. In other words, precipitation of PSA-GNPs after EndoNF-DM-mediated cross-linking generated a decolouration of the solution which could be quantified as a decrease in the absorbance at 525 nm. This one-pot biosensing strategy allowed the sensitive and specific detection of a glycosyltransferase-catalysed bond formation, in label-free glycosyl donor conditions, which could be further applied as a high-throughput GNP-based assay for the detection of enzymatic activity of other bond-forming enzymes.

GNPs have also been employed for the inhibition and detection of the influenza virus. This virus has two types of surface glycoproteins: hemagglutinin (HA) that recognises and binds sialic acids present on the surface of host cells, and neuraminidase (NA) that releases new-forming viruses from infected cells.[98] Haag and collaborators employed 2 nm and 14 nm gold nanoparticles functionalised with a sialic acid-terminated glycerol-based dendron to study their interactions with the viral fusion protein HA either on complete virions (influenza A strain X31) or isolated and purified HA trimers.[99] TEM and cryogenic transmission electron microscopy (cryo-TEM) were used to directly visualise the binding of nanoparticles to virosomes or isolated HA trimers. After pre-incubation of the virus with sialylated NPs of 14 nm size, the viral infection of adherent MDCK (Madin–Darby canine kidney) cells was inhibited by 40%, as determined by immunostaining.

On the other hand, 2 nm GNPs did not show any significant inhibition. These results were also confirmed by measuring the virus-induced agglutination of red blood cells (hemagglutination inhibition assay, HAI). The lowest concentration of 14 nm GNP that prevents agglutination was in the nanomolar range. This work does not make use of colorimetric detection, but it was a breakthrough in this field and inspired other work which use colorimetric-based assays with GNPs. For example, Russell and collaborators used GNPs to discriminate between human and avian strains of influenza virus. Gold nanoparticles (16 nm diameter) functionalised with varying ratios of a thiolated trivalent α-2,6-thio-linked sialic acid derivative [*S*-α-Neu*p*5Ac-(2→6)-β-D-Gal*p*-(1→)] and a thiolated tetraethylene glycol (TEG) ligand were used in agglutination assays for the detection of human influenza virus X31 (H3N2) (Figure 5.7).[100] A sialic acid/TEG ratio of 25/75 was found to be optimum for this purpose, confirming that the percentage of

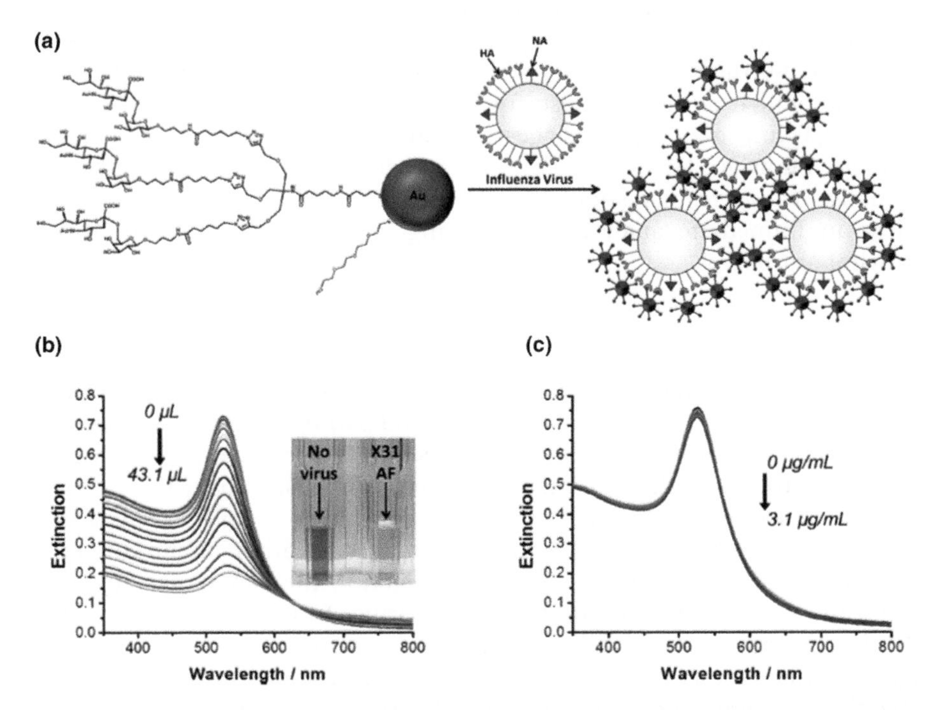

Figure 5.7 (a) Schematic representation of gold nanoparticles functionalised with a trivalent sialic acid-containing disaccharide [*S*-α-Neu*p*5Ac-(2→6)-β-D-Gal*p*-(1→)] and thiol-ending tetraethylene glycol, and their aggregation in the presence of influenza virus due to the interaction with hemagglutinin (HA) at the viral surface. (b) UV/Vis spectra of GNP bearing a sialic acid/TEG ratio of 25/75 after the addition of increasing volumes of human influenza virus X31 from influenza allantoic fluid (AF), *e.g.* X31 AF (H3N2); *inset*: cuvettes containing the GNPs before (*left*) and following (*right*) addition of X31 AF (43.1 μL). (c) UV/Vis spectra of the same GNP after the addition of increasing concentrations of avian virus RG14 (H5N1) (from 0 to 3.1 μg mL^{-1}).[100]

"active ligand" at the gold surface is extremely important, depending on the protein counterpart under study.[38] Importantly, the glyconanoparticles were able to discriminate between the human (α-2,6 binding) and avian (α-2,3 binding) RG14 (H5N1) influenza virus, highlighting the binding specificity of the trivalent α-2,6-thio-linked sialic acid ligand. Related to this, GNPs (13 nm gold core diameter) carrying terminal sialic acid moieties with an α-2,3 or α-2,6 linkage to the galactosyl core of β-lactoside analogues (sialyl lactose, sialyl lactosamine, 6-sulfated sialyl lactosamine and sialyl Lewis^X) have been recently used in similar colorimetric assays for the analysis of influenza virus receptor specificity.[101] The interactions between different HAs with the GNPs were screened. HAs from human-adapted viruses (09H1, 18H1, 68H3) prefer GNPs bearing α-2,6-linked sialic acid, while those from avian-adapted viruses (vieH5, qinH5, sheH5) showed a better affinity for GNPs bearing α-2,3-linked sialic acid. For the H7 haemagglutinin (generated from an emerging human-infecting H7N9 A/Anhui/1/2013 virus, and thus named anhH7) this assay revealed a dual human/avian α-2,6/α-2,3 specificity. Further analysis of a whole virus strain of the emerging H7N9 avian flu, which is infective also for humans due to its capability to bind both α-2,3- and α-2,6-types, suggested that anhH7 HA could be an intermediate in avian H7 HA evolution towards H7N9 virus adaptation to humans. These results show how GNP-based assays of this kind may allow an immediate analysis of emerging viruses without the need for antiviral antibodies.

5.3.1.2 *Other Techniques*

The detection of anti-glycan antibodies in serum is of great interest for the evaluation of carbohydrate-based vaccines and pathogen infection as well as for the detection of biomarkers in diseases like cancer.[102] A fast and highly sensitive serology method for biomarker detection based on GNPs has been recently described.[103] In this work, commercially available ELISA plates were directly coated with GNPs carrying different carbohydrate antigens. Anti-carbohydrate antibodies could be detected in the nanomolar range by performing an ELISA based on the GNP coating. In fact, the purified anti-HIV human monoclonal antibody 2G12 and serum from mice immunised against *Streptococcus pneumoniae* type 14 (see Section 5.4.1.2.1) showed specific and high binding to the GNPs bearing high mannose-type oligosaccharides and the repeating unit of the capsular polysaccharide of the bacterium, respectively. Moreover, the GNPs were employed in a solid phase assay to profile the carbohydrate binding of human cells and to profile lectin affinities on a highly multivalent surface.

Nishimura demonstrated that GNP-assisted MALDI-TOF MS can facilitate the analysis of glycosphingolipids (GSLs) at the surface of living cells.[104] Aminooxy-functionalised gold nanoparticles were used to capture GSLs extracted from cellular matrices and previously subjected to ozonolysis in order to transform the double bond in the ceramide moieties of natural GSLs into a reactive aldehyde. Subsequent MS-based structural analysis revealed the molecular diversity of GSLs in biological samples (adult and embryonic mouse brain gangliosides as well as mouse B16 melanoma cells) in a straightforward way. The application of a GNP-based "glyco-blotting" concept to the use of a technique like MS can help in the individuation of disease biomarkers and in advanced structural glycosphingolipidomics.

All these examples evidence the opportunity to design GNP-based assays for biosensing and biolabelling. The working conditions should facilitate the use of GNPs as nano-biotools for *in vitro* diagnostics.

5.3.2 GNPs for Molecular Imaging

Magnetic resonance imaging (MRI), X-ray computed tomography (CT), positron emission tomography (PET) and ultrasound (US) are broadly used in medical imaging, especially for diagnostics. Spatial resolution and sensitivity are two main issues and, usually, techniques with high sensitivity lack good spatial resolution and *vice versa*. Contrast agents are often used to enhance image contrast and to supply functional information. In this context, the design of nanotechnology-based contrast agents may offer some advantages with respect to classical imaging probes, especially in terms of multimodality (*e.g.* combining multiple imaging modalities by using the same molecular probe) and targeting.[2,5,105] In this section we will give some highlights on the application of gold glyconanoparticles in MRI, optical imaging, PET and other imaging techniques.

5.3.2.1 MRI

Superparamagnetic iron oxide-based nanoparticles (SPIOs), which are approved MRI contrast agents, have been broadly used as platforms to attach biomolecules.[5] Examples of glycan functionalisation of SPIOs have appeared in the literature.[41,64] The most interesting examples are the sLeX-functionalised SPIOs by Davis and collaborators[106] (selectins targeting for *in vivo* imaging of cerebral inflammation in animal models of multiple sclerosis and stroke) and monosaccharide-functionalised iron oxide/silica core/shell magnetic nanoparticles by Huang[107] (use of a glyco-code to decipher *in vitro* the protein biomarkers of cancer cells). Our group has dedicated efforts to the development of glyconanoparticles based on metal-doped ferrites/gold core/shell (XFe_2O_4@Au, X = Fe, Mn or Co) nanometric nuclei which are superparamagnetic systems and allow thiol chemistry at the gold surface.[108]

Another way to transform gold nanoparticles in magnetic probes for MRI is to coat the gold core with paramagnetic complexes.[109,110] By a combination of thiol-ending glycosides and Gd(III) complexes, our group developed paramagnetic Gd-based gold glyconanoparticles (Gd-GNPs).[111] Different sugar derivatives (β-D-glucose, β-D-galactose and β-lactose) and gadolinium complexes (Gd:DO3A conjugates) were simultaneously tailored on the same gold nanoplatform in order to obtain biocompatible paramagnetic GNPs with high relaxivity (contrast enhancement), depending on the nature of the sugar and the relative distance between the paramagnetic ion and the carbohydrate, among other factors. These novel Gd-GNPs were used *in vivo* as MRI probes imaging glioma (generated with GL261 tumour cells) in mice. At the same Gd concentration, Gd-GNPs coated with glucose were able to enhance the contrast in the tumour zones better than the small molecule Dotarem® (DOTA:Gd), a T_1 contrast agent in clinical use.

By using a different synthetic protocol, Gd-GNPs bearing simple monosaccharides (glucose, mannose or galactose) and paramagnetic Gd(III) complexes were also prepared as biocompatible and targeted MRI probes for unveiling cellular receptors by *in vitro* studies.[112] The multivalent presentation of carbohydrates on the Gd-GNPs enhances the avidity of sugars for GBPs at the cell surface and increases the local concentration of the paramagnetic probes. A significant T_1 reduction (brighter images) upon incubation of cells with Gd-GNPs has been achieved, depending on the sugar coating and the type of cell. Burkitt lymphoma cells Raji− or Raji+ (*e.g.* cells that do not express or over-express lectin DC-SIGN, respectively), murine glioma cells GL261, and hepatocytes HepG2 were tested. Quantification of the percentage change of T_1 (eqn (5.1)) was also obtained from the T_1 maps of control cell pellets and those of Gd-GNP incubated cells (Figure 5.8):

$$\%\ \Delta T_1 = (T_{1\ \text{cells}} - T_{1\ \text{cells+Gd-GNPs}}) \times 100 / T_{1\ \text{cells}} \tag{5.1}$$

The preferential binding of Man-GNP to Raji− and Raji+ can be explained by the presence of a mannose-binding receptor,[113] and of mannose-binding lectin DC-SIGN[60] in the case of Raji+ cells. Monosaccharides with an equatorial 4-hydroxyl group (*e.g.* mannose and glucose) preferentially bind to mannose receptors with respect to galactose, which has an axial 4-hydroxyl group.[114] This may also explain why Glc-GNP induces a similar ΔT_1 (70%) in Raji−, Raji+ and GL261 cells that also over-express GLUT transporters,[115] while the reduction of T_1 in HepG2 cells with Glc-GNPs is only 37%. Hepatocytes HepG2 seem to take up preferably

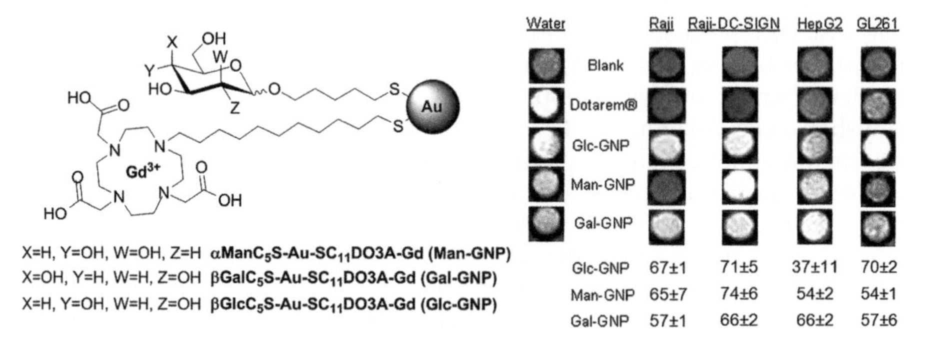

Figure 5.8 T_1-weighted images of fixed cells at 11.7 T and the percentage change of T_1 (% ΔT_1) after incubation with Dotarem® or Gd-GNP at [Gd] = 130 μM.[112]

Gal-GNP (ΔT_1 66%), probably due to the presence of the β-galactose/*N*-acetylgalactosamine binding lectin, named the asialoglycoprotein receptor.[116] These results stress the benefits of using carbohydrates for targeted cellular labelling with stable gold nanoparticles.

Biodistribution assays of Gd-GNPs in mice demonstrated that, 24 hours after intravenous administration, Glc-GNP accumulated almost 10 times more in brain tumour than normal brain[117] and that gold preferentially accumulates in liver and spleen, in agreement with other studies with gold nanoparticles.[12] As we previously commented, the administration of gold nanoparticles into the body is promising, especially for sporadic uses as is the case of diagnostics. However, (nano)toxicity concerns have to be carefully addressed. For example, regarding the interaction of gold NPs with the central nervous system, Kotov and collaborators have recently reported on the increased excitability of neurons after incubation with 5 nm and 40 nm gold NPs *in vitro.*[118] This effect could lead to neurological diseases.

5.3.2.2 GNPs as Luminescent Probes

As in the case of MRI, luminescent nanoparticles for optical imaging have been mainly obtained by modification of the NP nuclei and several efforts have been especially dedicated to semiconductor elements for the creation of quantum dots (QDs).[119] Some interesting examples of glycoside-functionalised QDs for *in vivo* imaging have also started to appear in the literature, such as the one reported by Seeberger's group that tested PEGylated QDs capped with mannose, galactose and galactosamine derivatives for specific liver sequestration.[120] It is important to know whether it is feasible to use these systems *in vivo*, both for reasons of poor light penetration in tissues (although these systems may be of help in open surgery) and for toxicity aspects to be clarified.[121] In the former case, great efforts are being made to develop luminescent probes which can work in the near-infrared window (NIR, 650–900 nm), where tissue auto-fluorescence and light absorption are low (light penetration is higher). In any case, *in vitro* cellular studies by optical imaging can afford precious information on GNP behaviour.[122–124]

The use of gold nanoparticles for optical imaging is usually associated with the insertion of fluorophores into the organic shell, as is the case of hyaluronic acid-GNPs proposed by Park for *in vivo* studies, although in this case a polysaccharide and not a defined glycoside is used.[125] The conjugation of luminescent dyes to gold nanoparticles is usually limited to very small gold nanostructures and/or to the use of long linkers in order to avoid fluorescent quenching.[126]

The study of GNP uptake by cells is a wide-open field in which the size/shape of gold nanoparticles, the charge of the organic shell and the type of carbohydrate can play a critical role to favour a way of interacting cells with respect to one another. Usually, things are complex and many phenomena occur at the same time and the route to NP internalisation is not an easy task to address. For example, it has been shown by microscopy that fluorescein-labelled GNPs bearing high mannose-type oligosaccharides can be taken up by different ways also involving DC-SIGN-mediated internalisation.[127]

In recent work,[128] the state of aggregation and dynamics of gold nanoparticles functionalised with glucose and fluorescently labelled with HiLyte Fluor647 could be studied by means of fluorescence correlation spectroscopy (FCS) in the intracellular environment of hepatocytes. This work shows that FCS can be used to visualise the presence of single GNPs or GNP aggregates following uptake and to estimate, locally, NP concentrations within the cell.

The role of carbohydrates in cellular uptake, the rate of internalisation and the routing of nanoparticles is thus of outstanding importance for GNP applications. Recently, de la Fuente and collaborators[129] reported on the difference in Vero cell line (monkey kidney epithelial cells) uptake of fluorescently labelled and magnetic GNPs, depending on the carbohydrate coating. Importantly, the authors demonstrated that a glucose coating can be used as a valid alternative to PEGylation of NPs and that glucose-coated NPs mainly follow a caveolar/lipid raft uptake, which ends in lysosomes after 12 hours of incubation in serum-free media. To work in serum-free media introduces another hot topic in nanomaterials and nanomedicine, which has been scarcely considered for glyconanoparticles but is of extreme importance for their *in vivo* applications. Non-specific binding of serum proteins to NPs may generate a "protein corona" that could critically affect the NP fate and function in biological fluids.[130] In this sense, the work of Dawson and collaborators evidence that blood plasma-derived coronas at a nanoparticle surface are quite long-lived and are "likely to be what the cell sees."[131] For this reason, the biofunctionalisation of gold nanoparticles has to properly face the prevention or selective adsorption of proteins.[132,133] Indirect evidence that carbohydrate coatings may avoid non-specific adsorption of proteins to gold nanoparticles is present in some papers discussed above. The experiments of GNP-based lectin sensing presented by Yan,[95] as well as those of GNP-ELISA for the detection of GBPs reported by us,[103] were performed in bovin serum albumin (BSA)-containing buffers. The presence of the globular protein BSA, a major component of fetal bovine serum, did not compromise the functional recognition properties of the tested GNPs.

In an effort to develop non-labelled fluorescent NPs as an alternative to potentially cytotoxic QDs, Chang and co-workers reported that Man-GNPs having a core diameter of 1.8 nm (mannose-protected Au nanodots, NDs) fluoresced at 545 nm, when excited at 375 nm.[134] Although the photoluminescence properties of gold nanoclusters is still under debate and is beyond the scope of this review, these Man-NDs could be successfully used for sensing mannose-binding lectin concanavalin A and *Escherichia coli*. Incubation of Man-NDs with *E. coli* yielded brightly fluorescent cell clusters, due to the multivalent interactions with mannose-binding receptors located on the bacterial pili.

5.3.2.3 Other Imaging Techniques

Gold nanoparticles coated with a blood–brain barrier (BBB) permeable neuropeptide, a glucose derivative and a chelator of the positron emitter ^{68}Ga for positron emission tomography (PET) studies were used *in vivo* for studying BBB permeability.[135] These small gold nanoparticles (2 nm) were stabilised by the glucose conjugate and were well-dispersible in buffer. The targeting functionality was in this case relegated to the leucine-enkephaline (Leu-Enk) neuropeptide for improving the BBB crossing. A biodistribution study in rats was performed for all ^{68}Ga-GNPs and showed different distribution patterns depending on the surface ligands. The presence of the neuropeptide on the nanoparticles (targeted ^{68}Ga-GNPs) improved brain accumulation three-fold compared to non-targeted ^{68}Ga-GNPs. This is a nice example where the gold nanoplatform can be used to insert different components, each one with a specific and complementary functionality in order to address a biological problem.

Gold-coated iron oxide nanoparticles of 6 nm size were recently proposed as mutimodal contrast agents for X-ray CT and ultrasound due to the gold coating, and T_2-weighted MRI due to the superparamagnetic core.[136] In order to ensure water dispersibility, a thiol-ending neoglucoconjugate with a spacer consisting of an aliphatic alkyl chain of 11 carbon atoms and a tetra(ethylene glycol) unit was anchored to the gold surface.

Although the role of carbohydrates in the design of GNPs reported in these examples is mainly to assist water solubility at physiological pH and to render the nanostructures more biocompatible, it can be foreseen that the targeting properties of carbohydrates may be of help for further development of multimodal imaging probes based on GNPs.

5.4 Nanotherapy

Anti-adhesion therapy, vaccine development, drug delivery and hyperthermia are the main research lines where gold glyconanoparticles may find application in the future. Regarding anti-adhesive therapy, the first demonstration that GNPs can work as anti-adhesion agents, depending on the carbohydrate coating (a β-D-lactoside or a β-D-glucopyranoside), was reported by Penadés and collaborators.[137] In this work, pre-incubation of lactose-GNPs with B16F10 melanoma cells before inoculation into mice prevented the development of lung metastasis *in vivo* by up to 70%. Recently, gold nanoparticles (~10 nm) covered with mannose-6-phosphate (M6P) analogues were prepared as anti-angiogenic agents.[138] Thiol-ending α- and β-D-mannosides with formal substitution of a phosphate group at position 6 of the mannopyranoside with azide or carboxylic acid groups were functionalised with different thiol-ending spacers *via* Huisgen cycloaddition and Julia reaction (β-D-mannosides) or thiol–ene reaction (α-D-mannosides), and then inserted onto the surface of 10 nm gold nanoparticles. All the tested GNPs showed anti-angiogenic activity *in vitro*, as determined by the avian chorioallantoic membrane assay. The GNPs bearing azide-substituted mannosides performed better than the carboxylic-substituted ones and the inhibitory effect of M6P derivatives on gold nanoparticles was higher than that of the corresponding monomers.

Most of the examples of gold glyconanoparticles in therapy are related to the generation of carbohydrate-based vaccines. In the next sections we will present in detail the main works in which GNPs have been used in the activation/modulation of the immune system. Some reports on GNPs in drug delivery have also appeared in the literature and an overview will be given in the last part of this review, including the use of gold nanoparticles as carriers of DNA-related biomolecules for gene therapy.[139] The application of different types of gold nanoparticles as carriers of commercially available antibiotics is also a promising strategy to develop antibacterial therapeutics.[140]

Finally, plasmonic gold nanoparticles can act as highly enhanced photo-absorbing agents to trigger laser near-infrared irradiation-induced hyperthermia.[12,141] Hyperthermia can provide an efficient anti-cancer therapy strategy, because the tumour tissue is exposed to temperatures high enough to induce cell damage and/or to make cancer cells more sensitive to the action of a drug or radiation. However, no examples of GNPs in hyperthermia have been reported, as far as we know.

5.4.1 Multivalent Glycoconjugates in Vaccine Development

Conjugate vaccines, *i.e.* conjugates of non- or poorly-immunogenic small molecules (haptens) to immunogenic carrier proteins, have re-emerged nowadays as alternatives to the use of antibiotics that can be complicated by the emergence of drug-resistant clones. As carbohydrates are poorly immunogenic (T-independent antigens), they are conjugated to carrier proteins (such as CRM_{197}) in order to elicit a specific and functional antibody response (glycoconjugate vaccines).[142] The seminal work on glycoconjugate vaccines was reported by Avery and Goebel in 1931. In this work, it was described for the first time that "type-specific antipneumococcus immunity has been induced in rabbits by immunisation with antigen prepared by combining a specific derivative of the capsular polysaccharide of type III pneumococcus with globulin from horse serum."[143] Surprising, in the same year (1931), organic colloidal particles were used as carriers for bacterial polysaccharides, trying to improve their immunogenic properties. The adsorption of anthrax polysaccharide to collodion particles (mainly consisting of nitrocellulose dispersed in ether/ethanol) was exploited to have an "adjuvant effect" in rabbit immunisation.[144] Although this seminal work lacks information on molecular mechanisms, it was shown for the first time that polysaccharides can be rendered immunogenic by adsorption on colloidal carriers.

In the last 80 years, glycoscientists have improved and optimised these seminal observations towards a better design of carbohydrate vaccines. In particular, much effort has been devoted to the individuation and synthesis of structurally defined antigenic motifs in order to avoid the use of purified microbial materials (such as polysaccharides), as well as to understanding antigen–antibody interactions at a molecular level.[145] Multidisciplinary approaches involving oligosaccharide synthesis, glycan array and computational methods for structure–activity correlation have been explored to rationally design and develop new carbohydrate vaccines. Multivalent and semi-synthetic glycoconjugates, mainly based on the conjugation of synthetic carbohydrate fragments to immunogenic proteins, have thus been prepared for stimulating the innate and adaptive immune systems.[146,147]

It has also to be considered that the immunogenic properties of protein carriers may result in epitope suppression, *e.g.* a strong immune response towards the carrier can lower the immune response towards the selected antigen.

Nanotechnology opens up the possibility to design a "minimalist approach" based on the concept of reduction to a minimum of the components to be tailored at the NP surface (the minimum antigen and the minimum immunogen, the better carrier/adjuvant). The combined use of the smallest repetitive and antigenic unit regarding (poly)saccharides or defined carbohydrate antigens of viral envelop glycoproteins, the consensus/immunogenic peptides in place of whole carrier proteins, and selected inorganic carriers (gold or silver nanostructures) may open new routes towards carbohydrate-based vaccines. Furthermore, gold or silver nanostructures are not immunogenic and have interesting properties both in terms of bioactivation (gold adjuvant; silver bactericide) and in terms of detection techniques (depending on the size and shape). The opportunity to immunise with very low quantities of nanoparticles, followed by only a few other booster doses, makes this field very promising for *in vivo* use. Last but not least, nanoplatforms potentially allow putting more than one antigen on the same construct (multiantigenic and fully synthetic vaccine candidates). In this part of the review we will describe the application of gold nanoparticles as carriers for carbohydrate-based vaccines (see Table 5.2) and as interesting tools to modulate immune responses. Here again the concept of mimicking pathogen presentation of carbohydrates at a gold surface (see Section 5.2) is applied.

5.4.1.1 GNPs as Carriers for Potential Anti-cancer Vaccines

Gold nanoparticles may be an alternative to protein carriers in vaccine development. The first steps to apply gold-based nanotechnology to carbohydrate vaccines were taken in cancer treatment. Penadés' group reported the preparation of different multifunctional GNPs (2 nm gold diameter) incorporating the disaccharide antigen sialyl-Tn [α-Neu*p*5Ac-(2→6)-α-D-Gal*p*-NAc-(1→)], the tetrasaccharide Lewis Y antigen {Le^{Y}, α-L-Fuc*p*-(1→2)-β-D-Gal*p*-(1→4)-[α-L-Fuc*p*-(1→3)]-β-D-Glc*p*NAc-(1→)} and a T cell helper peptide (the amino acid sequence 89–105 from tetanus toxoid) on the same platform.[148] Ten different GNPs were prepared in which glucose was the major component

Table 5.2 GNPs as synthetic carbohydrate-based vaccine candidates.

Glycosyl antigen	Neoglycoconjugates at GNP surface	Observations	Ref.
Sialyl-Tn [α-Neu*p*5Ac-(2→6)-α-D-Gal*p*NAc-(1→)] and Le^{Y} {α-L-Fuc*p*-(1→2)-β-D-Gal*p*-(1→4)-[α-L-Fuc*p*-(1→3)]-β-D-Glc*p*NAc-(1→)}	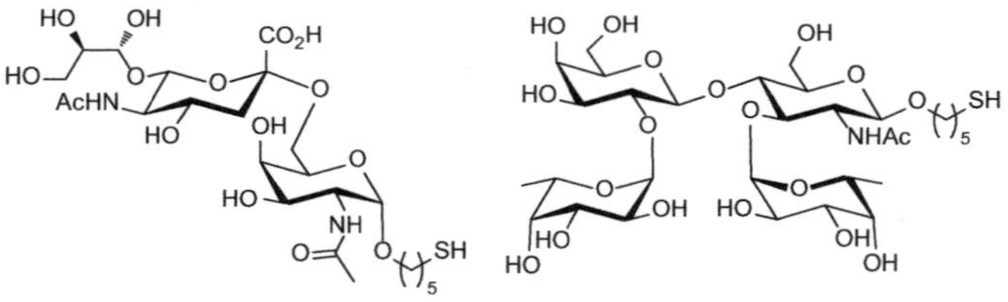	Au-NPs coated with four different components including the disaccharide antigen sialyl-Tn, the tetrasaccharide Lewis Y antigen, a glucoside as inner component and a T cell helper peptide were prepared	148
Thomsen–Friedenreich (TF) tumour-associated carbohydrate antigen	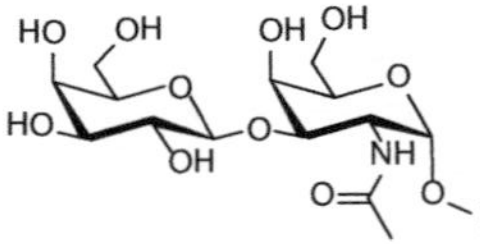 MUC4 peptide conjugated to a thiol ending linker	Au-NPs coated with mucin MUC4 peptides bearing one or more copies of the TF disaccharide, C3d-p28 peptide as an adjuvant and an amphiphilic linker elicited IgGs against the glycopeptides after mice immunization	150
High mannose-type tetrasaccharide α-D-Man*p*-(1→2)-α-D-Man*p*-(1→2)-α-D-Man*p*-(1→3)-α-D-Man*p*-(1→)	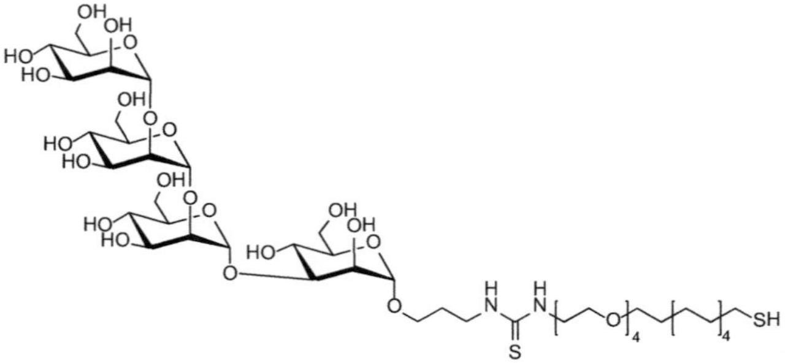	Au-NPs coated with a tetramannoside antigen related to HIV gp120 and a glucoside as inner component showed high avidity for 2G12 antibody and inhibited 2G12-mediated neutralization of HIV infection of selected cells	162

Analogues of the repeating unit of *Neisseria meningitidis* type A capsular polysaccharide (CPS)	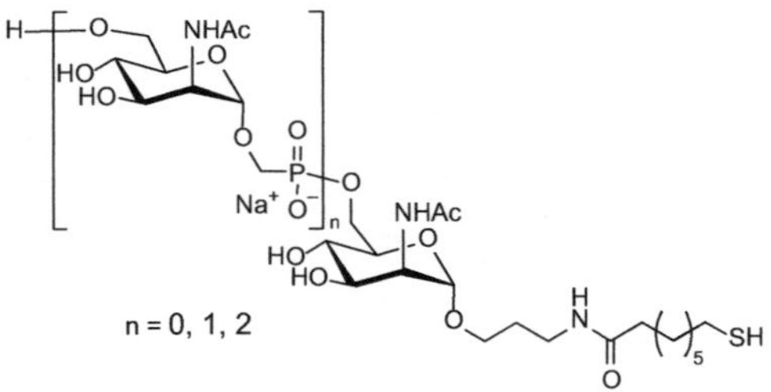 n = 0, 1, 2	Au-NPs 100% coated with mannosaminidine antigens competed with immune serum (obtained from mice vaccinated with anti-meningococcal A + C + W135 + Y polysaccharide vaccine) for the binding to Men A CPS in ELISA assays	155
Analogues of the repeating unit of *Streptococcus pneumoniae* type 14 capsular polysaccharide	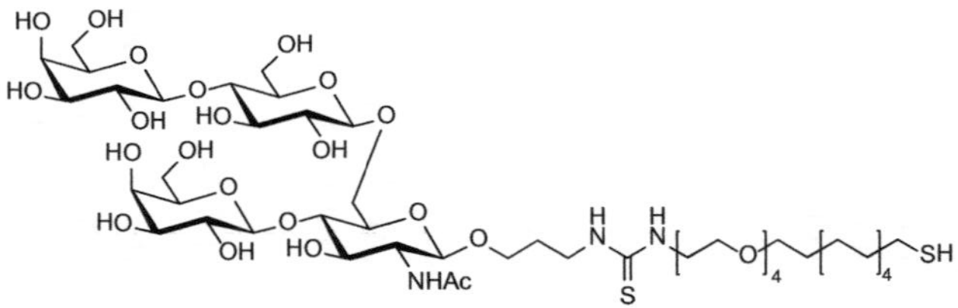	Au-NPs coated with tetrasaccharide antigen, $OVA_{323-339}$ immunogenic peptide and a glucoside as inner component elicited in mice specific and functional IgGs against the capsular polysaccharide of *S. pneumoniae* type 14	152
Tumour-associated Tn antigen (α-D-Gal*p*NAc)	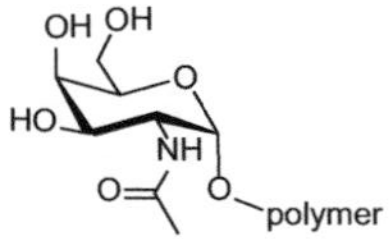	Au-NPs coated with RAFT polymers bearing dithioester end groups and multiple copies of the Tn antigen generated antibodies in rabbits that recognize natural Tn glycans and mammalian-mucin glycoproteins. These GNPs do not contain immunogenic protein material	151

and all of them contained a low density of T cell helper peptide (~3% of the total monolayer). The proportion of the two carbohydrate epitopes (sialyl-Tn and Lewis Y) on the GNPs ranged from 3% to 30%.

Barchi *et al.* reported on gold nanoparticles functionalised with the Thomsen–Friedenreich (TF) tumour-associated disaccharide antigen [β-D-Gal*p*-(1→3)-α-D-Gal*p*NAc-(1→)] and a model glycopeptide designed to mimic the TF antigen [Galβ-(1→3)-GalNAc-α1-OSer/Thr] in order to elicit a strong cytotoxic T lymphocyte response against the TF disaccharide.[149] Later, the same group developed more complex gold nanoparticles carrying simultaneously tumour-associated glycopeptides and immunological adjuvants.[150] In this latter work, the GNPs incorporated two immune elements: the molecular adjuvant C3d-p28 peptide conjugated to a thiol-ending linker and a MUC4 glycopeptide conjugated to a thiol-ending linker and containing four glycosylation sites for the insertion of one or more copies of the TF disaccharide. To maintain bioavailability and modulate the density of the immunological ligand on the GNPs, an amphiphilic thiol-ending linker was used. After mice immunisation, some of these GNPs triggered the production of IgGs that recognised the glycopeptide on the GNPs in a weak but significant manner and without the use of additional adjuvants admixed with the vaccine preparation.

Recently, the synthesis of "multicopy-multivalent" gold nanoparticles covered with the tumour-associated Tn antigen (α-D-Gal*p*NAc) has been proposed by Davis and collaborators.[151] Homopolymers bearing the Tn antigen and dithioester end groups were prepared by reversible addition – fragmentation chain transfer polymerisation (RAFT) and then used to protect gold nanoparticles. This new peptide-free platform showed a significant immune response in mice, with good indications on the generated antibodies' capacity in recognizing natural Tn antigen.

5.4.1.2 GNPs as Anti-bacterial Vaccine Candidates

5.4.1.2.1 ***Streptococcus pneumoniae*****.** An example of a glyconanotechnology-based approach *en route* to anti-*Streptococcus pneumoniae* type-14 carbohydrate vaccines has been recently presented.[152] *S. pneumoniae* is a leading cause of invasive respiratory tract infections in both children and the elderly.[153] Gold nanoparticles were explored as carriers of a synthetic branched tetrasaccharide antigen {β-D-Gal*p*-(1→4)-β-D-Glc*p*-(1→6)-[β-D-Gal*p*-(1→4)]-β-D-Glc*p*NAc-(1→)} related to the *S. pneumoniae* type-14 capsular polysaccharide (Pn14PS). In order to elicit B cell-mediated immune responses, low amounts (~5%) of the T cell-stimulating $OVA_{323-339}$ peptide were also tailored at the gold surface. Last but not least, a glucoside with a short aliphatic linker was chosen as an inner component in order to ensure water dispersibility and to make the tetrasaccharide antigen protrude above the organic shell of the GNPs. These GNPs were capable of inducing specific IgG antibodies against native Pn14PS. One of the most interesting results described in this work was related to the co-presence of the $OVA_{323-339}$

peptide and the tetrasaccharide on the same gold nanoparticles. This was in fact a key requisite to trigger the induction of the anti-Pn14PS IgGs. In addition, the molar ratio between tetrasaccharide, $OVA_{323-339}$ and glucose on the gold nanoparticles was critical for an optimal immune response. Importantly, $OVA_{323-339}$ on the nanoparticles did not lead to antibodies against the peptide, avoiding the risk of epitope suppression, and anti-glucose antibodies were not detected. The evoked IgGs were also able to coat heat-inactivated fluorescein isothiocyanate-labelled *S. pneumoniae* type 14 and make this bacterium susceptible to the action of human polymorphonuclear leukocytes. This work is one of the first examples of fully synthetic gold nanoparticle-based carbohydrate vaccines able to evoke functional IgGs *in vivo*.

5.4.1.2.2 *Neisseria meningitidis.* The development of glycoconjugate meningococcal vaccines is of current interest due to the fact that *Neisseria meningitidis* is still a serious health danger all over the world.[154] Scrimin and collaborators reported the use of gold nanoparticles as a platform to present synthetic analogues of the capsular polysaccharide (CPS) repeating unit of serogroup A *N. meningitidis* (MenA) in a multivalent way.[155] The native phosphodiester bridge connecting the glycosidic units of MenA CPS was substituted with a phosphonate as a hydrolytically more stable moiety in the mono-, di- and trisaccharide analogues (Figure 5.9). GNPs having a gold diameter of 2, 3.5 or 5 nm and bearing the monosaccharide analogue were tested in competitive ELISA experiments to inhibit the binding of

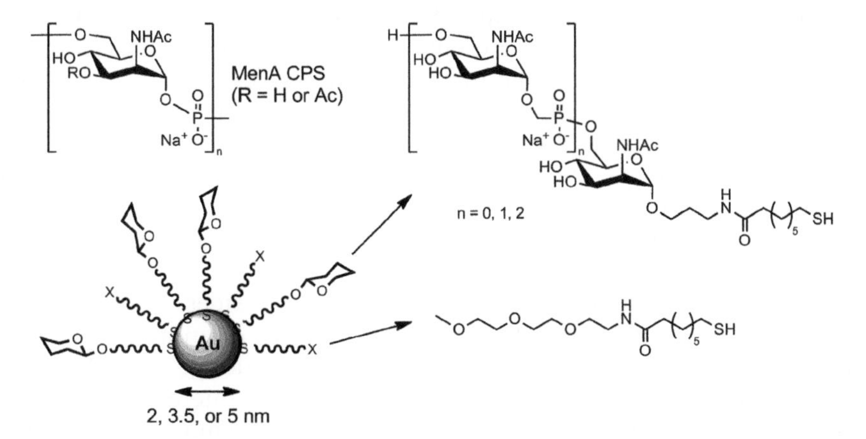

Figure 5.9 Repeating unit of the capsular polysaccharide of type A *Neisseria mengitidis* (MenA CPS) and gold nanoparticles bearing thiol-ending saccharide analogues of the repeating unit. GNPs (2, 3.5 and 5 nm) bearing different proportions (100%, 75% and 50%) of the monosaccharide analogue with respect to the amphiphilic thiol-ending ligand depicted here were prepared to study the influence of gold size and carbohydrate density in anti-MenA antibody binding. 5 nm GNPs 100%-coated with the di- and trisaccharide analogues were also prepared for comparison.

MenA CPS to mouse polyclonal antibodies specific for this bacterium. The 5 nm-sized GNPs were the most efficient inhibitors. Regarding the 5 nm NPs, a change of the monosaccharide analogue loadings from 100% to 75% and 50% with respect to a tri(ethylene glycol) thiol-ending ligand caused a decrease of inhibition capabilities, providing evidence of the multivalent nature of the interaction with the antibodies. Thus, 5 nm NPs 100%-loaded with the di- and trisaccharide analogues were also tested. No significant improvement in binding inhibition was found, indicating that the antibodies do not recognise the number of saccharides per thiolated unit but instead the terminal glycosidic unit. All the GNPs were two orders of magnitude more potent than the corresponding monovalent saccharides.

Not only the size of gold nanoparticles, but also their shape can influence the immunological response. The use of gold nanoparticles as a platform for vaccines against West Nile virus has been recently described and the results confirmed that the efficiency of gold nanoparticles as vaccine adjuvants/carriers can be increased by modulation of their sizes and shapes.[156] This recent finding further stresses the opportunity to use gold nanoparticles as vaccine carriers and it is predicted that more systematic studies with glyconanoparticles will be performed in this direction.

5.4.1.3 GNPs for HIV Vaccine Design

The development of a carbohydrate-based vaccine against HIV is a great challenge in the field of glycoconjugate vaccines. Several studies have been carried out to disclose the molecular mechanisms of antibody recognition of the high mannosides clusters present at the HIV envelope surface.[157] The cluster-type presentation of high mannose-type N-glycans at the viral gp120 surface evokes an immune response able to raise specific anti-carbohydrate IgG antibodies, such as the broadly neutralizing antibody 2G12.[158] The multivalent nature of the 2G12–oligomannoside interactions inspired scientists to multimerise oligomannosides and/or their partial structures onto different scaffolds in order to mimic the natural presentation of gp120 glycans.[159]

Gold nanoparticles functionalised with oligomannosides (*manno*-GNPs)[61] can offer an alternative scaffold in the development of an anti-HIV carbohydrate-based vaccine. In order to rationalise the design of gold nanoparticles covered with synthetic oligomannosides, the interactions between 2G12 and selected oligomannosides were studied at the molecular level also by our group by saturation transfer difference NMR spectroscopy (STD-NMR) and transferred NOE in isotropic solution.[160,161] In good agreement with other work, it was found that linear tri- and tetramannosides containing the α-D-Man*p*-(1→2)-α-D-Man*p*-(1→2)-α-D-Man*p*-(1→) moiety were the strongest 2G12 binders, among the analysed series of ligands. This structural information was a key starting point for the design of synthetic multivalent GNPs for the development of HIV vaccine candidates. *manno*-GNPs carrying different amounts of linear tri- and tetramannosides were able to bind 2G12 with high affinity and interfere with 2G12/gp120 binding, as determined by SPR-based biosensors and STD-NMR.[162] In addition, cellular neutralisation assays demonstrated that GNPs coated with the linear tetramannoside α-D-Man*p*-(1→2)-α-D-Man*p*-(1→2)-α-D-Man*p*-(1→3)-α-D-Man*p*-(1→) were able to inhibit the 2G12-mediated neutralisation of a replication-competent HIV-1 in the micromolar range. These results provided fundamental evidence for further exploration of gold nanoparticles as carriers for anti-HIV carbohydrate-based vaccines.

5.4.1.4 Immune Modulation

To improve the effectiveness of glycoconjugates in inducing adaptive immune responses, the behaviour of multivalent carbohydrate systems has been also studied *in vitro* for triggering the innate part of the human immune system. Metallic nanoparticles (5–10 nm) coated with tiopronine[163] or Cys-modified peptides[164] have been described to trigger macrophage activation *in vitro*, demonstrating that it is possible to modulate cell-mediated immune responses with suitable nanotools. Few examples have been presented with gold nanoparticles bearing carbohydrate antigens. To better explore the potentiality of GNPs as carriers for carbohydrate-based meningococcal vaccines, their ability to induce T cell responses was determined in human autologous monocytes–T cell co-cultures.[165] Gold nanoparticles of ∼2 and 5 nm coated with sugars related to the capsular polysaccharide of serogroup A of the *Neisseria meningitidis* bacterium (Figure 5.10a) were co-incubated with monocytes and the cross-talk with autologous T cells was studied. The 5 nm GNPs functionalised with a disaccharide analogue of MenA CPS significantly induced human T cell proliferation (Figure 5.10b) and IL-2 release. The corresponding GNPs of ∼2 nm showed a weaker but significant effect. The free disaccharide, tri(ethylene glycol)-functionalised GNPs and GNPs functionalised with a monosaccharide analogue failed to induce a significant T cell proliferation at all concentrations tested.

A parallel set of experiments with murine cell lines was also performed in this study. The 5 nm GNPs carrying the mono- or disaccharide analogue of *N. meningitidis* showed similar effects in activating a mouse monocyte/macrophage cell line (RAW 264.7). These GNPs were able to induce mouse macrophage differentiation into dendritic-like cells, as shown by morphological transformation characterised by multiple prominent cytoplasmic processes. In addition, these GNPs induced the up-regulation of MHC II on macrophages, a phenomenon that is involved in antigen presentation to T lymphocytes. Also in this part of the work, ∼2 nm GNPs showed a weaker effect than the corresponding 5 nm GNPs. These combined results, especially for the GNP dimension effect, can contribute to understanding the interactions between carbohydrate-multivalent nanotools and the immune system for designing nanoparticle-based vaccines.

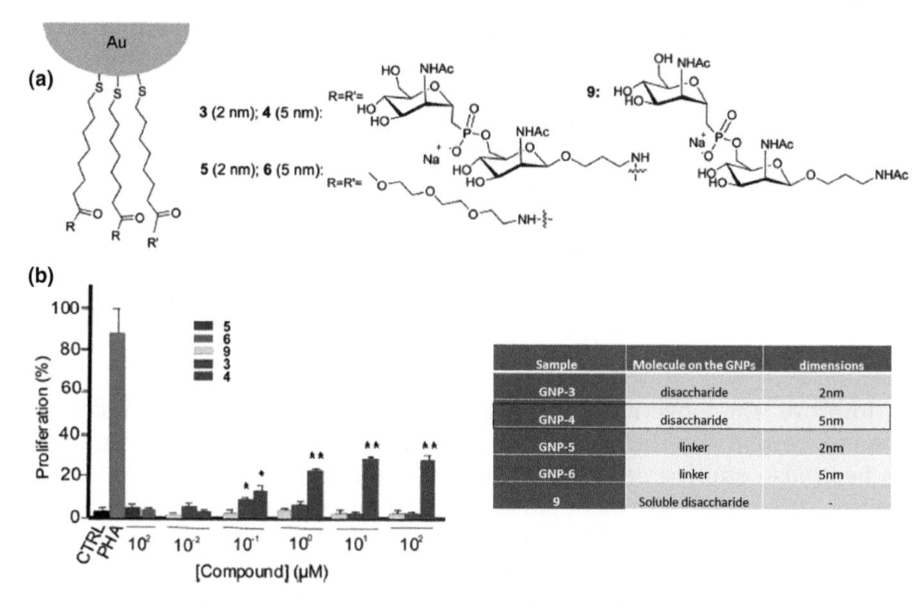

Sample	Molecule on the GNPs	dimensions
GNP-3	disaccharide	2nm
GNP-4	disaccharide	5nm
GNP-5	linker	2nm
GNP-6	linker	5nm
9	Soluble disaccharide	-

Figure 5.10 (a) Structural representation of GNPs and thiol-ending conjugates tailored at a gold surface. (b) Effects of the different samples on T cell proliferation, as measured by fluorescence assisted cell sorting (FACS) and expressed as mean percentages of proliferated T cells. Disaccharide-functionalised GNP-3 (2 nm gold diameter) and especially GNP-4 (5 nm gold diameter) induced a significant T cell proliferation, the latter in a concentration-dependent way. Tri(ethylene glycol)-functionalised GNP-5 and GNP-6 and monovalent disaccharide 9 did not trigger T cell proliferation. Compound-untreated cells (CTRL) were used as negative controls. Cells treated with phytohaemagglutinin (PHA) were considered as positive controls. $*p<0.05$; $**p<0.01$ *vs.* CTRL.[165]

Glycoconjugates containing terminal galactofuranoside (Gal*f*) residues are found in many microorganisms, including trypanosomatids, fungi and bacteria.[166] The study of the specific role of Gal*f* with different cells of the host immune system is complicated by the presence of many other molecules on the surface of these pathogens. To clarify the role of β-D-galactofuranosides in modulating the immune system, Gal*f*-coated gold nanoparticles (Figure 5.11a) were prepared and their interactions with human dendritic cells were studied.[167] It was found that human dendritic cells (DCs) bind to Gal*f*-GNPs *via* a C-type lectin and Gal*f*-GNPs are capable of eliciting a pro-inflammatory response in DCs, as demonstrated by the up-regulation of surface co-stimulatory markers such as CD86, CD80 and HLA-DR and secretion of pro-inflammatory cytokines (IL-6 and TNF-α) (Figure 5.11b). In addition, the pro-inflammatory response was also related to the Gal*f* loading on the GNPs. In this study, soluble Gal*f* was not able to induce these kinds of immune responses. These data suggest that Gal*f* is a pathogen molecular pattern that is recognised by the human innate immune system when suitably presented. The key information coming from this work

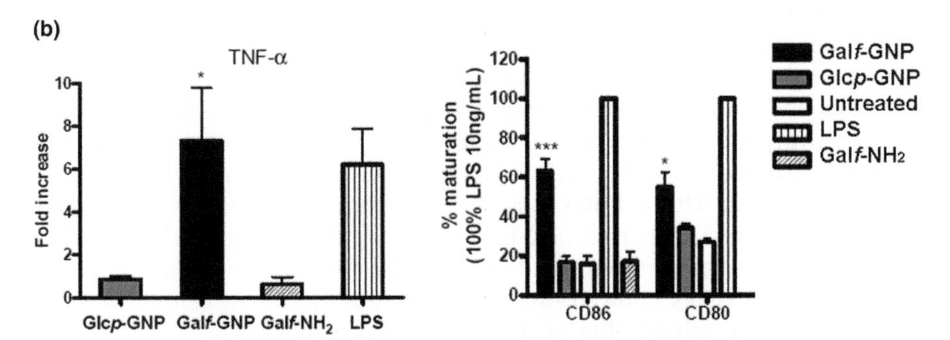

Figure 5.11 (a) Structure of galactofuranose-coated GNPs (Gal*f*-GNPs); ~2 nm gold nanoparticles carrying approximately 90% β-D-Glc*p* and 10% β-D-Gal*f* thiol-ending neoglycoconjugates. (b) *Left*: TNF-α production by human dendritic cells (DCs) after stimulation with Glc*p*- or Gal*f*-GNPs and soluble Gal*f*-NH_2, as measured by ELISA (the fold increase over the untreated DCs for three donors is represented); lipopolysaccharide (LPS) was used as positive control. *Right*: Surface expression of the maturation markers CD86 and CD80 by FACS analysis. DCs from different donors were untreated, or incubated for 16 h with Gal*f*-GNPs (10 μg mL^{-1}) or LPS (10 ng mL^{-1}). The percentage of maturation was normalised, with the LPS-treated DCs set at 100%. For CD86, the results of six donors are shown; for CD80, the results for three donors are shown. Error bars depict SEM, and asterisks indicate significant differences between Gal*f*-GNPs and Glc*p*-GNPs (*$p<0.05$; ***$p<0.001$).[167]

was that multivalent GNPs are an interesting tool to better understand the role of pathogen-derived carbohydrates in host–pathogen interactions.

All these studies may help the design and synthesis of carbohydrate-based vaccines, thanks to the ability of some carbohydrates in triggering the human innate immune system, acting like adjuvants.

5.4.2 Drug and Gene Delivery

The delivery of drugs or genetic material by means of gold nanoparticles is a flourishing field[12] and several strategies have been developed to attach drugs to gold nanoparticles.[168] The most outstanding examples towards the generation of non-viral transfection vectors and gene therapy come from the groups of Mirkin[139] and Rotello.[169]

The application of gold nanoparticles coated with carbohydrates for non-viral gene delivery has attracted interest in recent years. The binding of a model DNA plasmid to gold GNPs that present α- or β-galactopyranosides and amino groups at their surface was studied by TEM and AFM.[170] β-Gal*p*-functionalised GNPs exhibited much less affinity for the plasmid than amino/α-Gal*p*-GNPs. This indicates that a change at the anomeric configuration can influence the binding to DNA, with implications for the potential use of GNPs as non-viral gene delivery systems.

Carbohydrate–oligonucleotide conjugates coupled to gold nanoparticles were prepared for oligonucleotide delivery to hepatic parenchymal cells.[171] Oligonucleotides were modified with different monosaccharides at the 3′-terminus and multimerised on gold nanoparticles. Galactose (Gal), glucose (Glc) or a mixture of Gal and Glc (Gal/Glc = 1/1) oligonucleotide conjugates were constructed and the cellular uptake into rat primary hepatocytes was investigated. The cellular uptake of oligonucleotide-conjugated gold nanoparticles was increased ~2-fold by the galactose modification. No internalisation enhancement was found for the GNPs carrying Gal/Glc. Glc–oligonucleotide GNPs showed only slightly higher levels of cellular uptake than that of unmodified oligonucleotide–gold nanoparticles.

Narain and colleagues prepared cationic glucopolymer-stabilised gold nanoparticles as gene delivery carriers.[172] These biocompatible and stable GNPs were complexed to the enhanced cyan fluorescent protein-encoded DNA plasmid (pECFP-N1) and the mechanism of uptake by Hela cells of the resulting plasmid–glyconanoparticles was studied by confocal microscopy. The transfection efficiency was comparable to the one obtained with the commercially available transfection reagent Lipofectamine® 2000 (a liposomal formulation). The choice of more complex and targeting carbohydrates may help in the development of efficient gene delivery systems.

Other examples of gold nanoparticles coated with carbohydrate-based materials, such as chitosan[173] and β-cyclodextrin,[174] as vectors for DNA binding, have appeared in the literature. Usually, polysaccharide-based glyconanoparticles have received much more attention in drug delivery with respect to hybrid GNPs. This is due to their high biocompatibility and low cytotoxicity and the possibility to encapsulate the drug/DNA inside the polymer-like matrix or to functionalise selected groups in the external part of the polysaccharide.

In the field of drug release, we have studied the anti-viral activity of novel GNPs loaded with anti-HIV drugs in clinical use.[175] The nucleoside reverse transcriptase inhibitors abacavir and lamivudine were conjugated to ~3 nm glucose-coated gold nanoparticles as thiol-ending esters. The choice of glucose as a component of the organic shell of these NPs was made on the basis of previous results indicating that glucose-coated gold nanoparticles are water-dispersible, non-cytotoxic to different cell lines at high concentrations and facilitate the capture of the nanoparticles by different types of cells.[112,128] Furthermore, glucose-bearing gold nanoparticles do not elicit any immune response in model animals.[103,152] The drug loadings of the GNPs were characterised by LC-MS and the pH-mediated release of the drug from the nanoparticles was studied. The anti-viral activity of these glyconanomaterials was tested *in vitro* by evaluating the replication of an HIV strain in TZM-bl infected cells. The proof-of-principle presented in this work aims at introducing GNPs as a new multifunctional drug delivery system in the therapy against HIV.

A nice example of drug loading in multifunctional NPs has been reported by Lin and co-workers.[176] Trivalent β-galactoside dendrons (targeting ligands for hepatocytes) and an anticancer drug (paclitaxel) were combined on fluorescent silicon oxide NPs. The high sugar loading also overcame the low water solubility of paclitaxel. This concept may be further translated to gold nanoparticles.

Before closing, it is worthwhile to dedicate a few words to drugs containing carbohydrate moieties, a category which includes antibiotics, antiviral compounds and chemotherapeutics, among others.[177] Conjugation of such kinds of molecule to gold nanoparticles is appealing for targeted drug delivery. Although these systems cannot be strictly defined as "glyconanoparticles", some examples may be of help to those readers who are interested in this topic. One of the first works on drugs bearing carbohydrate components conjugated to gold nanoparticles was reported by Xu and collaborators.[178] The conjugation of the antibiotic vancomycin to gold nanoparticles led to enhanced *in vitro* antibacterial activities against vancomycin-resistant enterococcocci. Macrolide antibiotics (erythromycin and its synthetic derivatives) consist of a macrocyclic moiety attached to deoxy sugars, such as cladinose and desosamine, which are essential for the therapeutic properties of the drug.[177] Thiol-PEGylated azithromycin, clarithromycin and a tricyclic ketolide were conjugated to gold nanorods (NRs) for targeting tumour-associated macrophages (TAM) that exhibit high intrinsic accumulation and infiltration into solid tumours.[179] The effects of NR-loaded TAMs and NIR laser exposure ($\lambda = 808$ nm) on cell viability in breast adenocarcinoma co-cultures was studied. TAM-dependent cytotoxicity towards cancer cells was selectively induced, with no significant photothermal ablation of treated TAMs in co-culture. These experiments suggest that drug-functionalised and anisotropic gold nanoparticles are promising candidates for targeted cancer drug delivery and photothermal therapy.

5.5 Conclusion

In this chapter, we have showed how carbohydrates and gold nanoparticles can be integrated in a sort of "glyconanotechnology" in order to create multivalent systems which can help in basic studies, but also in solving biomedical problems where carbohydrate-based interactions are involved. In particular, we focused on gold glyconanoparticles (GNPs) as multivalent and synthetic nanotools which can find potential application as anti-pathogenic agents or in imaging-based diagnostics and therapy, including nanovaccines. While the translation of GNPs to the clinic is easy to foresee in biosensors (following the steps taken with DNA–gold NPs), much more has to be done to clarify the effectiveness and safety for systemic and routine administration in humans. On the other hand, GNPs have a great potential for *in vivo* applications which consist in external administration and/or sporadic use, such as molecular imaging and vaccines. Although "gold glyconanotechnology" is relatively young, it can certainly contribute to advances in the design of carbohydrate-based drugs and the development of multifunctional and multimodal systems which could exert one or several specific biological functions (targeting, delivery, therapeutic), together with the opportunity of being monitored by imaging techniques. In the field of therapeutics, it can also be foreseen that the use of carbohydrate-coated anisotropic gold nanoparticles will soon find a prominent place in glycoscience. In conclusion, GNPs can be added to the list of carbohydrate-based multivalent systems at the disposal of glycoscientists to address biological problems.

Acknowledgements

This work was supported by the Spanish Government (Grant CTQ2011-27268) and the Basque Government (Etortek Program IE11-301 and Project PI2012-46). M.M. acknowledges COST action CM1102. Prof. Soledad Penadés is gratefully acknowledged for her support throughout the last eight years.

References

1. *Nanotechnology*, ed. V. Vogel, Wiley-VCH Verlag GmbH & Co. KGaA, Weinheim, 2009, vol. 5.
2. D.-E. Lee, H. Koo, I.-C. Sun, J. H. Ryu, K. Kim and I. C. Kwon, *Chem. Soc. Rev.*, 2012, **41**, 2656.
3. T. L. Doane and C. Burda, *Chem. Soc. Rev.*, 2012, **41**, 2885.
4. N. L. Rosi and C. A. Mirkin, *Chem. Rev.*, 2005, **105**, 1547.
5. Y.-W. Jun, J.-W. Seo and J. Cheon, *Acc. Chem. Res.*, 2008, **41**, 179.
6. R. Weissleder, M. Nahrendorf and M. J. Pittet, *Nat. Mater.*, 2014, **13**, 125.
7. R. Tel-Vered, O. Yehezkeli and I. Willner, *Adv. Exp. Med. Biol.*, 2012, **733**, 1.
8. C.-C. You, A. Chompoosor and V. M. Rotello, *Nano Today*, 2007, **2**, 34.
9. B. Pelaz, G. Charron, C. Pfeiffer, Y. Zhao, J. M. de la Fuente, X.-J. Liang, W. J. Parak and P. del Pino, *Small*, 2013, **9**, 1573.
10. S. Su, X. Zuo, D. Pan, H. Pei, L. Wang, C. Fan and W. Huang, *Nanoscale*, 2013, **5**, 2589.
11. L. Dykman and N. Khlebtsov, *Chem. Soc. Rev.*, 2012, **41**, 2256.
12. E. Boisselier and D. Astruc, *Chem. Soc. Rev.*, 2009, **38**, 1759.
13. E. C. Dreaden, A. M. Alkilany, X. Huang, C. J. Murphy and M. A. El-Sayed, *Chem. Soc. Rev.*, 2012, **41**, 2740.
14. B. Duncan, C. Kim and V. M. Rotello, *J. Control. Release*, 2010, **148**, 122.
15. J. C. Love, L. A. Estroff, J. K. Kriebel, R. G. Nuzzo and G. M. Whitesides, *Chem. Rev.*, 2005, **105**, 1103.

16. K. Saha, S. S. Agasti, C. Kim, X. Li and V. M. Rotello, *Chem. Rev.*, 2012, **112**, 2739.
17. C. J. Murphy, A. M. Gole, J. W. Stone, P. N. Sisco, A. M. Alkilany, E. C. Goldsmith and S. C. Baxter, *Acc. Chem. Res.*, 2008, **41**, 1721.
18. D. A. Giljohann, D. S. Seferos, W. L. Daniel, M. D. Massich, P. C. Patel and C. A. Mirkin, *Angew. Chem., Int. Ed.*, 2010, **49**, 3280.
19. M. Grzelczak, J. Pérez-Juste, P. Mulvaney and L. M. Liz-Marzán, *Chem. Soc. Rev.*, 2008, **37**, 1783.
20. A. Varki and N. Sharon, in *Essentials of Glycobiology*, ed. A. Varki, R. D. Cummings, J. D. Esko, H. H. Freeze, P. Stanley, C. R. Bertozzi, G. W. Hart and M. E. Etzler, Cold Spring Harbor, New York, 2nd edn, 2009, ch. 1, pp. 1–22.
21. R. A. Dwek, *Chem. Rev.*, 1996, **96**, 683.
22. H. J. Gabius, H. C. Siebert, S. André, J. Jiménez-Barbero and H. Rüdiger, *ChemBioChem*, 2004, **5**, 740.
23. Y. C. Lee and R. T. Lee, *Acc. Chem. Res.*, 1995, **28**, 321.
24. S. I. Hakomori, *Pure Appl. Chem.*, 1991, **63**, 473.
25. M. Mammen, S.-K. Choi and G. M. Whitesides, *Angew. Chem., Int. Ed.*, 1998, **37**, 2755.
26. J. E. Hudak and C. R. Bertozzi, *Chem. Biol.*, 2014, **21**, 16.
27. J. M. de la Fuente and S. Penadés, *Glycoconjugate J.*, 2004, **21**, 149.
28. N. C. Reichardt, M. Martín-Lomas and S. Penadés, *Chem. Soc. Rev.*, 2013, **42**, 4358.
29. The acronym GNP is often used in the literature to indicate Gold NanoParticles. However, for simplicity and for maintaining the terminology of the seminal works, we will use the GNP acronym for Gold GlycoNanoParticles all through this article.
30. I. García, M. Marradi and S. Penadés, *Nanomedicine*, 2010, **5**, 777.
31. J. M. de la Fuente, A. G. Barrientos, T. C. Rojas, J. Rojo, J. Cañada, A. Fernández and S. Penadés, *Angew. Chem., Int. Ed.*, 2001, **40**, 2258.
32. I. Eggens, B. Fenderson, T. Toyokuni, B. Dean, M. Stroud and S.-I. Hakomori, *J. Biol. Chem.*, 1989, **264**, 9476.
33. G. N. Misevic, J. Finne and M. M. Burger, *J. Biol. Chem.*, 1987, **262**, 5870.
34. A. Carvalho de Souza and J. P. Kamerling, *Methods Enzymol.*, 2006, **417**, 221.
35. J. I. Santos, A. Carvalho de Souza, F. J. Cañada, S. Martín-Santamaría, J. P. Kamerling and J. Jiménez-Barbero, *ChemBioChem*, 2009, **10**, 511.
36. A. J. Reynolds, A. H. Haines and D. A. Russell, *Langmuir*, 2006, **22**, 1156.
37. M. Marradi, M. Martín-Lomas and S. Penadés, *Adv. Carbohydr. Chem. Biochem.*, 2010, **64**, 211.
38. M. Marradi, F. Chiodo, I. Garcia and S. Penadés, *Chem. Soc. Rev.*, 2013, **42**, 4728.
39. A. G. Barrientos, J. M. de la Fuente, T. C. Rojas, A. Fernandez and S. Penades, *Chem. Eur. J.*, 2003, **9**, 1909.

40. M. Marradi, F. Chiodo, I. García and S. Penadés, in *Synthesis and Biological Applications of Glycoconjugates*, ed. O. Renaudet and N. Spinelli, Bentham Science Publishers Ltd., 2011, ch. 10, pp. 164–202.
41. K. El-Boubbou and X. Huang, *Curr. Med. Chem.*, 2011, **18**, 2060.
42. G. Huang, F. Cheng, X. Chen, D. Peng, X. Hu and G. Liang, *Curr. Pharm. Des.*, 2013, **19**, 2454.
43. R. Sunasee and R. Narain, *Macromol. Biosci.*, 2013, **13**, 9.
44. T. K. Dam and C. F. Brewer, *Adv. Carbohydr. Chem. Biochem.*, 2010, **63**, 139.
45. A. Varki, M. E. Etzler, R. D. Cummings and J. D. Esko, in *Essentials of Glycobiology*, ed. A. Varki, R. D. Cummings, J. D. Esko, H. H. Freeze, P. Stanley, C. R. Bertozzi, G. W. Hart and M. E. Etzler, Cold Spring Harbor, New York, 2nd edn, 2009, ch. 26, pp. 375–386.
46. H. Lis and N. Sharon, *Chem. Rev.*, 1998, **98**, 637.
47. R. J. Pieters, *Org. Biomol. Chem.*, 2009, 7, 2013.
48. L. L. Kiessling and N. L. Pohl, *Chem. Biol.*, 1996, **3**, 71.
49. J. J. Lundquist and E. J. Toone, *Chem. Rev.*, 2002, **102**, 555.
50. L. L. Kiessling, J. E. Gestwicki and L. E. Strong, *Curr. Opin. Chem. Biol.*, 2000, **4**, 696.
51. M. Ambrosi, N. R. Cameron and B. G. Davis, *Org. Biomol. Chem.*, 2005, **3**, 1593.
52. H. Otsuka, Y. Akiyama, Y. Nagasaki and K. Kataoka, *J. Am. Chem. Soc.*, 2001, **123**, 8226.
53. G. L. Nicolson and J. Blaustein, *Biochim. Biophys. Acta, Biomembr.*, 1972, **266**, 543.
54. N. Bogdan, R. Roy and Mario Morin, *RSC Adv.*, 2012, **2**, 985.
55. X. Wang, E. Matei, L. Deng, O. Ramström, A. M. Gronenborn and M. Yan, *Chem. Commun.*, 2011, **47**, 8620.
56. A. Imberty and A. Varrot, *Curr. Opin. Struct. Biol.*, 2008, **18**, 567.
57. Y. M. Chabre and R. Roy, *Adv. Carbohydr. Chem. Biochem.*, 2010, **63**, 165.
58. A. Bernardi, J. Jiménez-Barbero, *et al.*, *Chem. Soc. Rev.*, 2013, **42**, 4709.
59. E. De Clercq, *Curr. Opin. Pharmacol.*, 2010, **10**, 507.
60. Y. van Kooyk and T. B. H. Geijtenbeek, *Nat. Rev. Immunol.*, 2003, **3**, 697.
61. O. Martinez-Avila, K. Hijazi, M. Marradi, C. Clavel, C. Campion, C. Kelly and S. Penades, *Chem. – Eur. J.*, 2009, **15**, 9874.
62. O. Martínez-Ávila, L. M. Bedoya, M. Marradi, C. Clavel, J. Alcami and S. Penadés, *ChemBioChem*, 2009, **10**, 1806.
63. A. Berzi, J. Reina, R. Ottria, I. Sutkeviciute, P. Antonazzo, M. Sánchez-Navarro, E. M. Chabrol, M. Biasin, D. Trabattoni, I. Cetin, J. Rojo, F. Fieschi, A. Bernardi and M. Clerici, *AIDS*, 2012, **26**, 127.
64. M. Marradi, I. García and Soledad Penadés, Progress in Molecular Biology and Translational Science, in *Nanoparticles in Translational Science and Medicine*, ed. A. Villaverde, Elsevier, Academic Press, Burlington, 2011, vol. 104, pp. 141–173.
65. L. Sihelníková and I. Tvaroška, *Chem. Pap.*, 2007, **61**, 237.

66. T. K. Lindhorst, in *Synthesis and Biological Applications of Glycoconjugates*, ed. O. Renaudet and N. Spinelli, Bentham Science Publishers Ltd., 2011, ch. 2, pp. 12–35.
67. C.-C. Lin, Y.-C. Yeh, C.-Y. Yang, C.-L. Chen, G.-F. Chen, A. C. Chen and Y.-C. Wu, *J. Am. Chem. Soc.*, 2002, **124**, 3508.
68. G. Cioci, E. P. Mitchell, C. Gautier, M. Wimmerova, D. Sudakevitz, S. Pérez, N. Gilboa-Garber and A. Imberty, *FEBS Lett.*, 2003, **555**, 297.
69. S. P. Diggle, R. E. Stacey, C. Dodd, M. Camara, P. Williams and K. Winzer, *Environ. Microbiol.*, 2006, **8**, 1095.
70. M. Reynolds, M. Marradi, A. Imberty, S. Penadés and S. Pérez, *Chem. – Eur. J.*, 2012, **18**, 4264.
71. J. R. Govan and V. Deretic, *Microbiol. Rev.*, 1996, **60**, 539.
72. E. Lameignere, T. C. Shiao, R. Roy, M. Wimmerova, F. Dubreuil, A. Varrot and A. Imberty, *Glycobiology*, 2010, **20**, 87.
73. M. Reynolds, M. Marradi, A. Imberty, S. Penadés and S. Pérez, *Glycoconj. J.*, 2013, **30**, 747.
74. P. I. Kitov, J. M. Sadowska, G. Mulvey, G. D. Armstrong, H. Ling, N. S. Pannu, R. J. Read and D. R. Bundle, *Nature*, 2000, **403**, 669.
75. Y. Y. Chien, M. D. Jan, A. K. Adak, H. C. Tzeng, Y. P. Lin, Y. J. Chen, K. T. Wang, C. T. Chen, C. C. Chen and C. C. Lin, *ChemBioChem*, 2008, **9**, 1100.
76. A. A Kulkarni, C. Fuller, H. Korman, A. A. Weiss and S. S. Iyer, *Bioconjugate Chem.*, 2010, **21**, 1486.
77. J. Balzarini and L. Van Damme, *Lancet*, 2007, **369**, 787.
78. P. Di Gianvincenzo, M. Marradi, O. M. Martinez-Avila, L. M. Bedoya, J. Alcamí and S. Penadés, *Bioorg. Med. Chem. Lett.*, 2010, **20**, 2718.
79. D. Baram-Pinto, S. Shukla, A. Gedanken and R. Sarid, *Small*, 2010, **6**, 1044.
80. J. Vonnemann, C. Sieben, C. Wolff, K. Ludwig, C. Böttcher, A. Herrmann and R. Haag, *Nanoscale*, 2014, **6**, 2353.
81. A. Varki, *Proc. Natl. Acad. Sci. U. S. A.*, 1994, **91**, 7390.
82. R. Hevey and C.-C. Ling, *Adv. Carbohydr. Chem. Biochem.*, 2013, **69**, 125.
83. M. Roskamp, S. Enders, F. Pfrengle, S. Yekta, V. Dekaris, J. Dernedde, H.-U. Reissig and S. Schlecht, *Org. Biomol. Chem.*, 2011, **9**, 7448.
84. A. G. Barrientos, J. M. de la Fuente, M. Jiménez, D. Solís, J. Cañada, M. Martín-Lomas and S. Penadés, *Carbohydr. Res.*, 2009, **344**, 1474.
85. M. B. Thygesen, J. Sauer and K. J. Jensen, *Chem. – Eur. J.*, 2009, **15**, 1649.
86. Y.-K. Lyu, K.-R. Lim, B.-Y. Lee, K. S. Kim and W.-Y. Lee, *Chem. Commun.*, 2008, 4771.
87. E. Mahon, Z. Mouline, M. Silion, A. Gilles, M. Pinteala and M. Barboiu, *Chem. Commun.*, 2013, **49**, 3004–3006.
88. D. Craig, J. Simpson, K. Faulds and D. Graham, *Chem. Commun.*, 2013, **49**, 30.
89. O. A. Loaiza, P. J. Lamas-Ardisana, E. Jubete, E. Ochoteco, I. Loinaz, G. Cabañero, I. García and S. Penadés, *Anal. Chem.*, 2011, **83**, 2987.

90. L. Ding, R. Qian, Y. Xue, W. Cheng and H. Ju, *Anal. Chem.*, 2010, **82**, 5804.
91. J. A. Creighton and D. G. Eadon, *J. Chem. Soc., Faraday Trans.*, 1991, **87**, 3881.
92. W. Zhao, M. A. Brook and Y. Li, *ChemBioChem*, 2008, **9**, 2363.
93. R. Elghanian, J. J. Storhoff, R. C. Mucic, R. L. Letsinger and C. A. Mirkin, *Science*, 1997, **277**, 1078.
94. L. Schofield, R. A. Field and D. A. Russell, *Anal. Chem.*, 2007, **79**, 1356.
95. H. S. Jayawardena, X. Wang and M. Yan, *Anal Chem.*, 2013, **85**, 10277.
96. S.-J. Richards, E. Fullam, G. S. Besra and M. I. Gibson, *J. Mater. Chem. B*, 2014, **2**, 1490.
97. C.-C. Yu, L.-D. Huang, D. H. Kwan, W. W. Wakarchuk, S. G. Withers and C.-C. Lin, *Chem. Commun.*, 2013, **49**, 10166.
98. S. J. Gamblin and J. J. Skehel, *J. Biol. Chem.*, 2010, **285**, 28403.
99. I. Papp, C. Sieben, K. Ludwig, M. Roskamp, C. Bottcher, S. Schlecht, A. Herrmann and R. Haag, *Small*, 2010, **6**, 2900.
100. M. J. Marín, A. Rashid, M. Rejzek, S. A. Fairhurst, S. A. Wharton, S. R. Martin, J. W. McCauley, T. Wileman, R. A. Field and D. A. Russell, *Org. Biomol. Chem.*, 2013, **11**, 7101.
101. J. Wei, L. Zheng, X. Lv, Y. Bi, W. Chen, W. Zhang, Y. Shi, L. Zhao, X. Sun, F. Wang, S. Cheng, J. Yan, W. Liu, X. Jiang, G. F. Gao and X. Li, *ACS Nano*, 2014, **8**, 4600.
102. O. Oyelaran, L. M. McShane, L. Dodd and J. C. Gildersleeve, *J. Proteome Res.*, 2009, **8**, 4301.
103. F. Chiodo, M. Marradi, B. Tefsen, H. Snippe, I. van Die and S. Penadés, *PLoS One*, 2013, **8**(8), e73027.
104. N. Nagahori, M. Abe and S.-I. Nishimura, *Biochemistry*, 2009, **48**, 583.
105. A. Louie, *Chem Rev.*, 2010, **110**, 3146.
106. S. I. van Kasteren, S. J. Campbell, S. Serres, D.C. Anthony, N. R. Sibson and B. G. Davis, *Proc. Natl. Acad. Sci. U. S. A.*, 2009, **106**, 18.
107. K. El-Boubbou, D. C. Zhu, C. Vasileiou, B. Borhan, D. Prosperi, W. Li and X. Huang, *J. Am. Chem. Soc.*, 2010, **132**, 4490.
108. J. Gallo, I. García, D. Padro, B. Arnáiz and S. Penadés, *J. Mater. Chem.*, 2010, **20**, 10010.
109. P.-J. Debouttière, S. Roux, F. Vocanson, C. Billotey, O. Beuf, A. Favre-Règuillon, Y. Lin, S. Pellet-Rostaing, R. Lamartine, P. Perriat and O. Tillement, *Adv. Funct. Mater.*, 2006, **16**, 2330.
110. L. Moriggi, C. Cannizaro, E. Dumas, C. R. Mayer, A. Ulianov and L. Helm, *J. Am. Chem. Soc.*, 2009, **131**, 10828.
111. M. Marradi, D. Alcántara, J. M. de la Fuente, M. L. García-Martín, S. Cerdán and S. Penadés, *Chem. Commun.*, 2009, 3922.
112. A. Irure, M. Marradi, B. Arnáiz, N. Genicio, D. Padro and S. Penadés, *Biomater. Sci.*, 2013, **1**, 658.
113. M. Kato, K. J. McDonald, S. Khan, I. L. Ross, S. Vuckovic, K. Chen, D. Munster, K. P. A. MacDonald and D. N. Hart, *Int. Immunol.*, 2006, **18**, 857.

114. S. V. Su, P. Hong, S. Baik, O. A. Negrete, K. B. Gurney and B. Lee, *J. Biol. Chem.*, 2004, **279**, 19122.
115. M. Mueckler, *Eur. J. Biochem.*, 1994, **219**, 713.
116. J. Lunney and G. Ashwell, *Proc. Natl. Acad. Sci. U. S. A.*, 1976, **73**, 341.
117. A. P. Candiota, M. Acosta, R. V. Simões, T. Delgado-Goñi, S. Lope-Piedrafita, A. Irure, M. Marradi, O. Bomati-Miguel, N. Miguel-Sancho, I. Abasolo, S. Schwartz Jr, J. Santamaria, S. Penadés and C. Arus, *J. Nanobiotechnol.*, 2014, **12**, 12.
118. S. Jung, M. Bang, B. S. Kim, S. Lee, N. A. Kotov, B. Kim and D. Jeon, *PLoS One*, 2014, **9**, e91360.
119. A. P. Alivisatos, *ACS Nano*, 2008, **2**, 1514.
120. R. Kikkeri, B. Lepenies, A. Adibekian, P. Laurino and P. H. Seeberger, *J. Am. Chem. Soc.*, 2009, **131**, 2110.
121. K.-T. Yong, W.-C. Law, R. Hu, L. Ye, L. Liu, M. T. Swiharte and P. N. Prasad, *Chem. Soc. Rev.*, 2013, **42**, 1236.
122. A. Robinson, J.-M. Fang, P.-T. Chou, K.-W. Liao, R.-M. Chu and S.-J. Lee, *ChemBioChem*, 2005, **6**, 1899.
123. A. Cambi, D. S. Lidke, D. J. Arndt-Jovin, C. G. Figdor and T. M. Jovin, *Nano Lett.*, 2007, **7**, 970.
124. D. Benito-Alifonso, S. Tremel, B. Hou, H. Lockyear, J. Mantell, D. J. Fermin, P. Verkade, M. Berry and M. C. Galan, *Angew. Chem., Int. Ed.*, 2014, **53**, 810.
125. H. Lee, K. Lee, I. K. Kim and T. G. Park, *Biomaterials*, 2008, **29**, 4709.
126. E. Dulkeith, M. Ringler, T. A. Klar, J. Feldmann, A. Muñoz Javier and W. J. Parak, *Nano Lett.*, 2005, **5**, 585.
127. B. Arnáiz, O. Martínez-Ávila, J. M. Falcon-Perez and S. Penadés, *Bioconjugate Chem.*, 2012, **23**, 814.
128. R. A. Murray, Y. Qiu, F. Chiodo, M. Marradi, S. Penadés and S. E. Moya, *Small*, 2014, 2602.
129. M. Moros, B. Hernáez, E. Garet, J. T. Dias, B. Sáez, V. Grazú, A. González-Fernández, C. Alonso and J. M. de la Fuente, *ACS Nano*, 2012, **6**, 1565.
130. P. del Pino, B. Pelaz, Q. Zhang, P. Maffre, G. U. Nienhaus and W. J. Parak, *Mater. Horiz.*, 2014, **1**, 301.
131. D. Walczyk, F. B. Bombelli, M. P. Monopoli, I. Lynch and K. A. Dawson, *J. Am. Chem. Soc.*, 2010, **132**, 5761 and references therein.
132. G. W. Doorley and C. K. Payne, *Chem. Commun.*, 2011, **47**, 466.
133. M. Chanana, P. Rivera_Gil, M. A. Correa-Duarte, L. M. Liz-Marzan and W. J. Parak, *Angew. Chem., Int. Ed.*, 2013, **52**, 4179.
134. C.-C. Huang, C.-T. Chen, Y.-C. Shiang, Z.-H. Lin and H.-T. Chang, *Anal. Chem.*, 2009, **81**, 875.
135. J. Frigell, I. Garcia, V. Gomez-Vallejo, J. Llop and S. Penadés, *J. Am. Chem. Soc.*, 2014, **136**, 449.
136. M. Carril, I. Fernandez, J. Rodriguez, I. Garcia and S. Penades, *Part. Part. Syst. Charact.*, 2014, **31**, 81.

137. J. Rojo, V. Díaz, J. M. de la Fuente, I. Segura, A. G. Barrientos, H. H. Riese, A. Bernade and S. Penadés, *ChemBioChem*, 2004, **5**, 291.
138. S. Combemale, J. N. Assam-Evoung, S. Houaidji, R. Bibi and V. Barragan-Montero, *Molecules*, 2014, **19**, 1120.
139. N. L. Rosi, D. A. Giljohann, C. S. Thaxton, A. K. R. Lytton-Jean, M. S. Han and C. A. Mirkin, *Science*, 2006, **312**, 1027.
140. Y. Zhao and X. Jiang, *Nanoscale*, 2013, **5**, 8340.
141. X. Huang, P. K. Jain, I. H. El-Sayed and M. A. El-Sayed, *Lasers Med. Sci.*, 2008, **23**, 217.
142. R. D. Astronomo and D. R. Burton, *Nat. Rev. Drug Discovery*, 2010, **9**, 308.
143. O. T. Avery and W. F Goebel, *J. Exp. Med.*, 1931, **54**, 437.
144. J. Zozaya, *Science*, 1931, **74**, 270–271.
145. C. Anish, B. Schumann, C. Lebev Pereira and P. H. Seeberger, *Chem. Biol.*, 2014, **21**, 38.
146. F. Berti and R. Adamo, *ACS Chem. Biol.*, 2013, **8**, 1653.
147. F. Peri, *Chem. Soc. Rev.*, 2013, **42**, 4543.
148. R. Ojeda, J. L. de Paz, A. G. Barrientos, M. Martín-Lomas and S. Penadés, *Carbohydr. Res.*, 2007, **342**, 448.
149. S. A. Svarovskya, Z. Szekely and J. J. Barchi Jr, *Tetrahedron: Asymmetry*, 2005, **16**, 587.
150. R. P. Brinãs, A. Sundgren, P. Sahoo, S. Morey, K. Rittenhouse-Olson, G. E. Wilding, W. Deng and J. J. Barchi Jr., *Bioconjugate Chem.*, 2012, **23**, 1513.
151. A. L. Parry, N. A. Clemson, J. Ellis, S. S. R. Bernhard, B. G. Davis and N. R. Cameron, *J. Am. Chem. Soc.*, 2013, **135**, 9362.
152. D. Safari, M. Marradi, F. Chiodo, H. A. Th. Dekker, Y. Shan, R. Adamo, S. Oscarson, G. T. Rijkers, M. Lahmann, J P. Kamerling, S. Penadés and H. Snippe, *Nanomedicine*, 2012, 7, 651.
153. H. Snippe, W. T. M. Jansen and J. P. Kamerling, Carbohydrate-Based Vaccines, in *ACS Symposium Series*, ed. R. Roy, American Chemical Society, Washington, DC, 2008, ch. 5, vol. 989, pp. 85–104.
154. R. Gasparini and D. Panatto, *Hum. Vaccines*, 2011, **7**, 170.
155. F. Manea, C. Bindoli, S. Fallarini, G. Lombardi, L. Polito, L. Lay, R. Bonomi, F. Mancin and P. Scrimin, *Adv. Mater.*, 2008, **20**, 4348.
156. K. Niikura, T. Matsunaga, T. Suzuki, S. Kobayashi, H. Yamaguchi, Y. Orba, A. Kawaguchi, H. Hasegawa, K. Kajino, T. Ninomiya, K. Ijiro and H. Sawa, *ACS Nano*, 2013, 7, 3926.
157. L. Kong, J.-P. Julien, D. Calarese, C. Scanlan, H.-K. Lee, P. Rudd, C.-H. Wong, R. A. Dwek, D. R. Burton and I. A. Wilson, *Glycobiology and Drug Design, ACS Symposium Series*, American Chemical Society, Washington, DC, 2012, ch. 7, vol. 1102, pp. 187–215.
158. D. A. Calarese, C. N. Scanlan, M. B. Zwick, S. Deechongkit, Y. Mimura, R. Kunert, P. Zhu, M. R. Wormald, R. L. Stanfield, K. H. Roux, J. W. Kelly, P. M. Rudd, R. A. Dwek, H. Katinger, D. R. Burton and I. A. Wilson, *Science*, 2003, **270**, 2065.

159. L. X. Wang., *Curr. Opin. Chem. Biol.*, 2013, **17**, 997 and references therein.
160. P. M. Enriquez-Navas, M. Marradi, D. Padro, J. Angulo and S. Penadés, *Chem. - Eur. J.*, 2011, **17**, 1547.
161. P. M. Enríquez-Navas, F. Chiodo, M. Marradi, J. Angulo and S. Penadés, *ChemBioChem*, 2012, **13**, 1357.
162. M. Marradi, P. Di Gianvincenzo, P. M. Enriquez-Navas, O. M. Martínez-Ávila, F. Chiodo, E. Yuste, J. Angulo and S. Penadés, *J. Mol. Biol.*, 2010, **410**, 798.
163. P. M. Castillo, J. L. Herrera, R. Fernandez-Montesinos, C. Caro, A. P. Zaderenko, J. A. Mejías and D. Pozo, *Nanomedicine*, 2008, **3**, 627.
164. N. G. Bastús, E. Sánchez-Tilló, S. Pujals, C. Farrera, M. J. Kogan, E. Giralt, A. Celada, J. Lloberas and V. Puntes, *Mol. Immunol.*, 2009, **46**, 743.
165. S. Fallarini, T. Paoletti, C. Orsi Battaglini, P. Ronchi, L. Lay, R. Bonomi, S. Jha, F. Mancin, P. Scrimin and G. Lombardi, *Nanoscale*, 2013, **5**, 390.
166. B. Tefsen, A. F. Ram, I. van Die and F. H. Routier, *Glycobiology*, 2012, **22**, 456.
167. F. Chiodo, M. Marradi, J. Park, A. F. Ram, S. Penadés, I. van Die and B. Tefsen, *ACS Chem. Biol.*, 2014, **9**, 383.
168. L. Vigderman and E.R. Zubarev, *Adv. Drug Deliv. Rev.*, 2013, **65**, 663.
169. S. Rana, A. Bajaj, R. Mout and V. M. Rotello, *Adv. Drug Deliv. Rev.*, 2012, **64**, 200.
170. P. Eaton, A. Ragusa, C. Clavel, C. T. Rojas, P. Graham, R. V. Durán and S. Penadés, *IEEE Trans. Nanobiosci.*, 2007, **6**, 309.
171. Y. Ikeda, D. Kubota and Y. Nagasaki, *Colloid. Polym. Sci.*, 2013, **291**, 2959.
172. M. Ahmed, Z. Deng, S. Liu, R. Lafrenie, A. Kumar and R. Narain, *Bioconjugate Chem.*, 2009, **20**, 2169.
173. X. Zhou, X. Zhang, X. Yu, X. Zha, Q. Fu, B. Liu, X. Wang, Y. Chen, Y. Chen, Y. Shan, Y. Jin, Y. Wu, J. Liu, W. Kong and J. Shen, *Biomaterials*, 2008, **29**, 111.
174. H. Wang, Y. Chen, X.-Y. Li and Y. Liu, *Mol. Pharmaceutics*, 2007, **4**, 189.
175. F. Chiodo, M. Marradi, J. Calvo, E. Yuste and S. Penadés, *Beilstein J. Org. Chem.*, 2014, **10**, 1339.
176. C.-H. Lai, T.-C. Chang, Y.-J. Chuang, D.-L. Tzou and C.-C. Lin, *Bioconjugate Chem.*, 2013, **24**, 1698.
177. A. A. Klyosov, Carbohydrate Drug Design, in *ACS Symposium Series*, ed. A. A. Klyosov, Z. J. Witczak and D. Platt, American Chemical Society, Washington, DC, 2006, ch. 1, vol. 932, pp. 2–24.
178. H. W. Gu, P. L. Ho, E. Tong, L. Wang and B. Xu, *Nano Lett.*, 2003, **3**, 1261.
179. E. C. Dreaden, S. C. Mwakwari, L. A. Austin, M. J. Kieffer, A. K. Oyelere and M. A. El-Sayed, *Small*, 2012, **8**, 2819.

CHAPTER 6

Boosting Humoral Immune Responses to Tumor-associated Carbohydrate Antigens with Virus-like Particles

ZHAOJUN YIN AND XUEFEI HUANG*

Department of Chemistry, Chemistry Building, 578 S. Shaw Lane, Michigan State University, East Lansing, MI 48824, USA
*Email: xuefei@chemistry.msu.edu

6.1 Introduction

Cancer immunotherapy that harnesses the body's own immune system to fight against cancer is emerging as a highly attractive approach for cancer treatment,[1,2] which was recently selected as "Breakthrough of the Year 2013" by *Science*.[3] The prophylactic cancer vaccines against human papillomavirus-associated cervical cancer and hepatitis B virus-induced liver cancer have been successfully developed,[4] while a dendritic cell-based therapeutic cancer vaccine targeting prostate cancer was approved by the FDA in 2010.[5] Another landmark of cancer immunotherapy is the discovery of blocking monoclonal antibodies (mAbs) against immune system checkpoint molecules such as human anti-cytotoxic T lymphocyte antigen-4 (CTLA-4) or programmed cell death 1 protein (PD1), which has shown very striking therapeutic effects on advanced melanoma.[6,7]

RSC Drug Discovery Series No. 43
Carbohydrates in Drug Design and Discovery
Edited by Jesús Jiménez-Barbero, F. Javier Cañada and Sonsoles Martín-Santamaría

Published by the Royal Society of Chemistry, www.rsc.org

Although the majority of efforts for cancer vaccine development have focused on peptide or protein antigens, the unique carbohydrate structures overexpressed by cancer cells (Figure 6.1), known as tumor-associated carbohydrate antigens (TACAs), are also recognized as promising targets.[8–11] TACAs are among the most frequent antigens found on a cancer cell's surface, which correlate strongly with cancer progression. For example, the expression level of sialyl Lewis^a (CA19-9 antigen) is associated with a high incidence of cancer recurrence and shortened survival time, as well as hepatic metastasis in colon and stomach cancer patients.[12] In addition, sialyl Lewis^a is the most frequently applied serum tumor mark for diagnosis of pancreatic cancer, with higher sensitivity than oncogene markers such as ras^k and p53.[13] For melanoma, patients having a high expression of ganglioside antigens such as GM3 have been found to have a high rate of metastasis and shortened survival.[14] Another TACA, the Tn antigen, is rarely found in normal tissues, but expressed on the surface of 70–90% of breast, lung, prostate, and pancreatic cancers.[15] High levels of Tn expression are significantly correlated with shortened disease-free interval and increased tumor metastasis in breast cancer patients.[16] In clinical studies, patients with elevated levels of naturally occurring autoantibodies against TACAs, such as GM2,[17,18] TF,[19] and STn[20] antigens, have been found to be associated with improved prognosis. Therefore, the generation of high titers of anti-TACA antibodies by active immunization can potentially be an effective approach towards carbohydrate-based cancer immunotherapy.

In order to elicit an antibody response, the antibody secreting B cells require two signals: the first is the interaction between the antigen and its B cell antigen receptor (BCR), and the second is the costimulation by cognate CD4$^+$ helper T cells through the immunological synapses (Figure 6.2). The BCR is composed of a transmembrane-bound immunoglobulin molecule for binding with extracellular antigens, and the Igα-Igβ heterodimer with immunoreceptor tyrosine-based activation motifs (ITAM). Each B cell displays up to 120 000 BCRs on its cell surface for antigen recognition. Engagement of BCR by specific antigens triggers tyrosine phosphorylation of the ITAMs and a series of downstream signaling, which lead to rapid uptake of the antigen complex, reorganization of intracellular MHC-II compartments, and elevated expression of costimulatory receptors such as CD80, CD40, and CD86.[21] Antigens internalized by BCR are subsequently processed by enzymes, and presented in complexes with MHC-II molecules to recruit specific helper T cells. Therefore, B cell activation is critically dependent on the interaction between antigens and BCR.

There are multiple challenges in inducing strong humoral responses against TACAs. The first difficulty lies in the relatively weak binding between a carbohydrate and its corresponding protein receptor, typically in range of 10^3 to 10^6 M^{-1}.[22] For a soluble monomeric antigen to activate B cells, its affinity with BCR needs be higher than 10^6 M^{-1}.[23] Furthermore, carbohydrates are normally not T cell epitopes.[24] Direct administration of TACAs can only elicit short-lived IgM or IgG3 antibodies. To generate strong and

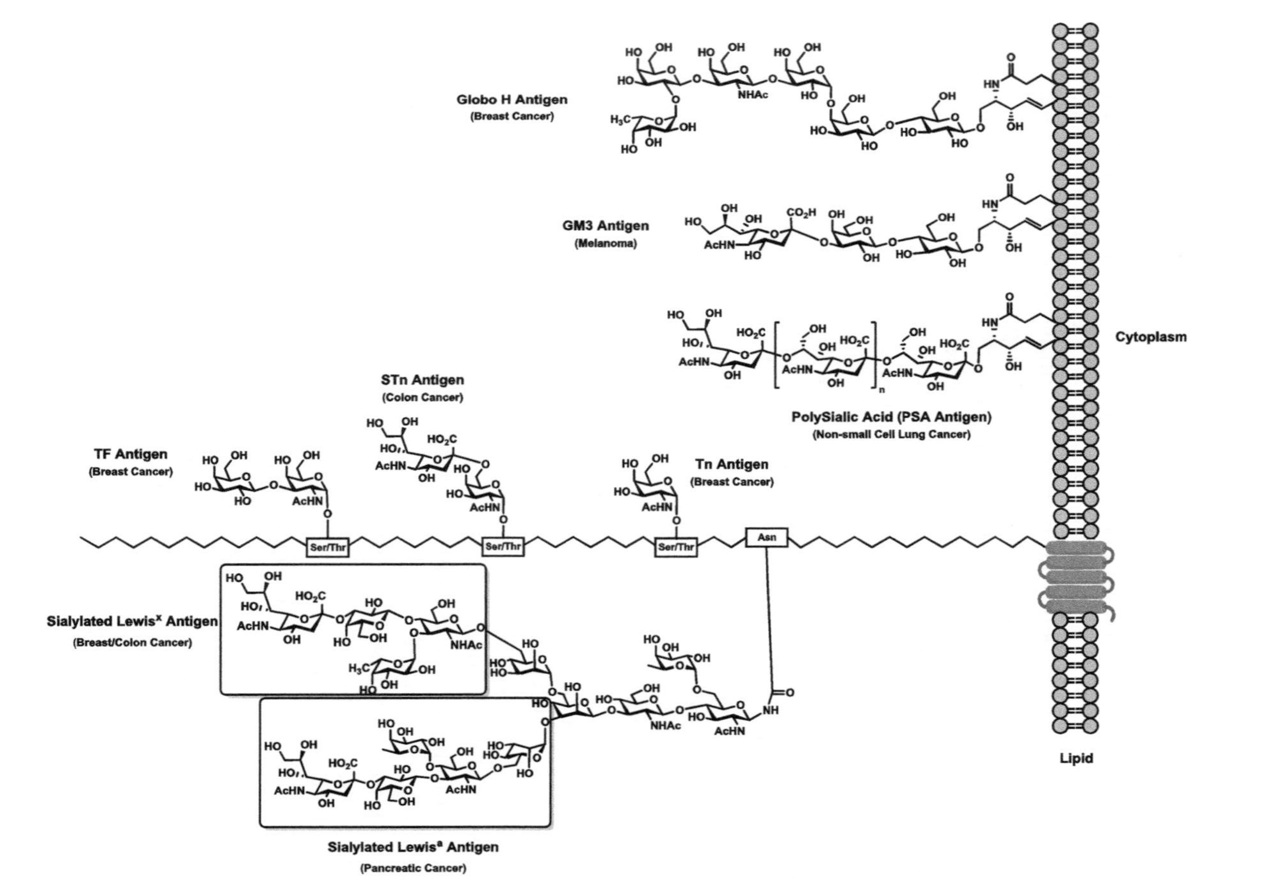

Figure 6.1 Illustration of representative tumor-associated carbohydrate antigens (TACAs). The names and chemical structures of carbohydrate antigens, as well as their expression in typical cancers, are shown. TACAs can be divided into two general classes: (1) glycolipid antigens such as Globo H, GM3, and PSA, which are linked to ceramide lipids and anchored to the lipid bilayer on the cell surface through hydrophobic interactions; (2) glycoprotein antigens such as STn, TF, Tn, and sialylated Lewis^{a}, which are linked to the protein backbone through the hydroxyl group of a serine or a threonine, or the amide group of an asparagine.

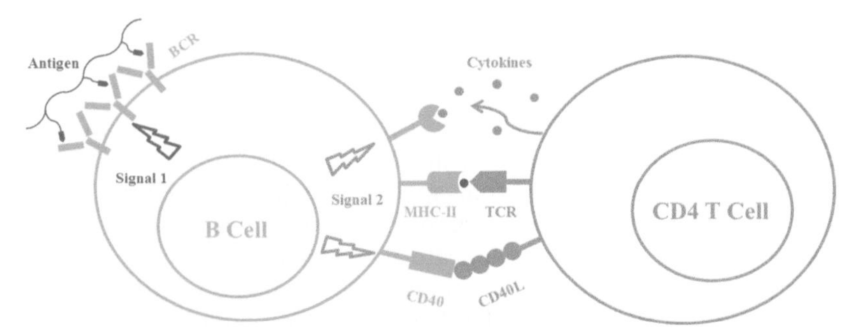

Figure 6.2 Activation of B cells. To secret high-affinity IgG antibodies, two signals are needed for B cell activation: B cell receptor (BCR) cross-linking by the antigen provides signal 1, and cognate interaction with $CD4^+$ T cells provides signal 2. In the case of polysaccharide antigens, only a strong signal 1 was provided by extensive cross-linking of BCR, which typically leads to production of low-affinity IgM antibodies.

long lasting immunity, B cell stimulation by helper T cells is crucial, which can lead to a switch of antibody isotype from IgM to high-affinity IgG, as well as induction of immunological memory.[25] An additional challenge in the development of TACA-based cancer vaccines is that TACAs are well tolerated by the immune system, as they are also expressed by normal cells, albeit at a low level resulting in low frequencies of TACA specific B cells. Therefore, enhancement of antigen affinity with BCR, incorporation of helper T cell epitopes as well as methods to break the immune tolerance are important factors to take into account in designing TACA-based vaccines.

Many innovative studies have been carried out to address the challenges discussed above, which include glycoengineering,[26] synthesis of novel adjuvants,[27,28] and development of delivery platforms such as dendrimers,[29] polysaccharides,[30] and nanoparticles,[31] as well as protein carriers.[32] Immunogenic protein carriers can provide multiple epitopes to promote helper T cell activation. Proteins such as tetanus toxoid (TT), cross-reactive materials (CRM), and keyhole limpet hemocyanin (KLH) have been conjugated with various TACAs to induce IgM and IgG responses in preclinical and clinical evaluations.[33] However, several limitations for these carriers exist. As they are widely used in many vaccine formulations, patients receiving the TACA conjugates may already have pre-existing anti-carrier antibodies due to prior immunization.[34] Such carrier-specific antibodies can suppress the immune responses to carbohydrate antigens in subsequent vaccinations.[35,36] Some carriers such as KLH are heavily glycosylated themselves.[37] The endogenous glycans on the carrier may compete with the desired TACA antigen for anti-carbohydrate antibody responses and interfere with the analysis of TACA-specific B cell responses. In addition, KLH can exit as multiple morphologies, including didecamer, multidecamers, or flexible tubules, depending on ion strength, protein concentration, and pH.[38] This renders it difficult to reduce batch-to-batch variability in vaccine preparations. The recent failures of KLH-STn[39] and KLH-GM2[40] vaccines in clinical trials indicate that further studies are needed for effective vaccine designs.

Recently, virus-like particles have been evaluated as a platform for carbohydrate-based anti-cancer vaccine development. Virus-like particles are composed of subunit proteins that spontaneously self-assemble into particles. They maintain authentic organization and conformation of native viruses without the ability to replicate in animals or humans, rendering them safe for handling and vaccination.[41] Compared with traditional protein carriers, virus-like particles have multiple potential advantages as vaccine carriers:[4]

1. Unique size: Virus-like particles are monodispersed in composition and size with a diameter typically ranging from 10 to 100 nm, which is in the size regimen for efficient trafficking to lymph nodes when administered subcutaneously.
2. Highly repetitive surface: Antigens can be displayed in a highly organized manner, which can efficiently cross-link BCR and lower the threshold for B cell activation.
3. Well-defined structure: The crystal structures of many virus-like particles are available at high resolution, and almost all virus-like particles are non-glycosylated.
4. Versatile modification: The virus-like particles can be mutated by manipulation of viral genomes to introduce reactive functional groups. In addition, the assembly–disassembly process can be controlled *in vitro* for many virus-like particles, which allows the packaging of immunostimulatory molecules such as CpG adjuvant into the particle to fine-tune the immune responses.[42]
5. Superior stability: Virus-like particles are more stable towards a broad range of pH, temperature, and solvents than standard proteins. For example, the cowpea mosaic virus (CPMV) particles can tolerate temperatures up to 60 °C for at least 1 hour and a pH range of 2–12 without being denatured.[43] This bodes well for chemical modifications of virus-like particles for antigen introduction.[44]
6. Easy production: Virus-like particles can be prepared in a large quantity. For example, 1–2 g CPMV can be obtained from 1 kg of infected plant leaves.[43] The particulate nature of virus-like particles enables facile purification by size-exclusion chromatography or density-gradient centrifugation.

In this chapter, we discuss how a highly ordered presentation of carbohydrate antigens by virus-like particles can potently enhance the carbohydrate recognition by B cells for subsequent B cell activation. We also compare the efficacy of several virus-like particles in eliciting anti-carbohydrate immune responses, and analyze the critical factors that help tailor the immune responses to carbohydrate antigens.

6.2 General Strategies for Conjugation of Carbohydrate Antigens with Virus-like Particles

In order to attach carbohydrate antigens onto virus-like particles, suitable bioconjugation chemistry needs to be developed. Syntheses of TACAs are very tedious and labor intensive. Thus, it is desirable that bioconjugation can be performed without resorting to a large excess of the highly valuable carbohydrate reagents, and that the unreacted antigens can be recovered and recycled.

There are two general approaches for bioconjugation. In the first method, the endogenous reactive groups on the external surface of the virus-like particles are targeted, with amines of lysine residues being the most common site for reaction. TACA derivatives containing activated esters such as *N*-hydroxysuccinimide (NHS) can be chemically synthesized, which react with virus-like particles containing amines on the external surface (Scheme 6.1a). Alternatively, virus-like particles can be treated with the NHS ester of pent-4-ynoic acid **1** to introduce alkynes onto the surface. This will be followed by the copper-promoted alkyne–azide cycloaddition reaction (CuAAC) with azide-functionalized TACAs (Scheme 6.1b). The CuAAC is highly advantageous for bioconjugation, as it is specific towards the alkyne and only requires a small excess of the azide ($\sim$ 2 equiv) for the reaction to complete.[45] Besides amines, innovative chemistry has been developed to functionalize the tyrosine residue.[46,47] The electron-rich phenol of tyrosine can react with a diazonium alkyne **2** through an electrophilic aromatic substitution reaction to introduce alkynes (Scheme 6.1c). Subsequent CuAAC reaction with azide-functionalized TACAs will produce TACA/virus-like particle conjugates.

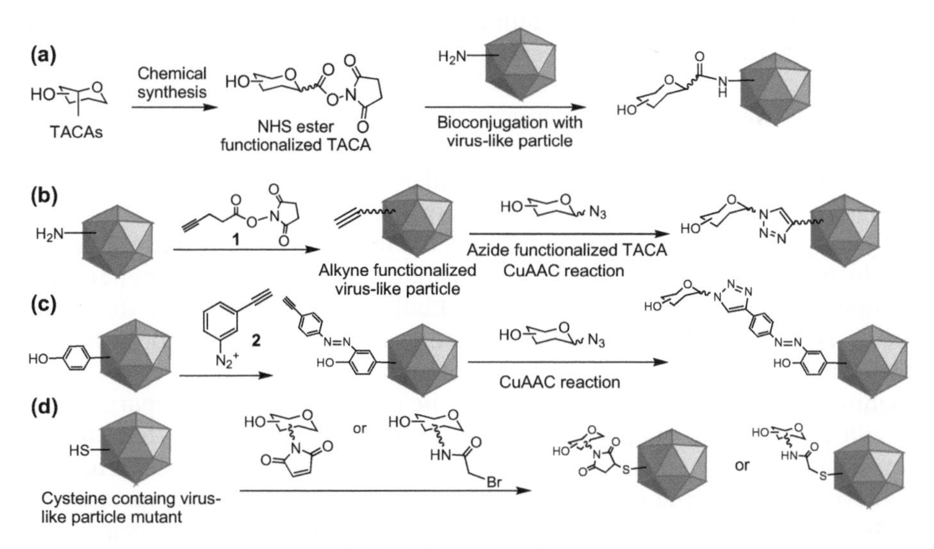

Scheme 6.1 General approach for bioconjugation of carbohydrate antigens with virus-like particles.

Besides targeting the endogenous functional groups, exogenous reactive sites can be introduced through site-directed mutagenesis. In virus-like particles lacking free cysteines, cysteine mutants can be created, which react with TACA derivatives bearing a thiol reactive moiety such as maleimide or bromoacetamide (Scheme 6.1d). Unnatural amino acid mutagenesis can also be utilized through expanded genetic codes to install the reactive site. For example, alkyne-containing L-homopropargylglycine (Hpg) can be inserted into virus-like particles,[48] which is followed by CuAAC reaction with azide-containing TACAs for bioconjugation.

Over the past decade, the CuAAC reaction[49] has become a powerful tool to prepare vaccine constructs.[48,50–56] However, a potential concern is that the resulting triazole linker is immunogenic, which can possibly compete with the TACAs for B cell recognition, thus suppressing anti-TACA humoral responses. Several studies have reported the generation of high antibody titers specific against the desired carbohydrate antigen by triazole-linked glyco-conjugates, suggesting that it is possible to use triazoles in vaccine development.[48,50–56] Further studies are needed to establish the impact of the triazole linker on anti-TACA antibody responses.

6.3 Multivalent Carbohydrate Display on Virus-like Particles for Enhancing Binding Avidity

As discussed above, the interaction of protein receptor and monomeric carbohydrate is typically of low affinity. Nature overcomes this problem by simultaneously engaging multiple copies of carbohydrates to enhance the ligand avidity. This phenomenon is commonly referred to as the cluster effect or multivalent effect.[57] Since the magnitude of B cell activation depends on the strength of interactions between BCR and antigen, multivalent displays of carbohydrate antigens on an optimal platform should strongly activate B cells by engaging BCRs and inducing antibody responses.

With their rigid, homogenous, and well-defined structures, virus-like particles are ideal scaffolds for multivalent presentation of carbohydrate antigens in an organized manner. Recently, Davis and co-workers successfully conjugated mannoside dendrimers to Hpg-incorporated bacteriophage Qβ in high yields.[58] The resulting virus-like glycodendrimer nanoparticles competitively blocked the binding of Ebola pseudotyped virus to C-type lectin DC-SIGN, which significantly reduced viral infection to Jurkat or dendritic cells at nanomolar to picomolar concentrations. It is worth mentioning that the apparent affinity of the glycodendrimer nanoparticles to DC-SIGN on a per mannose basis was 860-fold higher than that by the unconjugated monomer, suggesting a dramatic avidity enhancement was achieved by clustering carbohydrates on the surface of virus-like particles.

As alkynes and amines have orthogonal reactivities, they can be selectively modified on the same virus-like particle to expand the diversity of the carbohydrates displayed.[59] The power of this approach has been demonstrated by Finn and co-workers. Using the alkyne-functionalized Qβ particles, they were able to introduce two different types of oligomannosides onto the particle at defined locations. These heterologous mannoside-bearing Qβ particles were able to bind with the HIV broad neutralizing antibody 2G12 with apparent nanomolar affinities, which were stronger than Qβ containing a single type of mannoside, suggesting a possible cooperating effect.[48]

6.4 Repetitive Antigen Display for Enhancement of B Cell Responses

Studies have been performed to elucidate the criteria for engaging BCRs to induce a strong antibody response. Using a series of polyacrylamide polymers with various lengths as the delivery platform, Dintzis *et al.* performed pioneering studies and revealed that a critical number of antigens ($>$20) organized in appropriate distances ($\sim$10 nm) was critical to activate B cells.[60,61] This threshold level was attributed to the crucial clustering of the minimum number of BCRs. Once the valency exceeded the threshold level, decreasing the spacing between the haptens (*i.e.*, increasing hapten density) resulted in an enhanced antibody response.

With DNP (2,4-dinitrophenyl) as the model antigen and synthetic polymer as the scaffold, the Kiessling group demonstrated that the generation of the IgM response *in vivo*, the extent of BCR redistribution into polarized caps, as well as the magnitude of intracellular Ca^{2+} concentration were well correlated with the valency of the antigen.[62] Vaccine constructs bearing less repetitive epitopes were much less effective in B cell activation. Similar phenomenon was also reported recently by Schamel *et al.*[63] These elegant studies illustrate the importance of antigen display for B cell activation.

Since the immune system responds strongly to viruses with highly repeating structures, it has been hypothesized that B cells can break tolerance to produce autoantibodies when self-antigens are organized in a repetitive manner. To test this, Zinkernagel and co-workers created a transgenic mouse strain that expressed the cell membrane-associated glycoprotein of vesicular stomatitis virus (VSV-G).[66,67] Immunization with VSV-G protein, recombinant vaccinia virus expressing VSV-G, or VSV-adsorbing cells did not elicit any antibodies against VSV-G, suggesting the strong tolerance by transgenic mice. In contrast, VSV-G specific IgG and IgM antibodies were generated by immunization with inactivated VSV, indicating that tolerance or unresponsiveness of a self-antigen could be overcome by presenting the antigen in a highly organized repetitive form (Figure 6.3).[64,65] A similar observation was reported by Chackerian and co-workers with transgenic mice expressing soluble hen egg lysozyme (HEL). Trivalent HEL induced very little IgG antibodies (titer $<$ 100), while HEL-conjugated virus-like particles led to more than 1000-fold increase in IgG responses (titer $\sim$ 100 000).[66] These results suggest that repetitiveness is not only an essential requirement for

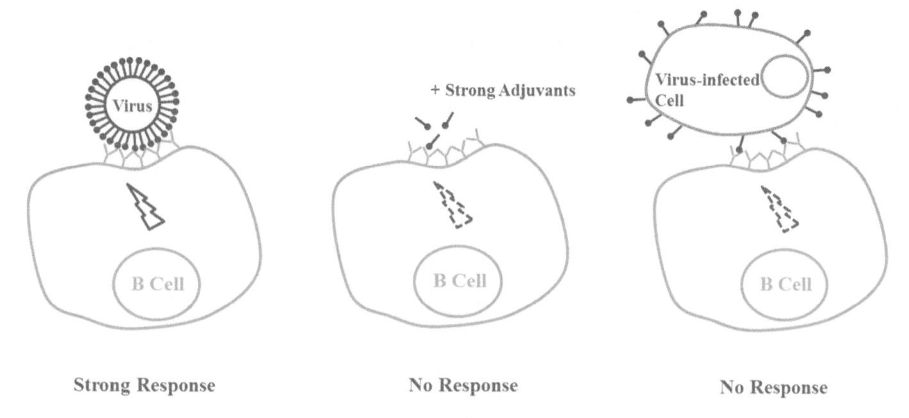

Figure 6.3 Effect of antigen pattern on B cell responses towards self-antigens. The inactivated virus could break tolerance and elicit strong B cell responses in transgenic mice expressing the coat protein of the virus. However, virus-infected cells or the coat protein itself even with strong adjuvants cannot generate any B cell immune responses.

cross-linking BCR to induce powerful T cell independent responses, but also a strong stimulus that can override B cell tolerance with the aid of helper T cells.

With the traditional protein carriers such as TT and KLH, it is impossible to precisely control the display patterns of antigens on the surface. In contrast, the well-defined, highly repetitive and rigid surface of virus-like particles, coupled with the polyvalency effect, renders virus-like particles an attractive platform to enhance antibody responses against TACAs. A variety of carbohydrate antigens were immobilized onto a virus-like particle, CPMV.[56] Chicken inoculated with these carbohydrate–CPMV conjugates produced strong and specific IgY antibodies. One direct comparison was carried out between KLH and CPMV conjugates with a carbohydrate antigen tri-LacNAc. Anti-tri-LacNAc antibodies elicited by the CPMV conjugate bound about 50 times stronger to antigens than those generated by the corresponding KLH construct. In another study, Tn antigen[67] was attached onto CPMV.[68] Previously, when monomeric Tn-KLH conjugates were evaluated in mice, no antibodies (IgG and IgM titers $\sim$0) were obtained, presumably because the monomeric form of Tn is an exceedingly weak antigen.[69] In contrast, the CPMV-Tn construct elicited good anti-Tn antibodies with an IgG titer of 10 500, which recognized Tn antigens displayed on the cancer cell surface. These results suggest the superiority of the virus-like particle platform.

6.5 Comparison of Virus-like Particles in Eliciting Anti-Carbohydrate Responses

Many virus-like particles are available.[4] To compare their efficiencies in TACA delivery, several virus-like particles have been investigated using Tn as the antigen.

CPMV is a plant virus that only infects a number of legume species, with particularly high titer in cowpea. The crystal structure shows that CPMV has a diameter of 31 nm with a single-stranded positive sense RNA inside. The capsid consists of 60 copies of an asymmetric unit including the S subunit (23.7 kDa) and L subunit (41.2 kDa),[70] with a capsid thickness of only 12 Å. As native CPMV does not contain any exposed cysteine on the exterior surface, a cysteine mutant was generated (S-CPMV) through mutagenesis (Figure 6.4),[71,72] which was subsequently conjugated with maleimide-functionalized Tn **3**, resulting in S-CPMV-Tn with on average 60 Tn per capsid (Scheme 6.2).

Similar to CPMV, tobacco mosaic virus (TMV) is a plant virus that can be produced on a gram scale from infected tobacco leaves.[46] In contrast to the icosahedron structure of CPMV, TMV displays a rod-like structure with a length of 300 nm and a diameter of 18 nm. The capsid of TMV is composed of 2130 identical coat proteins, and each coat protein is encoded by 159 amino acids with a molecular weight of 17 kDa. As there were no lysine

Figure 6.4 Illustration of surface-exposed thiol groups in the cysteine mutant of CPMV, in which residues 28 (threonine) and 102 (threonine) in the large subunit were replaced by cysteine. The structure was reconstructed based on the crystal structure of native CPMV.

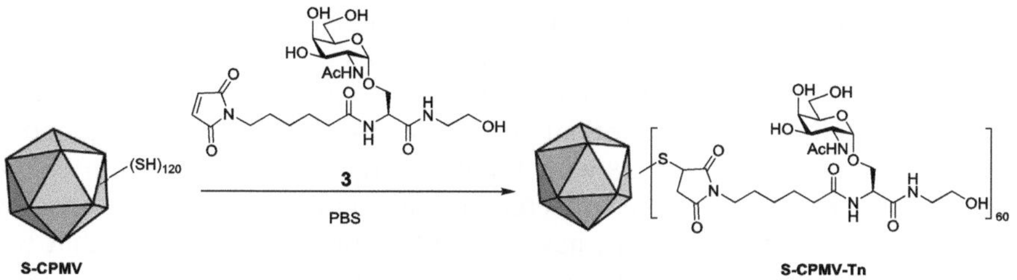

S-CPMV $(SH)_{120}$ → **3**, PBS → S-CPMV-Tn

Scheme 6.2 Synthesis of the S-CPMV-Tn vaccine.

Figure 6.5 Illustration of surface-exposed tyrosine residues in TMV. The side chain of the tyrosine residues is shown in the stick model. This crystal structure only contains 34 subunits, while the native TMV is composed of 2130 subunits.

residues on the external surface of TMV, tyrosine 139 was functionalized by the diazo alkyne **2** to introduce an alkyne onto TMV. CuAAC reaction with Tn-azide **4** introduced over 2000 copies of Tn onto each capsid (Figure 6.5 and Scheme 6.3).[53]

Another interesting virus-like particle is bacteriophage Qβ, which is obtained by recombinant expression in *E. coli*. During the assembly within *E. coli*, the RNA of host cells is packaged into Qβ and contributes almost 25% of the total mass of the Qβ particle. The icosahedron capsid of Qβ is made of 180 coat proteins (14 kDa each), with a diameter of 30 nm. There are four reactive amino groups available for each coat protein, including three lysine residues (K2, K13, and K16) and the N-terminus (Figure 6.6).[48] Compared with CPMV and TMV, Qβ displays a more dense arrangement of accessible amino groups, with the distance between neighboring amines around 4 nm. Native Qβ could be functionalized with the NHS ester of pent-4-ynoic acid **1** and the resulting alkyne Qβ reacted with Tn-N3 **5** through CuAAC to introduce an average of 340 Tn antigens per particle (Scheme 6.4).

Mice were immunized with S-CPMV-Tn, TMV-Tn, and Qβ-Tn at the same dose of Tn following an identical immunization protocol (subcutaneous injection on days 0, 14, and 28 with Freund's adjuvant). One week after the final boost, sera samples were collected and antibody titers were analyzed by enzyme linked immunosorbent assay (ELISA). The results showed that the Qβ-Tn vaccine induced the highest anti-Tn IgG titer (average titer 263 000) compared to S-CPMV-Tn and TMV-Tn (IgG titers of 10 500 and 2500,

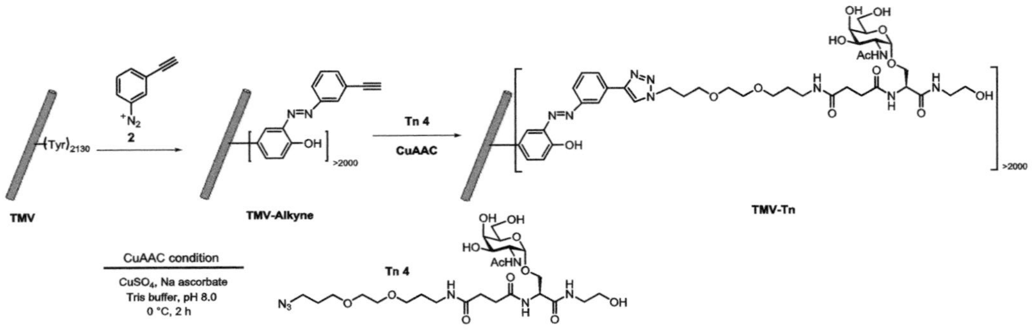

Scheme 6.3 Synthesis of the TMV-Tn vaccine.

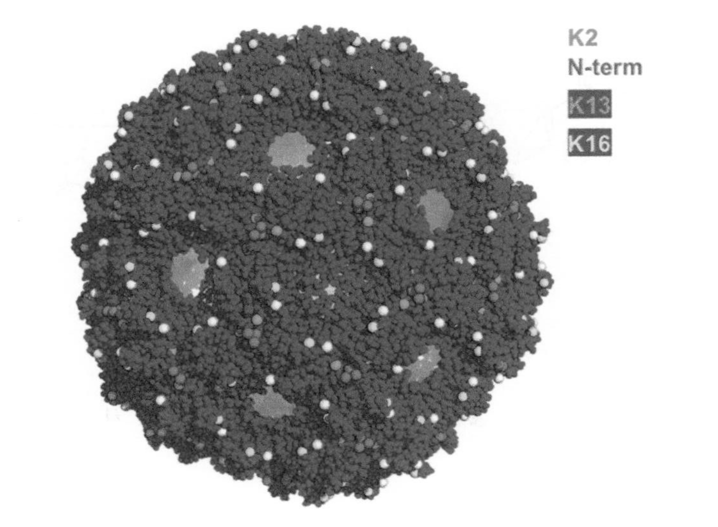

Figure 6.6 Illustration of surface-exposed amino groups on a wild-type Qβ particle.

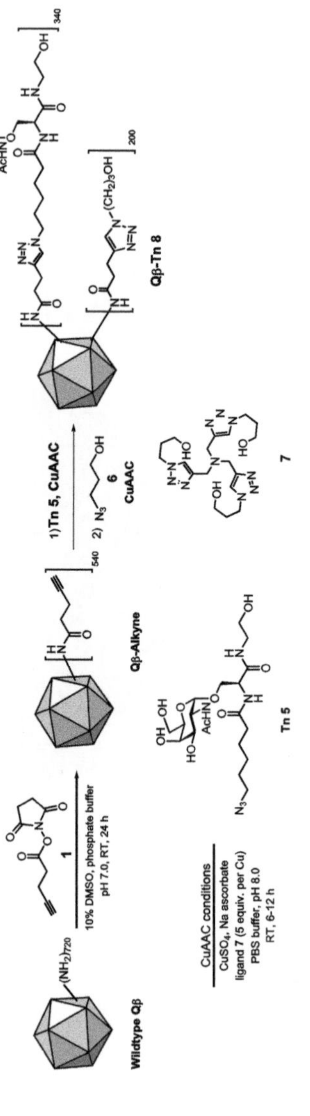

Scheme 6.4 Synthesis of the Qβ-Tn vaccine.

respectively). These results suggest that the Qβ particle is more superior in inducing anti-Tn responses than CPMV and TMV particles.

6.6 Influence of Local Density of Antigens on Anti-Carbohydrate Responses

To better understand the impact of antigen density on the immune potentiating effect of Qβ, three Qβ-Tn vaccines **8–10** were prepared with decreasing Tn antigen density (Scheme 6.5). Considering the possible influence of Qβ, two sets of immunization experiments were carried out. First, three groups of mice (groups 1–3) were immunized with Qβ-Tn vaccines **10**, **9**, and **8**, respectively, with the same amount of Qβ capsid (Table 6.1). The results showed that the group 3 mice, which received Qβ-Tn **8** with the highest level of Tn antigen, generated the highest IgG antibody responses (Figure 6.7, left). This indicated that higher levels of Tn antigen (~1 μg) were needed to elicit significant IgG responses when equal amounts of Qβ were present.

In the next experiment, the amount of Tn antigen was kept constant (1 μg) for another three groups of mice (Table 6.1, groups 4–6) immunized with Qβ-Tn **10**, **9**, or **8**. A control Qβ particle without Tn was added to groups 5 and 6 to render the total amount of Qβ particle the same (75 μg) for all three groups. Analysis of the post-immune sera demonstrated that only group 6 generated significant IgG antibody responses, suggesting that only the vaccine with high local density of antigen was able to induce IgG responses (Figure 6.7, left).

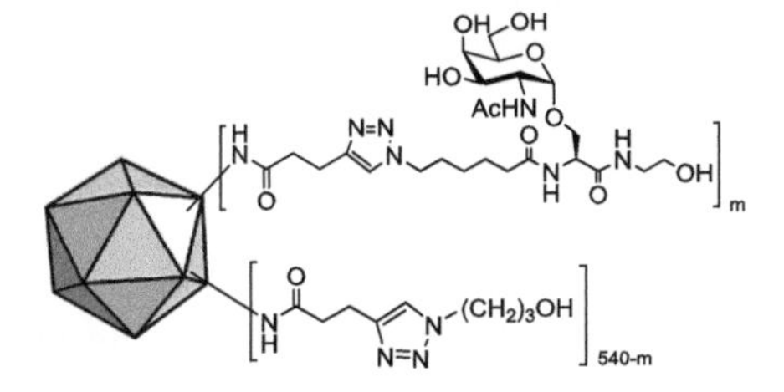

Qβ-Tn	m
8	340
9	150
10	78

Scheme 6.5 Qβ-Tn vaccines with different densities of Tn antigen.

Table 6.1 The amounts of Tn and Qβ in six groups of mice for deciphering the importance of antigen display patterns.

Group	1	2	3	4	5	6
Particles used	L (particle)	M	H	L (Tn)	M + N	H + N
Vaccine construct	**10**	**9**	**8**	**10**	**9** + Qβ	**8** + Qβ
Tn occupancy[a] (%)	23	44	100	23	44 + 0	100 + 0
Total Tn[b] (μg)	0.23	0.44	1	1	1	1
Total Qβ virion[b] (μg)	18	18	18	75	75	75

[a]Tn occupancy in Qβ-Tn **8**, which displays 340 Tn antigens, was set as 100%.
[b]Amount for each injection in each mouse.

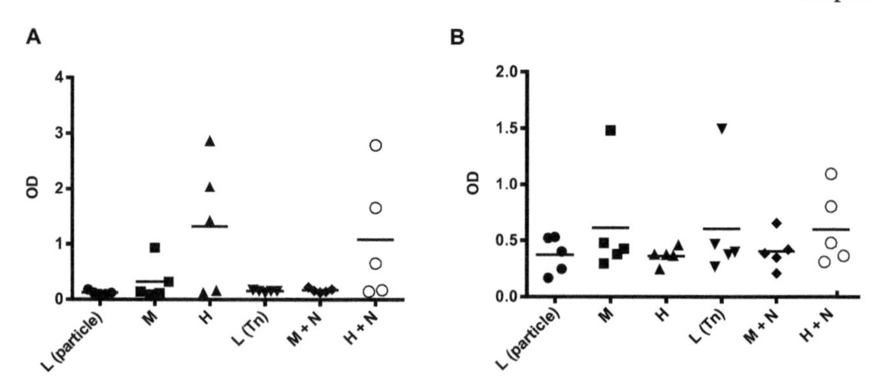

Figure 6.7 IgG (*left*) and IgM (*right*) responses induced by Qβ conjugated with variable Tn antigen densities. The data shown were measured at a serum dilution of 1 : 6400.

Interestingly, the anti-Tn IgM titers were similar in all groups, regardless of the density of Tn on the Qβ particle or the amount of particle administered (Figure 6.7, right). This is presumably because even the Qβ-Tn **10** with the lowest Tn density still contained 78 copies of Tn per capsid. With the spherical shape of the Qβ particle, it is estimated that half of the antigens, approximately 39 Tn, can engage BCRs, which is above the threshold number (20) predicted by Dintzis *et al.*[60,61] Therefore, substantial differences in IgM responses were not observed within these groups. One possible reason for enhanced IgG responses with the high antigen density construct is that stronger BCR cross-linking resulting from higher antigen density renders B cells more responsive to the co-stimulatory signals from helper T cells. This can lead to more facile IgG class switching, affinity maturation, and/or B cell proliferation.

6.7 Conclusions

During the past several years, virus-like particles have been explored as a new generation of carrier to improve the immune responses against TACAs. Virus-like particles have many potential advantages over the traditional protein carriers, including a well-defined structure, the possibility for an organized high-density display of antigens, and stability towards a wide range of chemical conditions. In several direct comparisons, virus-like particle–TACA conjugates elicited much stronger TACA-specific IgG antibodies than the traditional protein carrier KLH. Among various virus-like particles, the bacteriophage Qβ particle has been shown to be superior in eliciting anti-TACA IgG antibodies. The local density of the antigen was found to be critical for strong antibody responses. Upon further development, virus-like particles, in particular the bacteriophage Qβ particle, will become a powerful platform for carbohydrate-based anti-cancer vaccines.

Acknowledgements

We are grateful to the National Cancer Institute (R01CA149451-01A1) for financial support of our work. We thank Prof. M.G. Finn (Georgia Institute of Technology), Prof. Qian Wang (University of South Carolina), Dr Jeffrey Gildersleeve (National Cancer Institute) and Prof. BenMohamed (University of California, Irvine) for collaborations in the work mentioned in this review. We also thank Sebastian Raschka (MSU) for assistance in analyzing crystal structures of virus-like particles.

References

1. B. Acres, S. Paul, H. Haegel-Kronenberger, B. Calmels and P. Squiban, *Curr. Opin. Mol. Ther.*, 2004, **6**, 40–47.
2. O. J. Finn, *Nat. Rev. Immunol.*, 2003, **3**, 630–641.
3. J. Couzin-Frankel, *Science*, 2013, **342**, 1432–1433.
4. A. Roldao, M. C. M. Mellado, L. R. Castilho, M. J. T. Carrondo and P. M. Alves, *Expert Rev. Vaccines*, 2010, **9**, 1149–1176.
5. M. A. Cheever and C. S. Higano, *Clin. Cancer Res.*, 2011, **17**, 3520–3526.
6. C. Kyi and M. A. Postow, *FEBS Lett.*, 2014, **588**, 368–376.
7. C. G. Drake, E. J. Lipson and J. R. Brahmer, *Nat. Rev. Clin. Oncol.*, 2014, **11**, 24–37.
8. Z. Yin and X. Huang, *J. Carbohydr. Chem.*, 2012, **31**, 143–186 and references cited therein.
9. T. Buskas, P. Thompson and G. J. Boons, *Chem. Commun.*, 2009, 5335–5349.
10. R. M. Wilson and S. J. Danishefsky, *J. Am. Chem. Soc.*, 2013, **135**, 14462–14472.
11. Z. W. Guo and Q. L. Wang, *Curr. Opin. Chem. Biol.*, 2009, **13**, 608–617.
12. M. Ugorski and A. Laskowska, *Acta Biochim. Pol.*, 2002, **49**, 303–311.
13. J. J. Liang, E. T. Kimchi, K. F. Staveley-O'Carroll and D. F. Tan, *Int. J. Clin. Exp. Pathol.*, 2009, **2**, 1–10.
14. P. Fredman, K. Hedberg and T. Brezicka, *BioDrugs*, 2003, **17**, 155–167.
15. T. Freire and E. Osinaga, *Inmunologia*, 2003, **22**, 27–38.
16. G. F. Springer, *J. Mol. Med.*, 1997, **75**, 594–602.
17. P. C. Jones, L. L. Sze, P. Y. Liu, D. L. Morton and R. F. Irie, *J. Natl. Cancer Inst.*, 1981, **66**, 249–254.
18. P. O. Livingston, G. Y. C. Wong, S. Adluri, Y. Tao, M. Padavan, R. Parente, C. Hanlon, M. J. Calves, F. Helling, G. Ritter, H. F. Oettgen and J. O. Lloyd, *J. Clin. Oncol.*, 1994, **12**, 1036–1044.
19. O. Kurtenkov, K. Klaamas, K. Rittenhouse-Olson, L. Vahter, B. Sergejev, L. Miljukhina and L. Shijapnikova, *Exp. Oncol.*, 2005, **27**, 136–140.

20. G. D. MacLean, M. A. Reddish, R. R. Koganty and B. M. Longenecker, *J. Immunother.*, 1996, **19**, 59–68.
21. Y. M. Kim, J. Y. J. Pan, G. A. Korbel, V. Peperzak, M. Boes and H. L. Ploegh, *Proc. Natl. Acad. Sci. U. S. A.*, 2006, **103**, 3327–3332.
22. P. H. Liang, S. K. Wang and C. H. Wong, *J. Am. Chem. Soc.*, 2007, **129**, 11177–11184.
23. F. D. Batista and M. S. Neuberger, *Immunity*, 1998, **8**, 751–759.
24. J. J. Mond, A. Lees and C. M. Snapper, *Annu. Rev. Immunol.*, 1995, **13**, 655–692.
25. L. J. McHeyzer-Williams and M. G. McHeyzer-Williams, *Annu. Rev. Immunol.*, 2005, **23**, 487–513.
26. L. Qiu, X. Gong, Q. L. Wang, J. Li, H. G. Hu, Q. Y. Wu, J. P. Zhang and Z. W. Guo, *Cancer Immunol. Immunother.*, 2012, **61**, 2045–2054.
27. Y. L. Huang, J. T. Hung, S. K. C. Cheung, H. Y. Lee, K. C. Chu, S. T. Li, Y. C. Lin, C. T. Ren, T. J. R. Cheng, T. L. Hsu, A. L. Yu, C. Y. Wu and C. H. Wong, *Proc. Natl. Acad. Sci. U. S. A.*, 2013, **110**, 2517–2522.
28. G. Ragupathi, J. R. Gardner, P. O. Livingston and D. Y. Gin, *Expert Rev. Vaccines*, 2011, **10**, 463–470.
29. R. Lo-Man, S. Vichier-Guerre, R. Perraut, E. Deriaud, V. Huteau, L. BenMohamed, O. M. Diop, P. O. Livingston, S. Bay and C. Leclerc, *Cancer Res.*, 2004, **64**, 4987–4994.
30. R. A. De Silva, Q. Wang, T. Chidley, D. K. Appulage and P. R. Andreana, *J. Am. Chem. Soc.*, 2009, **131**, 9622–9623.
31. R. P. Brinas, A. Sundgren, P. Sahoo, S. Morey, K. Rittenhouse-Olson, G. E. Wilding, W. Deng and J. J. Barchi, *Bioconjugate Chem.*, 2012, **23**, 1513–1523.
32. M. E. Pichichero, *Hum. Vaccines Immunother.*, 2013, **9**, 2505–2523.
33. P. O. Livingston, *Semin. Cancer Biol.*, 1995, **6**, 357–366.
34. J. A. Englund, E. L. Anderson, G. F. Reed, M. D. Decker, K. M. Edwards, M. E. Pichichero, M. C. Steinhoff, M. B. Rennels, A. Deforest and B. D. Meade, *Pediatrics*, 1995, **96**, 580–584.
35. J. A. Chabalgoity, B. Villarealramos, C. M. A. Khan, S. N. Chatfield, R. D. Dehormaeche and C. E. Hormaeche, *Infect. Immun.*, 1995, **63**, 2564–2569.
36. M. P. Schutze, C. Leclerc, M. Jolivet, F. Audibert and L. Chedid, *J. Immunol.*, 1985, **135**, 2319–2322.
37. S. G. Bi, M. Baum and D. Morse, *J. Immunol.*, 2012, **188**, 5216.
38. J. R. Harris and J. Markl, *Micron*, 1999, **30**, 597–623.
39. D. Miles, H. Roche, M. Martin, T. J. Perren, D. A. Cameron, J. Glaspy, D. Dodwell, J. Parker, J. Mayordomo, A. Tres, J. L. Murray, N. K. Ibrahim and G. Theratope Study, *Oncologist*, 2011, **16**, 1092–1100.
40. A. M. M. Eggermont, S. Suciu, P. Rutkowski, J. Marsden, M. Santinami, P. Corrie, S. Aamdal, P. A. Ascierto, P. M. Patel, W. H. Kruit, L. Bastholt, L. Borgognoni, M. G. Bernengo, N. Davidson, L. Polders, M. Praet and A. Spatz, *J. Clin. Oncol.*, 2013, **31**, 3831–3837.
41. A. Zeltins, *Mol. Biotechnol.*, 2013, **53**, 92–107.

42. T. Storni, C. Ruedl, K. Schwarz, R. A. Schwendener, W. A. Renner and M. F. Bachmann, *J. Immunol.*, 2004, **172**, 1777–1785.
43. Q. Wang, E. Kaltgrad, T. W. Lin, J. E. Johnson and M. G. Finn, *Chem. Biol.*, 2002, **9**, 805–811.
44. E. Jonczyk, M. Klak, R. Miedzybrodzki and A. Gorski, *Folia Microbiol.*, 2011, **56**, 191–200.
45. V. Hong, S. I. Presolski, C. Ma and M. G. Finn, *Angew. Chem., Int. Ed.*, 2009, **48**, 9879–9883.
46. T. L. Schlick, Z. B. Ding, E. W. Kovacs and M. B. Francis, *J. Am. Chem. Soc.*, 2005, **127**, 3718–3723.
47. M. A. Bruckman, G. Kaur, L. A. Lee, F. Xie, J. Sepulveda, R. Breitenkamp, X. Zhang, M. Joralemon, T. P. Russell, T. Emrick and Q. Wang, *ChemBioChem*, 2008, **9**, 519–523.
48. R. D. Astronomo, E. Kaltgrad, A. K. Udit, S. K. Wang, K. J. Doores, C. Y. Huang, R. Pantophlet, J. C. Paulson, C. H. Wong, M. G. Finn and D. R. Burton, *Chem. Biol.*, 2010, **17**, 357–370.
49. H. C. Kolb and K. B. Sharpless, *Drug Discovery Today*, 2003, **8**, 1128–1137.
50. H. Cai, Z. Y. Sun, M. S. Chen, Y. F. Zhao, H. Kunz and Y. M. Li, *Angew. Chem., Int. Ed.*, 2014, **53**, 1699–1703.
51. Q. Y. Hu, M. Allan, R. Adamo, D. Quinn, H. L. Zhai, G. X. Wu, K. Clark, J. Zhou, S. Ortiz, B. Wang, E. Danieli, S. Crotti, M. Tontini, G. Brogioni and F. Berti, *Chem. Sci.*, 2013, **4**, 3827–3832.
52. Z. Yin, M. Comellas-Aragones, S. Chowdhury, P. Bentley, K. Kaczanowska, L. BenMohamed, J. C. Gildersleeve, M. G. Finn and X. Huang, *ACS Chem. Biol.*, 2013, **8**, 1253–1262.
53. Z. Yin, H. G. Nguyen, S. Chowdhury, P. Bentley, M. A. Bruckman, A. Miermont, J. C. Gildersleeve, Q. Wang and X. Huang, *Bioconjugate Chem.*, 2012, **23**, 1694–1703.
54. Q. L. Wang, Z. F. Zhou, S. C. Tang and Z. W. Guo, *ACS Chem. Biol.*, 2012, **7**, 235–240.
55. T. Lipinski, T. Luu, P. I. Kitov, A. Szpacenko and D. R. Bundle, *Glycoconjugate J.*, 2011, **28**, 149–164.
56. E. Kaltgrad, S. Sen Gupta, S. Punna, C. Y. Huang, A. Chang, C. H. Wong, M. G. Finn and O. Blixt, *ChemBioChem*, 2007, **8**, 1455–1462.
57. C. Fasting, C. A. Schalley, M. Weber, O. Seitz, S. Hecht, B. Koksch, J. Dernedde, C. Graf, E. W. Knapp and R. Haag, *Angew. Chem., Int. Ed.*, 2012, **51**, 10472–10498.
58. R. Ribeiro-Viana, M. Sanchez-Navarro, J. Luczkowiak, J. R. Koeppe, R. Delgado, J. Rojo and B. G. Davis, *Nat. Commun.*, 2012, **3**, 1303.
59. E. Strable, D. E. Prasuhn, A. K. Udit, S. Brown, A. J. Link, J. T. Ngo, G. Lander, J. Quispe, C. S. Potter, B. Carragher, D. A. Tirrell and M. G. Finn, *Bioconjugate Chem.*, 2008, **19**, 866–875.
60. H. M. Dintzis, R. Z. Dintzis and B. Vogelstein, *Proc. Natl. Acad. Sci. U. S. A.*, 1976, **73**, 3671–3675.
61. R. Z. Dintzis, M. H. Middleton and H. M. Dintzis, *J. Immunol.*, 1985, **135**, 423–427.

62. E. B. Puffer, J. K. Pontrello, J. J. Hollenbeck, J. A. Kink and L. L. Kiessling, *ACS Chem. Biol.*, 2007, **2**, 252–262.
63. S. Minguet, E. P. Dopfer and W. W. A. Schamel, *Int. Immunol.*, 2010, **22**, 205–212.
64. R. M. Zinkernagel, S. Cooper, J. Chambers, R. A. Lazzarini, H. Hengartner and H. Arnheiter, *Nature*, 1990, **345**, 68–71.
65. M. F. Bachmann, U. H. Rohrer, T. M. Kundig, K. Burki, H. Hengartner and R. M. Zinkernagel, *Science*, 1993, **262**, 1448–1451.
66. B. Chackerian, M. R. Durfee and J. T. Schiller, *J. Immunol.*, 2008, **180**, 5816–5825.
67. G. F. Springer, *Science*, 1984, **224**, 1198–1206.
68. A. Miermont, H. Barnhill, E. Strable, X. W. Lu, K. A. Wall, Q. Wang, M. G. Finn and X. Huang, *Chem. – Eur. J.*, 2008, **14**, 4939–4947.
69. E. Kagan, G. Ragupathi, S. S. Yi, C. A. Reis, J. Gildersleeve, D. Kahne, H. Clausen, S. J. Danishefsky and P. O. Livingston, *Cancer Immunol. Immunother.*, 2005, **54**, 424–430.
70. T. W. Lin, Z. G. Chen, R. Usha, C. V. Stauffacher, J. B. Dai, T. Schmidt and J. E. Johnson, *Virology*, 1999, **265**, 20–34.
71. A. S. Blum, C. M. Soto, C. D. Wilson, J. D. Cole, M. Kim, B. Gnade, A. Chatterji, W. F. Ochoa, T. Lin, J. E. Johnson and B. R. Ratna, *Nano Lett.*, 2004, **4**, 867–870.
72. T. Lin, *J. Mater. Chem.*, 2006, **16**, 3673–3681.

CHAPTER 7

Synthetic Glycosylated Ether Glycerolipids as Anticancer Agents

GILBERT ARTHUR,*[a] FRANK SCHWEIZER[b] AND MAKANJUOLA OGUNSINA[b]

[a] Department of Biochemistry and Medical Genetics, University of Manitoba, 754 Bannatyne Avenue, Winnipeg, Manitoba, Canada R3E 0J9;
[b] Department of Chemistry and Medical Microbiology, University of Manitoba, 460 Parker Building, Winnipeg, Manitoba, Canada R3T 2N2
*Email: Gilbert.Arthur@umanitoba.ca

7.1 Introduction

Cancer is one of the leading causes of death worldwide in both developed and developing countries and afflicts people from all socio-economic strata of the population. According to the WHO world cancer factsheet, in 2008 there were 12.7 million new cases of cancer and 7.6 million deaths worldwide and it is projected that if current trends remain unchanged there will be 22 million cases in 2030 with 11.4 million deaths.[1] These numbers are astonishing in light of the resources and efforts that have targeted the disease and the large number of anticancer drugs that have been approved for clinical use. The inescapable conclusion from these observations is that the drugs have not been as effective as desired. Indeed, there are certain cancers such as pancreatic cancer for which there are no effective drugs and, furthermore, there are no effective treatments for advanced stage cancers

RSC Drug Discovery Series No. 43
Carbohydrates in Drug Design and Discovery
Edited by Jesús Jiménez-Barbero, F. Javier Cañada and Sonsoles Martín-Santamaría

Published by the Royal Society of Chemistry, www.rsc.org

that have metastasized. While our inability to detect the cancer early is responsible for a large proportion of the deaths, this is by no means the sole reason for the lack of efficacy of cancer treating drugs. The fact is, cancer cells have devised several strategies to avoid being killed by these drugs and tumors have become drug resistant. The cells can express specific proteins that actively extrude the drugs from cells, they may have enhanced mechanisms to repair DNA destroyed by DNA-targeting drugs, and the cells may overexpress proteins that promote cell survival and/or mutated proteins of the cell death pathway, apoptosis, and thus thwart the death mechanism activated by the drugs to kill the cells.[2–4] Given the fact that most anticancer drugs kill cells by inducing the apoptotic pathway, resistance to apoptotic cell death is a major reason for our inability to kill cancer cells.[5] As a result, the cancer continues its growth and spreads to secondary organs and ultimately kills the patient. One of the confounding issues in cancer treatment has been the problem of relapse; initial treatment may cause the tumor mass to shrink and become undetectable, but over time the cancer reappears in an even more aggressive form which resists drug treatment.[6] A hypothesis that is gaining ground has been recently proposed to explain this phenomenon. This hypothesis, the cancer stem cell (CSC) hypothesis, posits that the tumors are initiated by progenitor or stem-like cells that have the ability to self-renew and also differentiate to form differentiated cells in the bulk tumor.[7] The CSC theory provides an explanation for the relapse or recurrence that often accompanies chemotherapy and radiotherapy. As a consequence of the intrinsic or induced mechanisms that resist cell death by apoptosis, and the ability of the CSCs to remain in Go phase, they survive chemotherapy and radiotherapy which kill the cells of the bulk tumor.[8–10] Their unlimited proliferation potential allows them to regenerate the tumor again with differentiated cells, including drug resistance types after cessation of treatment, thus causing the recurrence of the tumor which is refractory to treatment. In addition to being implicated in the development of drug resistance, CSCs have also been implicated as the driving force underpinning metastases and hence successful cancer treatment would require the ability to kill and eliminate the CSCs.[11,12] Unfortunately, very few compounds have been reported that kill CSCs.[13] From the above, it is quite clear that novel cancer treatments are required to have a significant impact on the current trends of cancer mortality. If resistance to apoptosis cell death is indeed responsible for the inefficacy of drugs, then one approach to overcome this impediment is to develop drugs that kill cells independent of apoptosis. Such compounds may be able to kill the CSCs. In addition, the development of compounds that kill CSCs will have a significant impact on preventing relapse of the disease.

Studies in our lab have indicated that a group of synthetic compounds dubbed glycosylated antitumor ether lipids (GAELs) are able to kill cancer cells by an apoptosis-independent mechanism and, furthermore, they are able to kill breast CSCs.[14–16] Such characteristics have identified GAELs as potential molecules for development into potent novel anticancer agents that could address some of the problems inherent in current drug treatments mentioned above. This chapter focuses on GAELs and describes their development to include synthesis and activity, as well as challenges that lie ahead in developing them into clinically useful agents.

7.2 Antitumor Ether Lipids

By convention, the term antitumor ether lipids (AELs) describes (denotes) a group of disparate synthetic compounds that share the common structural feature of being ether lipids with functional ability to kill cancer cell lines. They are synthetic compounds and do not exist in nature and they are primarily classified into three groups: alkyllysophospholids (ALPs), alkylphospholipids (APLs), and glycosylated antitumor ether lipids (GAELs) (Figure 7.1).[17] The first class of AELs developed belong to the group of ALPs. ALPs are phospholipid-like compounds with a long alkyl chain at the *sn*-1 position of the glycerol moiety, a short-chain ether group at the *sn*-2 position, and a phosphobase at the *sn*-3 position of glycerol. The ALPs were initially developed as stable analogues of lysophosphatidylcholine (LPC), by insertion of the two ether bonds, for studies on the immunomodulatory properties of LPC in animals.[18] They are typified by edelfosine, compound **1** (see Figure 7.2 for the structures of all AELs discussed herein). ALPs were subsequently discovered to possess antiproliferative and cytotoxic activity *in vitro* and *in vivo.*[18–21] Several analogues with modifications at the *sn*-2 position have been synthesized in the quest to develop the compounds as anticancer agents for clinical use.[13] Several characteristics of the compounds have garnered interest. AELs do not interact with DNA and are not mutagenic;[18–20] they appear to be more active against cancer cells relative to normal cells *in vitro* and *in vivo*[19] and could therefore kill cancer cells selectively; their effects on cells is independent of the p53 status of the cell[22] and they can be administered intravenously or orally. Edelfosine, which is the standard by which other ALPs are assessed, has undergone clinical studies but is not as yet used to treat solid tumors due to gastrointestinal toxicity.[23] Several studies have demonstrated that ALPs, including edelfosine, kill cells by initiating events that ultimately result in cell death by caspase-dependent apoptosis.[22,24–26]

Alkylphospholipids (APLs) represent a second subgroup of AELs. They lack a glycerol backbone and the alkyl group is directly esterified to the phosphobase, generally phosphocholine. The prototype of this subclass is

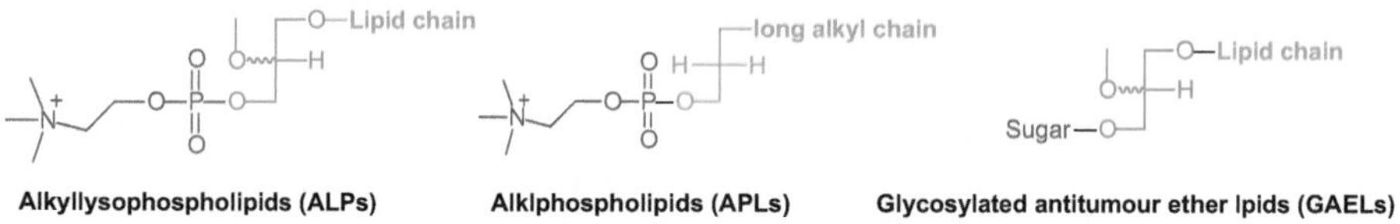

Figure 7.1 Subclasses of antitumor ether lipids (AELs).

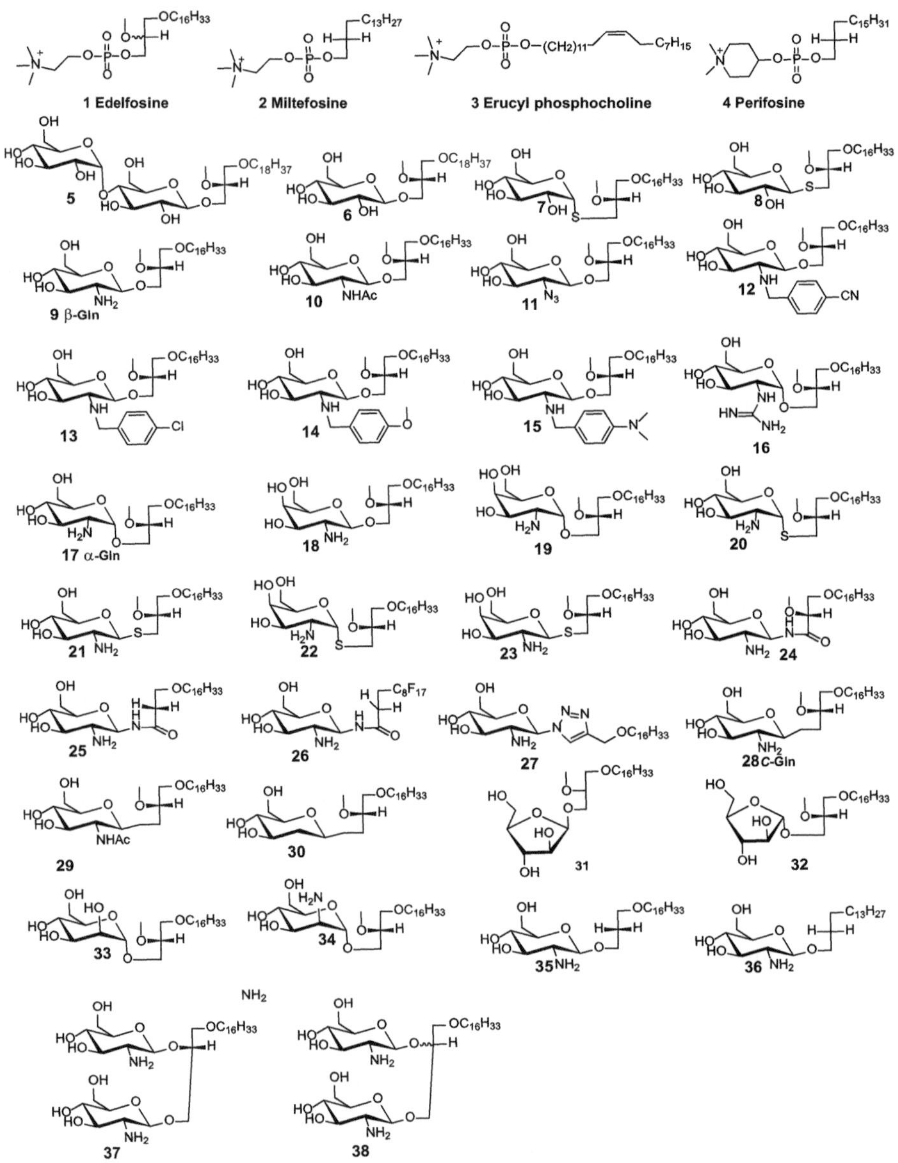

Figure 7.2 Structures of AELs with compound numbers discussed herein.

miltefosine, compound **2**, which has undergone clinical trials and is in use as a topical treatment for skin metastases in breast cancer.[23] Novel analogues of miltefosine with longer alkyl chains (C_{22}), namely eurycylphosphocholine, compound **3**, and a compound where the phosphocholine moiety was replaced with a heterocyclic piperidine group, perifosine **4**, have been synthesized and shown to possess promising anticancer activity *in vitro* and *in vivo.*[27] APLs perturb several processes involved in cell survival and growth and ultimately result in cell death *via* apoptosis. The possible mechanism of action utilized by these compounds has recently been reviewed.[23]

Glycosylated antitumor ether lipids (GAELs), the compounds that are the focus of this chapter, represent the third subclass of AELs. GAELs differ from the other two AEL subclasses by the absence of the phosphobase found in ALPs or APLs, and its replacement by a sugar moiety attached to the glycerol at the *sn*-3 position. The reason for our focus on developing GAELs as clinical anticancer drugs is because their mode of action is independent of apoptosis, and is thus distinct from the mode of action of ALPs and APLs as well as conventional chemotherapeutic agents. As alluded to above, cancer cells, including CSCs, have developed several strategies to escape cell death by apoptosis. Consequently, they survive exposure to apoptosis-inducing drugs and subsequently repopulate the tumor. Because GAELs do not kill cancer cells by apoptosis, they have the potential of by-passing the mechanisms utilized by cancer cells to avoid death by apoptosis and could conceivably kill CSCs. They could potentially be used in combination therapy with apoptosis-inducing compounds to completely eliminate the tumor cells.

7.3 Development of GAELs as Anticancer Agents

The anticancer properties of a compound include its cytostatic, cytotoxic, immunomodulatory, and antimetastatic properties. As far as GAELs are concerned, the end point for developing these compounds thus far has been based on their cytostatic and cytotoxic properties.

Even though compounds that can be classified as GAELs were synthesized as far back as 1986,[28] their development as anticancer agents has lagged behind that of the ALPs. In hindsight, this is not surprising given the inferior potency of the early GAEL analogues, compounds **5** and **6**,[28,29] relative to the ALPs, and edelfosine in particular. Also, with the assumption that the mechanisms of action of all AELs were similar, it was logical to focus developmental efforts on the more active compounds like edelfosine. It was the discovery of the novel apoptosis-independent mechanism used by GAELs to kill cells[15,16] that has provided the impetus to focus on developing clinically useful GAELs. Ultimately, if GAELs are to be effective clinical drugs they would have to deliver anticancer activity *in vivo* with minimal side effects. This necessitates the development of metabolically stable GAELs with potent activity against drug sensitive and drug resistant cancer cells, including CSCs. The exhibition of selective activity against cancer cells relative to normal cells *in vivo* would also go a long way towards minimizing side effects.

Until recently, development of GAELs has not been systematic. Very limited structural activity studies were conducted with these compounds to determine the pharmacophore and identify changes that would improve on the activity.

7.3.1 History of the Development of GAELs as Anticancer Agents

The thioglycosides 7 and **8** were synthesized by Bittman and associates for evaluation of their anticancer activity on murine cell lines as well as the human HL60 leukemia cell line.[30] The rationale for making these thio analogues was the expectation that substitution of the *O*-glycosidic linkage by an *S*-glycosidic linkage would make the molecule more lipophilic.[31,32] Overall, the results of the study showed that the α-linked thioglycolipid 7 was about 1.5 times more active in inhibiting the incorporation of [^{3}H]thymidine into mouse cell lines than the β-linked analogue **8**. However, when the activity of the compounds was assessed against human epithelial cells (T84, MCF-7, A549, and A427 cells), no significant differences were found in the activity of the α-*S*- and β-*S*-analogues.[33] Furthermore, the compounds were antiproliferative and not toxic to the epithelial cells. The activity of the α-thioglycoside 7 against HL60 cells was similar to that of compound **6** with either *R*- or *S*-configuration at the C-2 or *sn*-2 position of the glycero backbone,[30] but none of these compounds (CC_{50} 12 μM) were as active as edelfosine, compound **1** (IC_{50} 2.5 μM).

The development of GAELs as potential anticancer drugs took a significant turn with the synthesis of compound **9** [1-*O*-hexadecyl-2-*O*-methyl-3-*O*-(2″-acetamido-2″-deoxy-α-D-glucopyranosyl)-*sn*-glycerol], denoted as Gln. In Gln, the neutral glucose moiety of reference compound **6** is replaced by a cationic 2-deoxy-2-glucosamine. The presence of the amino function resulted in a significant increase in cytotoxic activity similar to or better than that of edelfosine (compound **1**).[34] Gln was initially synthesized using 2-acetamido-2-deoxy-3,4,6-tri-*O*-acetyl-α-D-glucopyranosyl chloride as glycoside donor and 1-*O*-hexdecyl-2-*O*-methyl-*sn*-glycerol as glycoside acceptor to give the acetate-protected Gln. The acetate protecting groups were cleaved using ethanolic potassium hydroxide to give Gln. A significant challenge in the synthesis of Gln was the saponification of the acetamido group, which resulted in a low yield of Gln. To avoid this difficulty, other glycosyl donors with different protecting groups that can easily be deprotected have been employed; these include glycoside donors where the amino substituent at position C-2 of glucosamine was protected as azide or phthalimide. The azido function can be reduced by catalytic hydrogenation or by trimethylphosphine in THF and the phthalimido protecting group can be removed by 50% ethylenediamine in butanol at 90 °C.[35,39,44] The advantage of using the azido function is that both α- and β-forms can be generated in one synthetic route because the azide is not a participating group during the glycosylation reaction. Gln demonstrated significant antiproliferative and toxic effects against epithelial cancer cell lines derived from prostate (DU145, PC3), breast (MCF7, MDA-MB 468, MDA-MB-231, JIMT-1, BT549, Hs578t), lung (A549, A427), kidney (A498), colon (T84), ovarian (OVCAR 3), cervix (HeLa), glioblastoma (U251), and pancreas (MiaPaCa2, BxPC3) and from neuroblastoma (SK-N-MC, SK-N-SH).[34,36] Complete loss of cell viability was obtained at a concentration of 15 μM of Gln **9** after 48 h incubation.[34,36]

Apart from demonstrating that Gln had activity comparable to edelfosine, the significant observation of the study was its effect on NIH OVCAR-3 cells.[34] NIH OVCAR-3 is an ovarian cancer cell line that is resistant to clinically relevant concentrations of adriamycin, melphalan, and cisplatin.[37,38] The effects of Gln on the growth and viability of OVCAR-3 cells were directly compared with the activities of edelfosine **1** and alkylphosphocholines such as hexadecylphosphocholine and erucylphosphocholine **3**.[34] Incubation of the cell lines with the compounds for 48 h gave CC_{50} values of 4 and 12 μM for Gln, whereas the CC_{50} values were 24 μM for edelfosine **1** and >30 μM for erucylphosphocholine **3** and hexadecylphosphocholine. OVCAR-3 cells were viable after incubation with 30 μM edelfosine for 48 h; all cells were dead after incubation with 15 μM of Gln. Thus, while the GAEL, Gln, was cytotoxic against this drug-resistant line, the other AEL classes were not. This study demonstrated and reinforced the clear differences in the activity profile of GAELs and the other AEL classes, suggesting dissimilar mechanisms of action.

7.3.2 Systematic Structure–Activity Studies on the Amino Function of Gln

Using Gln as a lead molecule for drug development, various parts of the molecule were identified for structural modifications (Figure 7.3), firstly to determine the pharmacophore and secondly to improve the potency of the compound. So studies were designed to investigate the effect of modifications on the amino substituent at the C-2 position of Gln, and modifications at the *sn*-2 position of the glycero moiety.

7.3.2.1 Effects of Converting the Amino Function to an Azido or Acetamido Group in Gln

The presence of a free NH_2 group at the C-2 position of glucose produced a compound that was as active and in some instances more active than

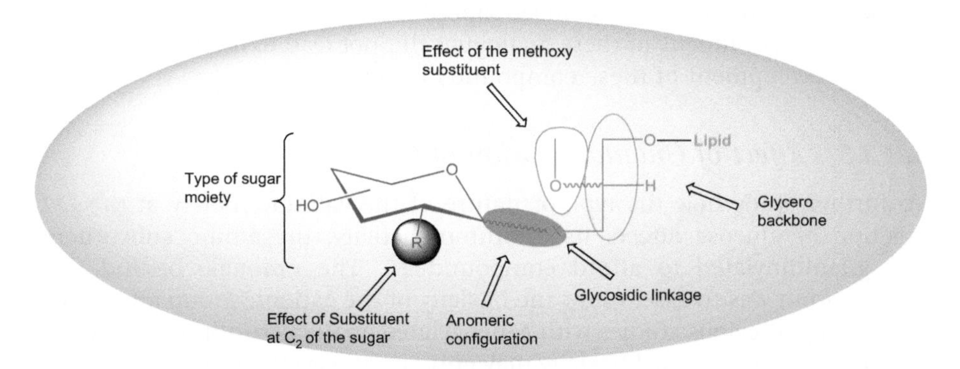

Figure 7.3 Schematic diagram illustrating parts of the glycosylated antitumor ether lipid (GAEL) molecule modified for structure–activity studies.

edelfosine **1**. To investigate the importance of the free amino substituent in Gln, a number of modifications were made to study their cytotoxic effects against cancer cell lines. *N*-Acetylation of the amino group in Gln produced compound **10** that resulted in significant loss of activity.[34] The CC_{50} values for most of the cell lines were three-fold higher with the *N*-acetyl analogue **10** relative to Gln. This result suggests that the cationic nature of the free amine may be crucial for activity. In spite of the loss in activity, it is worth pointing out that the acetamido analogue was more active than edelfosine **1** against the drug-resistant OVCAR-3 cell line.[34] The importance of the free amino group in Gln was reinforced with the synthesis of an analogue **11** bearing a neutral azido group in place of a cationic amino group.[39] The CC_{50} values against a panel of cancer cell lines were threefold higher than values observed with Gln.

7.3.2.2 Effects of N-*Benzylation*

Benzylation of the amino function in Gln would create lipophilic analogues of Gln **9**, which would be expected to enhance its accumulation in the cells. Increased accumulation of the compounds would theoretically lead to enhanced activity relative to Gln **9**. To investigate this hypothesis, *N*-benzylated analogues of Gln with different electron-withdrawing (*p*-cyanobenzylamino **12**, *p*-chlorobenzylamino **13**) or different electron-releasing groups (*p*-methoxybenzylamino, *p*-dimethylaminobenzylamino) were synthesized. Synthesis was achieved by reductive amination of Gln with corresponding *p*-substituted benzaldehydes.[36]

A comparison of the effects of the compounds on the viability of epithelial cancer cell lines (DU145, MiaPaCa2, U251, HeLa) relative to the parental Gln **9** revealed the following. The benzylamino analogues with electron-withdrawing groups were significantly less active than Gln **9**. The analogue with the electron-releasing substituent, *p*-methoxybenzylamine **14**, was as active as the parent compound **9**, while *p*-dimethylaminobenzylamino analogue **15** was slightly less active.[36] The results from the study did not support the notion that enhanced lipophilicity would lead to increased activity. As lipophilicity of these analogues did not lead to increased activity, further development of these compounds is not warranted.

7.3.2.3 Effect of Guanidinylation of Gln

To further study how the type or nature of the cationic moiety at the C-2 position of glucose affects its antitumor activity, the amino substituent was guanidinylated to afford compound **16**. The rationale behind this modification was to assess how the basicity of the cationic group affects the cytotoxicity. Previous studies with *N*-benzylated analogues indicated that Gln analogues with increased basicity may enhance the cytotoxic effects. As **16** is expected to be a much stronger base, we anticipated enhanced antitumor effects of **16** when compared to Gln. Contrary to this hypothesis, when the guanidylated analogue **16** was screened against a panel of human epithelial cancer cell lines, it was not cytotoxic to any of the cell lines at concentrations of 30 μM, the highest concentration tested.[39] The lack of cytotoxic activity by guanidylated Gln **16** may be the result of a steric effect and reemphasizes the importance of free amine at the C-2 position for anticancer activity.

In summary, our findings indicate that the free amine at the C-2 position of glucose appears to be very important for anticancer activity. Bulky substituents like *N*-benzyl or *N*-guanidinyl and abolishing the cationic charge (azide, *N*-acyl) reduced the cytotoxic effect. It is currently unknown how critical the position of the amino group in Gln is to the anticancer effect.

7.3.3 Effect of the Anomeric Linkage on Anticancer Properties in Gln

7.3.3.1 Effect of the Configuration of Anomeric Glycosidic Linkage on Activity

When the anticancer activity against murine cancer cell lines of compound 7 with an α-*S*-glycosidic linkage was compared to that of compound **8** with a β-*S*-glycosidic linkage, the α-anomer was more active than the β-anomer.[30] However, no differences in activity were observed when the compounds were tested against human epithelial cancer cell lines.[33] From these results, no generalizations can be made with respect to the activities of the anomers. In any case, the activities of these compounds were quite weak and unlikely to be of any clinical use.

Following the discovery that the activity of Gln against cancer cell lines was similar or greater than edelfosine **1**, depending on the cell line, the effect of the anomeric linkage on the activity of this molecule was investigated. The α-anomer of Gln, compound **17**, was synthesized.[39] Comparison of the activities of the α-anomer **17** and Gln against five epithelial cancer cell lines (JIMT-1, BT549, DU145, PC3, and MiaPaCa2) revealed that the α-anomer was consistently more active than the β-anomer, although the differences were slight and varied with the cell line (0.5-fold to 2-fold).[39] These results, together with results obtained with GAEL analogues **18** and **19** bearing sugars other than glucose, indicate that the α-anomeric compounds are more active than their corresponding β-anomers.[14] In this case, the α-anomer **19** was four times more potent than its β-anomer **18**. However, a caveat seems to be whether the sugars are in the pyranosyl or furanosyl forms.

7.3.3.2 Effect of Type or Nature of Glycosidic Linkage on Gln Activity

One of the desirable features of AELs is their purported metabolic stability as a consequence of the ether bonds at the C-1 and C-2 positions of the glycero moiety that resist metabolism. Thus ALPs cannot be metabolized by phospholipases A1 or A2 and are also not susceptible to acylation processes. ALPs have a phosphodiester bond at the C-3 position that is potentially susceptible to hydrolysis by phospholipase C-type enzymes, but studies have revealed that there is little hydrolysis of the molecule in cells or *in vivo*[40–42] and these compounds have a long half-life.

GAELs are also expected to be metabolically inert at the C-1 and C-2 positions of the glycerol moiety due to the presence of ether bonds. The *O*-glycosidic bond at the C-3 position of the glycerol moiety is likely to be susceptible to glycosidases *in vivo*. Although this has not been demonstrated experimentally, indirect evidence suggests that this is the case. We were unable to determine the maximum tolerability levels of Gln in Rag2M mice. The animals tolerated Gln **9** levels of 500 mg kg^{-1} given orally or 100 mg kg^{-1} given intravenously without any adverse effects.[43] In light of the activity shown *in vitro*, this observation is consistent with rapid glucosidase hydrolysis of the compound which prevents attainment of therapeutic levels. Not surprising, there were no effects on the growth of JIMT1 or MDA-MB-231 xenograft growing in female rats treated with 100 mg kg^{-1} of Gln given intravenously. Thus one goal of developing clinically useful GAELs would be to fashion GAELs with a metabolically stable bond at the C-3 position of the glycerol moiety. This would be required prior to any *in vivo* animal studies.

The problem of glycosidase hydrolases of GAELs is not a unique one to GAELs but is a general one that has bedeviled the development of glycosidic drugs. Several approaches have therefore been developed to generate metabolically stable analogues of *O*-glycosidic compounds. They involve the replacement of the *O*-glycosidic bond with *S*-, *N*-, or *C*-glycosidic bonds.[14,39,44,45] While the bonds formed are metabolically stable and would not be susceptible to glycosidase hydrolysis, the change may have an effect on biological activity which cannot be predicted and hence has to be determined empirically. Some of these metabolically stable compounds have been synthesized and are discussed below.

7.3.3.2.1 Effects of Thioglycosidic Linkages on Stability and Activity of GAELs. As discussed above, GAELs with an amine at the C-2 position of the sugar moiety have proved to be the most active compounds to date. Both β- and α-glucosamine-derived GAELs, compounds **9** and **17**, were active whereas only the α-galactosamine-derived GAEL **19** was active, while the β-galactosamine analogue **18** showed little activity[14] (Section 7.3.4.2). We synthesized the thio analogues of α- and β-glucosamine-derived GAELs, compounds **20** and **21**, respectively, as well as the α- and β-thio analogues of galactosamine-derived GAELs, compounds **22** and **23**, respectively.[14]

Comparison of the activity of the glucosamine-derived *O*-glycosidic linked **9** and **17** and the *S*-glycosidic linked **20** and **21** against the viability of five epithelial cell lines (BT549, JIMT1, PC3, DU145, and MiaPACa2) revealed that replacement of the *O*-glycosidic linkage in α- and β-anomers by the

S-glycosidic linkage resulted in two- to threefold loss of activity with either anomer. When the activity of the most active GAEL to date, the α-galactosamine-derived anomer **19**, was compared to its *S*-glycosidic linked analogue, the latter was three- to fivefold less active than the *O*-glycosidic linked analogue.[14] The β-*S*-linked anomer **23** was as inactive as the β-*O*-linked analogue **18**. From these results, the replacement of the glycosidase-susceptible *O*-glycosidic linkage with the glycosidase-resistant *S*-glycosidic linkage in amino sugar-derived GAELs leads to loss of activity. It is worth mentioning that thioglycosides **7** and **8** were as active as compound **6** with an *O*-glycosidic linkage; this may be an indication that replacement of the *O*-glycosidic linkage by the *S*-linkage had not impacted the activity of these non-cationic compounds (Section 7.3.1).[30] Thus possible interaction between the amine substituent at the C-2 position of the sugar, which confers a cationic nature on Gln, and the *S*-glycosidic linkage may be responsible for the differences observed between these two groups of GAEL molecules. In any case, as compound **6** and its analogues were not particularly active relative to edelfosine **1** and glucosamine-derived analogues **9** and **17** or galactosamine-derived analogues **18** and **19**, their development as potential drugs has not been pursued.

7.3.3.2.2 **Effect of *N*-Glycosidic Linkage on Anticancer Activity.** Another means of developing metabolically stable analogues of GAELs is to convert the *O*-glycosidic bond to a glycosylamide or glycosyltriazole linkage, as these bonds are resistant to cleavage by glycosidases. Synthesis of the glycosylamide analogues **24–26** was achieved by coupling a protected glucosylamine to a carboxylic acid derivative of the glycerol lipid alcohol using TBTU as the coupling reagent, whereas the synthesis of the glycosyltriazole **27** was achieved *via* click chemistry from glucosyl azide.[44] Unfortunately, these N-linked analogues were two- to fourfold less active than Gln **9**, depending on the cancer cell line.

7.3.3.2.3 **Effect of *C*-Glycosidic Linkage on Stability and Anticancer Activity of GAELs.** A successful approach for obtaining metabolically stable analogues of glycosidase-sensitive compounds is the replacement of *O*-glycosidic linkage with a *C*-glycosidic analogue. *C*-Glycosidic linkages are not susceptible to hydrolysis by glycosidases and are expected to have an enhanced half-life, which may extend the bioavailability of the compound relative to the *O*-glycosidic analogue. Because only minimal conformational differences are expected between the *O*-linked and *C*-linked GAEL compounds, they often maintain the characteristics of the original compound *O*-glycosidic linkage.[45] The synthesis of *C*-glycosides is an arduous and rather involved, and is not a very popular undertaking.

Analogue **28** of Gln with a *C*-glycosidic linkage was synthesized.[45] Subsequent to the synthesis, the antiproliferative and toxic effects of *C*-Gln against nine epithelial cancer cell lines were assessed and compared to Gln **9**.[45] The activity of the *N*-acetylated analogue of *C*-glycoside **29** was also investigated. Glucosamine-derived GAELs with *C*- and *O*-glycosidic linkages were equally active against epithelial cancer cell lines derived from a variety of tissues. The CC_{50} values for the *C*-glycoside/*O*-glycoside for the following cell lines are shown in parenthesis: SK-N-MC (CC_{50}, 4.1/4.1 μM), SK-N-SH (CC_{50}, 4.1/3.8 μM), DU145 (CC_{50}, 7.9/6.5 μM), MCF7 (CC_{50}, 8.1/8.0 μM), MDA-MB 468 (CC_{50}, 9.0/7.0 μM), MDA-MB-231 (CC_{50}, 9.1/7.1 μM), BT549 (CC_{50}, 8.9/6.5 μM), Hs578T (CC_{50}, 5.1/3.1 μM), and A498 (CC_{50}, 8.5/6.9 μM). Thus, there were no significant differences in the CC_{50} values of the two analogues. Compound **29** and the 2-deoxy analogue **30** with *C*-glycosidic linkages were also synthesized and their *in vitro* anticancer activities were compared to that of **28**. The 2-deoxyglucose-derived analogue **30** was three to five times less active than the corresponding glucosamine-derived analogue **28**, and the 2-*N*-acetamido-2-deoxy analogue **29** showed very little activity. The result of the activity studies of the non-cationic *C*-linked analogue is similar to their *O*-linked analogues.[46] This further supports the fact that a free amino group or maybe other small cationic substituents on GAELs is important for their anticancer activity.

Since Gln with a *C*-glycosidic linkage was as potent as the *O*-analogue, one can infer that the *O*-glycosidic bond is unnecessary for activity. While we expect the *C*-linked GAEL **28** to be more stable and thus have a longer half-life and extended bioavailability, it should be pointed out that these postulates have not yet been tested experimentally in cells or animals. In light of this, the *C*-linked analogues of GAELs will potentially be a better choice for preclinical studies, as unlike the *O*-glycosides they will not be susceptible to *in vivo* hydrolysis by glycosidases.

7.3.4 Effect of Sugar Moiety on Cytotoxicity of GAELs

The early GAELs synthesized contained the disaccharides maltose or lactose as the sugar core.[28,29] These compounds were not very cytotoxic, but recent studies have revealed that they are potent inhibitors of cell migration and could be developed into anti-metastatic compounds.[47] Even though most of the discussion above has focused on glucose-bearing GAELs, it is not an indication that the carbohydrate moiety of GAELs is limited to glucose. Indeed, in light of the available sugars that can be incorporated into GAELs, it comes as no surprise that GAELs bearing sugars other than glucose have been explored.

7.3.4.1 Other Non-Cationic GAELs

A variety of furanoside-based non-cationic GAELs have been studied. For instance, D-arabinose-derived GAELs, compound **31** with the β-*O*-anomeric configuration and compound **32** with the α-*O*-anomeric configuration, were synthesized by Bittman and associates[48] and their activity was compared to edelfosine **1**. Compound **31** was as active as edelfosine in killing a number of cancer cell lines, including A549, MCF-7, and MCF-7-Adr. In contrast, the α-anomer **32** exhibited little activity. In this instance, the α-anomer was much less active than the β-anomer, which contrasts with what was observed with pyranoside-based glucosamine analogues previously discussed. Whether this observation is limited to D-arabinose-based furanosides or a general feature of furanosides is unknown. Besides GAELs **5–8** and **31** and **32**, another non-cationic α-*manno*-configured GAEL **33** has also been synthesized and its activity was compared to the arabinose-containing GAELs **31** and **32** and edelfosine **1**.[48] Compound **33** was three to four times less active compared to edelfosine and the arabinose analogue. Because the corresponding β-anomer of the mannose analogue was not synthesized, it is unclear whether the lack of activity was due to the anomeric configuration or due to the sugar itself.

7.3.4.2 Other Amino-Sugar-Derived GAELs

As mentioned earlier, replacement of the glucose moiety in compound **6** by glucosamine enhances the antitumor properties effect that resulted in the discovery of Gln with activity comparable to or better than edelfosine **1**, depending on cell line. To investigate whether analogues of Gln that contain different sugar configurations possess antitumor activities, GAEL analogues were synthesized with mannosamine to give **34** or galactosamine compounds **18** and **19** and their activities were compared to glucosamine-linked GAEL **9**.[14] The chemistry involved in this synthesis of the mannosamine-derived GAEL **34** and the galactosamine-derived GAELs **18** and **19** followed a similar strategy as described earlier for Gln (Section 7.3.1).

The effects of Gln and other aminosugar-derived GAELs, including compounds **17**, **18**, **19**, and **34**, on the viability of BT-474, BT549, JIMT-1 (breast cancer cells), PC3, DU145 (prostate cancer cells), and pancreas MiaPaCa2 (pancreas cancer cells) cell lines were assessed.[14] The α- galactosamine-derived analogue **19** was the most active compound, followed by the α-glucosamine-derived analogue **17** and then the β-glucosamine-derived analogue **9**. Surprisingly, the β-galactosamine-derived analogue **18** was four- to fivefold less active than the α-anomer **19**. The α-mannosamine-derived analogue **34** was the least active of the compounds. Since the β-mannosamine GAEL has not been synthesized yet, we cannot say with any certainty that GAELs with mannose are inherently inactive because of the sugar.

7.3.5 Effect of Changes to the Glycerol Moiety on Cytotoxicity of GAELs

Changes in the glycerol moiety can have profound effects on AEL cytotoxic activity. For example, when the methoxy group at the C-2 position of edelfosine was replaced by a carbamate moiety, novel analogues were obtained that showed different activity profiles than edelfosine. These compounds preferentially killed prostate cancer cells.[49]

To investigate how essential the methoxy substituent is for activity in GAELs, compound **35**, an analogue of Gln **9** which lacked the methoxy group at the C-2 position of the glycerol backbone, was synthesized.[39] Comparison of the activity with Gln **9** revealed very little to no difference in the CC_{50} values observed for JIMT-1, BT549, DU145, PC3, and MiaPaCa2 cell lines. Thus the methoxy group *per se* is not essential for activity. In another study, the glycerol moiety was eliminated and the glucosamine group was attached to the long-chain alkyl group to produce **36**, a molecule analogous to the alkylphosphocholines represented by miltefosine **2**. This compound was about 1.3- to 2-fold less active than Gln **9** or the methoxy-less compound **35**, depending on the cell line.[39]

A GAEL analogue in which the methoxy group was replaced by a glucosamine to yield a diglycosylated molecule has also been synthesized.[44] Compound **37** with the *R*-configuration and **38**, a diastereomeric mixture with *R*/*S* configuration at the C-2 position of the glycerol moiety, were synthesized to evaluate the effect of the stereochemistry and replacement of methoxy with a glucosamine moiety at this position. Both forms of the compound were at least threefold less active than Gln **9**, showing that substituting the methoxy with a glucosamine moiety reduced the activity. This loss of activity could be due to steric effects or decreased lipophilicity, which can reduce the absorption of the drug into the cell. Comparison of the activity of **37** and **38** showed no difference. This is an indication that the stereochemistry at this position does not greatly affect the anticancer activity. Overall, it would appear that there is not a strict requirement for the methoxy group but a glycerol moiety is essential.

7.4 Selectivity

One of the highly touted characteristics of AELs is their ability to kill cancer or transformed cells at concentrations that do not affect normal cells.[18,19] Clearly, such selectivity would be highly desirable as it would be expected to have lower side-effects when the compounds are used in patients. This selectivity has been demonstrated for ALPs in a number of studies.[50–54] The greater sensitivity of leukemia cells to edelfosine **1** relative to normal bone marrow cells, neutrophils, leukocytes, or macrophages[50–54] underpins the use of edelfosine to purge bone marrow cells of cancer cells in autologous transplantation.[51] Edelfosine was also able to kill Swiss 3T3 fibroblasts transformed with SV40 virus, but not the untransformed cells.[55] It is worth pointing out that some studies with edelfosine reported little discrimination between transformed cells and parental cells. Thus, *v-sis* NIH 3T3 cells were threefold more resistant to edelfosine than the parental cell line,[56] and transformation of H184A, a human breast epithelial cell line, with the *v-erb* oncogene did not make the cells more susceptible to a variety of AELs.[57] The reasons for these disparate results *in vitro* are unclear. The basis for the selectivity is currently unknown and whether this is a universal phenomenon remains to be established.

We have investigated the selectivity of GAELs with respect to normal and transformed cells. In these studies the toxic effects of Gln **9** on parental NIH 3T3 cells and tumorigenic offspring generated by transformation of the parental cells with mos, A-raf, fes and src oncogenes were compared.[58] Transformation by these oncogenes renders the cells tumorigenic and metastatic.[58] Incubation of the cells with varying concentrations of Gln **9** showed that the compound killed all the cells following a 48 h incubation with 10 μM or higher concentrations of the compound.[59] The CC_{50} values were similar between the parental and transformed cells. Thus the studies showed no significant selectivity in the toxic effects of the GAEL between the normal and transformed cells.[59]

In another study, we investigated the effects of β- and α-Gln, **9** and **17**, and β- and α- galactosamine-derived analogues **18** and **19** on the viability of normal human epithelial cells from breast and prostate. Because these primary cells are grown in serum-free media, the protein content of the media was normalized to that of 10% FBS-containing media used to grow the cancer cell lines by adding BSA. As GAELs are lipids they associate with proteins and thus the effective concentration available to interact with the cells depends on the protein content of the medium. The results are displayed in Table 7.1. The CC_{50} for the α-Gln **17** and α-galactosamine-derived

Table 7.1 Effect of GAELs on the viability of human normal breast or prostate epithelial cells.[a]

Dose μM	α-Gln **17**	β-Gln **9**	α-Galn **19**	β-Galn **18**
A. Human mammary epithelial cells.				
0	100	100	100	100
2	74.2 ± 4.23	101.42 ± 10.28	79.24 ± 8.93	101.00 ± 4.27
4	28.44 ± 7.01	81.30 ± 18.7	41.31 ± 3.90	89.65 ± 9.06
6	1.1 ± 0.75	83.94 ± 11.19	1.20 ± 0.28	86.84 ± 11.77
9	0.66 ± 0.24	54.66 ± 7.20	0.96 ± 0.42	90.46 ± 6.37
12	0.7 ± 0.15	15.22 ± 5.6	0.98 ± 0.43	81.80 ± 8.20
15	0.74 ± 0.23	1.00 ± 0.14	1.11 ± 0.52	73.58 ± 2.86
20	1.08 ± 0.20	0.94 ± 0.31	1.3 ± 0.27	45.98 ± 3.65
B. Human prostate epithelial cells.				
0	100	100	100	100
2	102.95 ± 3.96	105.53 ± 4.71	86.64 ± 4.86	112.97 ± 2.9
4	56.31 ± 5.07	94.28 ± 8.25	57.02 ± 7.55	95.53 ± 10.5
6	11.12 ± 0.83	84.4 ± 7.54	9.99 ± 5.97	103.36 ± 8.14
9	0	69.56 ± 5.11	0	83.46 ± 16.43
12	0	32.43 ± 7.34	0	90.85 ± 10.42
15	0	5.57 ± 8.14	0	85.52 ± 10.39
20	0	0	0	65.21 ± 7.8

[a]The primary cells were obtained from Lonza (Allendale, NJ, USA) and grown in their respective recommended defined-serum-free medium. Cells were dispersed into 96 well plates and the compounds were added after 24 h. The cells were incubated for another 48 h and the viability of the cells was determined by the MTS assay. The results are expressed as the % of viable cells relative to the controls. The data show the means ± standard deviation of six independent determinations.

analogue **19** were between 3–4 μM, while β-Gln **9** had a CC_{50} of 10 μM and the β- galactosamine-derived analogue **18** was 20 μM for the HMEC. The corresponding values for PrEC were around 4 μM for α-anomers **17** and **19**, 10 μM for β-Gln **9**, and >20 μM for the β-galactosamine-derived analogue **18**. While we cannot directly compare the results for the normal primary cells with the cancer cell lines, the results nevertheless show that, in the *in vitro* setting, the normal cells are not less sensitive to the GAEL than the cancer cell lines.

The results to date indicate very little selectivity of GAELs for cancer or transformed cells *in vitro*. However, we cannot simply extrapolate and conclude that GAELs will show little selectivity *in vivo*. A whole host of factors would go into making that determination. For example, the extent of metabolism *in vivo* or selective accumulation in tumor *versus* normal tissues could greatly affect any differential effects of the compounds on normal and cancer tissue *in vivo*. It is worth pointing that, in spite of the demonstrated selectivity of edelfosine for tumor cells relative to cancer cells, a major factor preventing its use in treating cancers is gastrointestinal toxicity.[23] This is presumably due to the effects of the compound on the normal epithelial cell of the gut. Thus, *in vitro* selectivity may not necessarily translate into *in vivo* selectivity and *vice versa*. Selectivity *in vivo* would have to be determined experimentally.

7.5 Toxicity of GAELs

There have been limited studies on the toxicity of GAELs. A single dose of 0.5 g kg^{-1} body weight of the non-cationic analogue **6** given intraperitoneally in NMRI mice was toxic to the animals within 24–48 h.[60] In contrast, the animals tolerated 0.25 g kg^{-1} of the glycolipid administered consecutively for five days.

We have investigated the toxicity of orally delivered and intravenously delivered Gln in Rag 2M mice. A maximum tolerability dose could not be attained at concentrations of 500 mg kg^{-1} oral dosage or 100 mg kg^{-1} given intravenously and there was no effect on the behavior or growth (body weight) of the animals and no abnormal features were found during necroscopy.[43]

7.6 Hemolysis and Thrombogenic Effects

One impediment to the use of edelfosine as an intravenously delivered compound is the induction of haemolysis due to its structural similarity to lysophosphatidylcholine, a well-known lytic agent.[17] This undesirable effect may be overcome by using liposomal formulations of the compound EECL50.[61,62] With respect to GAELs, a maltosyl GAEL analogue **5**, that was not cytotoxic against cells growing in culture, caused hemolysis in *in vitro* studies with red blood cells suspended in PBS.[63] As binding of the GAELs to serum proteins is expected *in vivo* to lead to decreased levels of free

GAEL available to interact with RBCs, the lack of protein in the hemolytic assay set-up suggests that the results cannot be extrapolated to the *in vivo* situation. As we reported, we were unable to attain a maximum tolerability dose for intravenously delivered Gln **9** in Rag 2M mice up to a concentration of 100 mg kg^{-1}. Since this may be due to the rapid hydrolysis of the compound, it is clearly not the best analogue to use to assess the toxicity of GAELs. The toxicity will have to be assessed with an active stable GAEL analogue, like the *C*-Gln **28**.

Another limitation to the use of AELs in cancer treatment could be their potential prothrombogenic effects due to the structural similarity with platelet activating factor (PAF), a well-known thrombogenic agent. A comparative study of the ability of **7**, **8**, and **9** to induce the aggregation of rabbit platelets, relative to PAF, revealed that while PAF induced aggregation at 10^{-13} M, the C-16 analogue of edelfosine induced aggregation at 10^{-7} M, whereas the GAELs induced aggregation at a concentration of 10^{-5} M.[30] The lack of pro-aggregatory effects of GAELs relative to ET-16-OCH_3 suggests that thrombogenesis may not be a concern if these compounds are given intravenously. Clearly, pharmacokinetic studies to determine the concentrations in the plasma would have to be undertaken, especially if metabolically stable long-lived analogues are developed.

7.7 Metabolism of GAELs

As far as we are aware, there has been only one study to investigate the metabolism of GAELs. The compound studied was **6**, which showed moderate antiproliferative activity but is not toxic to human epithelial cancer cell lines.[29] A radioactive analogue of **6**, *rac*-1-*O*-[1′-^{14}C]octadecyl 2-*O*-methylglycero-β-D-glucopyranoside, was synthesized and its metabolism was investigated in Ehrlich ascites cells.[43] The results showed that 90% of compound **6** was taken up into the cells within 90 min. The compound was hydrolyzed by glycosidases to form *rac*-1-*O*-[1′-^{14}C]octadecyl-2-*O*-methylglycerol. After 24 h, 96% of the radiolabel was recovered as dialkylglycerol and 4% as 1-*O*-octadecyl-2-*O*-methyl-3-acylglycerol, indicating acylation of the diglyceride at the C-3 position. The rate of glycosidase formation of the diglyceride was estimated to be 70 pmol $(10^6 \text{ cells})^{-1}$ h^{-1} in the cells.

There have been no metabolic studies on the amino-substituted sugar GAELs for the simple reason that labeled analogues have not been synthesized to permit such studies. It is an open question whether the activity of the compounds is due to the parental molecule or a metabolite. The activity of ALPs has in fact been shown to be due to the parent molecule.[42,43,64] While our *in vivo* data suggest that Gln **9** is metabolized in animals, this metabolism is likely to be the reason for the lack of activity observed in animals. Interestingly, the β-*O*-glycosidic bond of alkyl glycosides was rapidly metabolized in the intestines and liver of mice.[83] We think it is unlikely that such hydrolysis occurs in the cell lines we have used to characterize the compounds. Since the activity of the *C*-glycosidic Gln **28** is similar to that of the *O*-analogue **9** and the *C*-glycoside is not susceptible to glycosidase action, we can infer that the activities of the compounds are likely due to the parental molecules.

7.8 Mechanism of Action of GAELs

7.8.1 Mechanism of GAEL-Induced Cell Death

As we have pointed out earlier, the attractive feature of GAELs that distinguishes them from other AEL classes and conventional anticancer agents is their ability to kill cancer cells *via* apoptosis-independent mechanism. That the mechanism used by GAELs to kill cells is distinct from the apoptotic mechanisms used by ALPs or APLs was evident from the differential cytotoxic effects the different classes had on several cell lines. A comparison of the cytotoxic effects of the GAEL subclasses against a drug-resistant ovarian cancer cell line, OVCAR-3, showed that GAELs were the only class that significantly inhibited the proliferation of the cells or killed them.[34] In studies with NIH 3T3 cells transformed with oncogenic protein kinases (mos, fes, raf, and src), edelfosine was unable to kill the transformed cell lines at concentrations that killed the wild-type NIH 3T3 cells. In contrast, GAELs killed both the wild-type and transformed cells at similar concentrations.[59]

More direct evidence for the fundamental differences between the mode of action of GAELs and ALPs was the demonstration that whereas GAELs were able to kill mouse embryonic fibroblasts (MEFs) devoid of key apoptotic components like caspase 3, caspase 9, and Apaf-1, which made them apoptotosis-incompetent, edelfosine was unable to kill these cells.[15] These studies showed that an active apoptosis pathway was required for manifestation of the cytotoxicity of ALPs but was not required for GAELs to kill the cells. We were also able to demonstrate that incubation of cells with GAELs did not result in the loss of mitochondria membrane permeability, the irreversible step in both caspase-dependent and caspase-independent apoptosis;[65] neither did we observe the release of cytochrome *c* to the cytosol, activation of caspase 9 or caspase 3, or cleavage of t-Bid.[15,16] Thus, in apoptosis-competent cells, incubation with GAELs did not induce apoptosis.

Although we have not fully elucidated the mechanisms *via* which GAELs kill cells, our studies have yielded insight into some events that may be involved. A striking observation of cells incubated with cytotoxic GAELs is the presence of large clear vacuoles that literally fill the cells (Figure 7.4). The vacuoles stain red with acridine orange, an indication of their acidic nature.

Studies with fluorescent markers, including lysotracker red, LAMP1, and lgp120, the rat homologue of LAMP1, indicate that the vacuoles have some lysosomal characteristics.[15] In addition to the lysosomal molecules, we also observed the presence of LC3-II, a key protein marker of autophagy, associated with the vacuoles. This observation suggested that the vacuoles could be autophagolysosomes and the GAEL was blocking their maturation.

Control

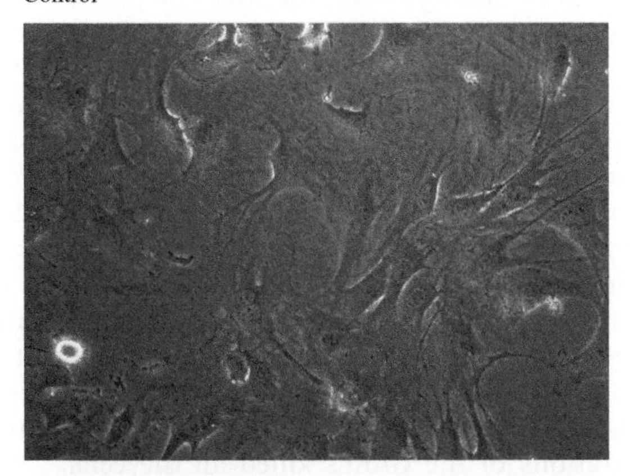

+ Gln **9** (5 μM) for 24 h

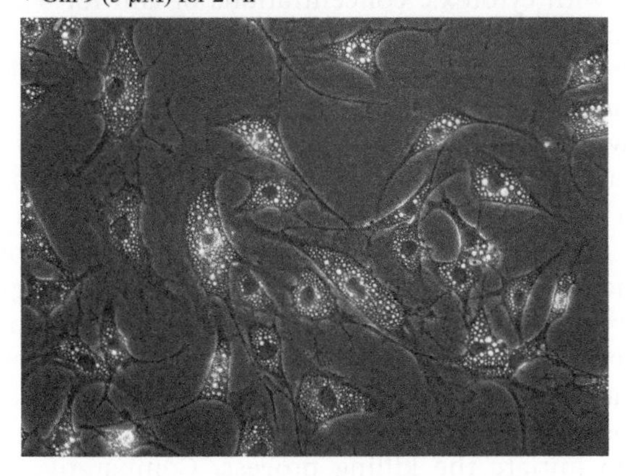

Figure 7.4 Generation of vacuoles by Gln **9** in ASK1 cells.

While this cannot be categorically ruled out, we believe this is unlikely because there was no distinction between the vacuoles formed in wild-type MEFs and ATG5-/-MEFs. The latter cells are unable to undergo autophagy as they cannot convert LC3-I to LC3-II.[66] We have also observed the formation of the vacuoles in both wild-type and Bif-/- cells, which are also autophagy incompetent.[67]

The isolation and characterization of the vacuoles would provide unequivocal knowledge of the composition of the vacuoles, which could lead to elucidation of their origin and biogenesis. Until then, studies conducted in our lab suggest that the vacuoles may be derived from perturbation of the endocytic pathway by the cytotoxic GAELs. When cells were incubated with GAELs in the presence or absence of β-methylcyclodextrin (MCD), which inhibits endocytosis,[36] no vacuoles were observed in the cells incubated with Gln **9** + MCD and, furthermore, the GAELs did not kill the cells. In contrast, in control cells with only Gln, the vacuoles were formed and the cells killed. The lack of LAV formation in MCD-treated cells suggests that

GAELs are taken up by endocytosis rather than by passive diffusion, as was previously proposed.[15] Unequivocal proof will have to await the synthesis of fluorescent-labeled analogues to allow visualization of the compounds in cells. If GAELs are taken up by endocytosis as our results suggest, then the large acidic vacuoles are likely generated by perturbation of the endocytic pathway that ultimately leads to lysosomal formation. To gain further insight into the events that may be transpiring, we manipulated the endocytic pathway by varying the incubation temperature of the cells.

At 22 °C, endocytic traffic proceeds from early endosomes to the endosomal carrier vesicles, but not into late endosomes/lysosomes.[68–71] When ATG5-/-MEFs were incubated with cytotoxic concentrations of Gln **9** at 20 °C, the cells generated vacuoles that were not as large as those observed at 37 °C, but, importantly, the cells did not die even though incubations at 37 °C with similar concentrations of the GAELs killed all the cells.[36] When the cells were incubated with cytotoxic concentrations of Gln for 4 h at 37 °C, and then were incubated for 20 h at 22 °C, the cells survived the Gln treatment. The results are consistent with the requirement of an active endocytosis pathway for manifestation of the cytotoxic effects of GAELs.

The current working hypothesis is that cells take up GAELs by endocytosis and the compound becomes incorporated in the early endosomes. During the maturation process that results in transformation of the early endosomes into late endosomes, GAELs initiate the vacuolarization process. The fusion and fission processes with components from the Golgi and ER occur that ultimately are responsible for the lysosomal-like characteristics that have been observed. The vacuoles formed from early endosomes may be regarded as immature in terms of their molecular composition, perhaps lacking the constituents or possessing them in inadequate quantities that are necessary to initiate the killing process. Comparative studies on the composition of the vacuoles isolated at 22 °C *versus* those from cells incubated at 37 °C would unequivocally validate or refute these proposals. Under the proposed scenario, association of the vesicles with autophagosomes, if it occurs, is an incidental event that has no direct role in the mechanism of action of GAELs.

Our studies have clearly linked the vacuolarization in cells with cell death. Thus, only compounds that generate the vacuoles kill the cells, while those that do not are unable to kill the cells. In addition, as reported above, in cells incubated with MCD, no vacuoles are formed and the cells do not die when incubated with cytotoxic concentrations of GAELs. The molecular events linking vacuole generation to cell death have yet to be elucidated, but there is evidence it may involve lysosomal hydrolases such as cathepsins. The acidic and lysosomal-like nature of the vacuoles makes this feasible. It has been demonstrated that incubation of cells with Gln results in the release of cathepsins B, D, and L into the cytosol of the cells.[16] This was assessed by extracting the cytosol of the cells under conditions that do not cause rupture of the lysosomes. The levels of cathepsins B, D, and L in the cytosolic fraction as measured by the enzyme activity and Western blotting for the proteins were greatly elevated in the cytosol of GAEL-treated cells. These observations suggest that Gln **9** induces lysosomal membrane permeability (LMP) in cells. When the cells were incubated with Gln **9** in the presence of pepstatin A, a cathepsin D inhibitor, Gln **9**-induced cell death was reduced by 40% relative to the controls with the GAELs alone.[16] The lack of complete protection may be due to a contribution by the other cathepsins in the death process. Unfortunately, it was not possible to assess the effects of inhibiting the other cathepsins as E64, a (cathepsin B) inhibitor, was toxic to the cells.[16] While the above results implicate cathepsin D in the GAEL-induced non-apoptotic cell death, the role of the other cathepsins, if any, remains to be demonstrated experimentally. The involvement of cathepsins in apoptosis has been previously reported and is thought to occur *via* conversion of Bid to t-Bid.[72] However, cathepsins released in response to GAELs do not induce apoptosis since t-Bid was not generated and neither was there a loss of MMP, cytochrome *c* release, or activation of caspses.[15,16] Our current hypothesis is that cathepsins and perhaps other lysosomal hydrolases are the effectors of GAEL-induced cell death and they achieve this *via* a non-apoptosis mechanism that has yet to be elucidated.

7.9 GAELs and Cancer Stem Cells

As discussed in the introduction, two major factors contributing to the high mortality of cancer are the problem of drug resistance to the current crop of chemotherapeutic agents and our inability to treat metastatic disease. Cancer stem cells (CSCs) have been implicated as the driving force underpinning drug resistance and metastases.[11,12] The CSC theories serves as a framework to guide the development of strategies to successfully eliminate tumors. The theory envisions CSCs as the driving force behind the growth and progression of the tumor. Accumulating evidence supports this. CSCs have now been reported in virtually all solid tumors.[73–78] Aggressive cancers that are refractory to treatment contain more CSCs and there is correlation between the presence of CSC markers with clinical progression and clinical outcome.[11,12,79,80] Elimination of CSCs is therefore crucial to successful cancer therapy, according to the CSC theory. This may be achieved by directly killing the CSCs, inducing differentiation with loss of CSC characteristics, or disrupting the niche signals they require for maintenance.

Since CSCs appear to be intrinsically resistant to apoptosis and radiotherapy, it is not surprising that current chemotherapeutic agents and radiotherapy are unable to prevent relapse, as their mode of action is to kill cells by inducing apoptosis. Agents that kill cells by an apoptosis-independent mechanism could be effective in killing CSCs and thus prevent recurrence of the tumor. As discussed above, GAELs are unique in killing cells *via* an apoptosis-independent mechanism and would be expected to be effective against cancer stem cells. We have tested this hypothesis by isolating breast cancer stem cells and comparing the effects of the apoptosis-inducing edelfosine with GAELs.

Breast CSCs were isolated from BT474 and JIMT1 breast cancer cell lines by staining the cells and sorting for aldehyde dehydrogenase-1 (ALDH1), a stem cell marker for breast CSCs, using the Aldefluor assay kit. The sorted cells were grown as tumorspheres in suspension in ultra-low adhesion tissue culture ware in mammocult medium.[14] Large tumorspheres were formed within 7 days.

The ability of the GAELs on the growth of CSCs to form spheroids was investigated. When GAELs were added to freshly sorted CSCs, at concentrations of 10 μM and above for 6 days, the cells were unable to form spheroids, whereas the controls formed spheroids.[14] In contrast, when the cells were incubated with edelfosine, some spheroid growth was observed.

During these studies we observed that when BT-474 cells were grown under conditions of constant shaking on a Nutator mixer, BT-474 cancer stem cells congregated to form a large compact sphere, the size of which increased over time. These spheres were subsequently incubated with *C*-Gln **28** (0, 20, 30 μM) for 7 days. Within 24 h of incubation with *C*-Gln, the compact nature of the spheroid was lost, the mass was loose, and single cells emanating from the mass were evident. After 96 h, the mass was very loose and amorphous, and chunks of material had been lost from the main mass (Figure 7.5). After 6 days there is almost complete disintegration of the mass

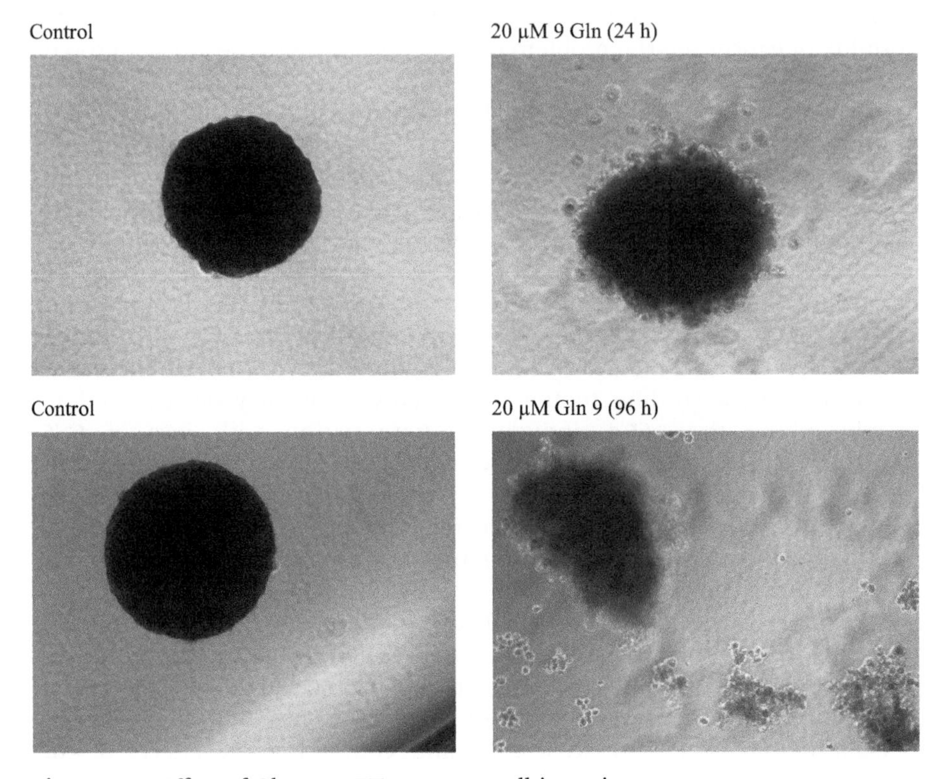

Figure 7.5 Effect of Gln **9** on BT-474 stem cell integrity.

and when the viability of the materials in the wells was tested using the MTS assay, the cells remaining in the wells that were incubated with 20 or 30 μM *C*-Gln **28** were not viable. Thus under these conditions of constant agitation of the CSCs, *C*-Gln **28** was able to disrupt the mass and kill the cells.

The effects of the most active GAELs synthesized to date, *i.e.* **9**, **17**, **19**, and **28**, on the growth and viability of BT-474 tumorspheres growing under static conditions have been investigated in comparison with the ALP edelfosine **1**.[14] The breast cancer stem cells were seeded into wells and incubated for 7 days to form tumorspheres which were subsequently incubated with the AELs for 6 days. In spheroids incubated with the GAELs, disintegration of the spheroids was apparent after 48 h and by the 6th day disintegration was complete. Assessment of the viability of the material at the end of the incubation revealed that viability of the cells was down to zero between 10 and 20 μM GAEL. In contrast, when the experiment was repeated with edelfosine **1**, the gold standard AEL, while some disintegration was observed, with 20 μM edelfosine the main mass remained relatively compact. At a concentration of 20 μM edelfosine, 70% of the cells were still viable. At 30 μM the viability was around 45% relative to controls. Thus at a concentration of 30 μM *in vitro*, edelfosine **1** kills most cancer cells lines and there would be no difference between the activity of GAELs and edelfosine. The fact that it is unable to kill the stem cells even at this concentration is significant as its mode of action is *via* apoptosis while GAELs kill by an apoptosis-independent mechanism.

In JIMT1 cells, Gln analogues **9**, **17**, and *C*-Gln **28** (10–30 μM) had a profound effect on the morphology and compactness of the spheroids (Figure 7.3). Virtual disintegration was observed after 6 days incubation with these compounds. In contrast, incubation with edelfosine **1** had minimal effect on the morphology of the spheroids (Figure 7.6). When the viability of the spheroids was assessed after 6 days incubation with the drugs, at 10 μM concentration the GAEL affected the viability of greater than 65% of the cells while at 20 or 30 μM the viability of the cells was completely lost. In contrast, 10 μM edelfosine affected the viability of 45% of the stem cells, and at 30 μM, 30% of the stem cells were still viable.

The above results clearly show that GAELs **9**, **17**, and **29** were able to cause the disintegration of CSC spheroids and the complete loss of cell viability. Under identical incubation conditions, the ALP, edelfosine **1**, was unable to achieve a similar effect on the morphology of the cancer stem spheroids and, more importantly, a significant proportion of the stem cells were still viable even at concentrations of 30 μM. These results are in line with our hypotheses that because GAELs kill cells by an apoptosis-independent pathway, they would be able to kill CSCs, whereas the apoptosis inducing edelfosine **1** is unable to kill the cells as CSCs resist cell death by apoptosis.

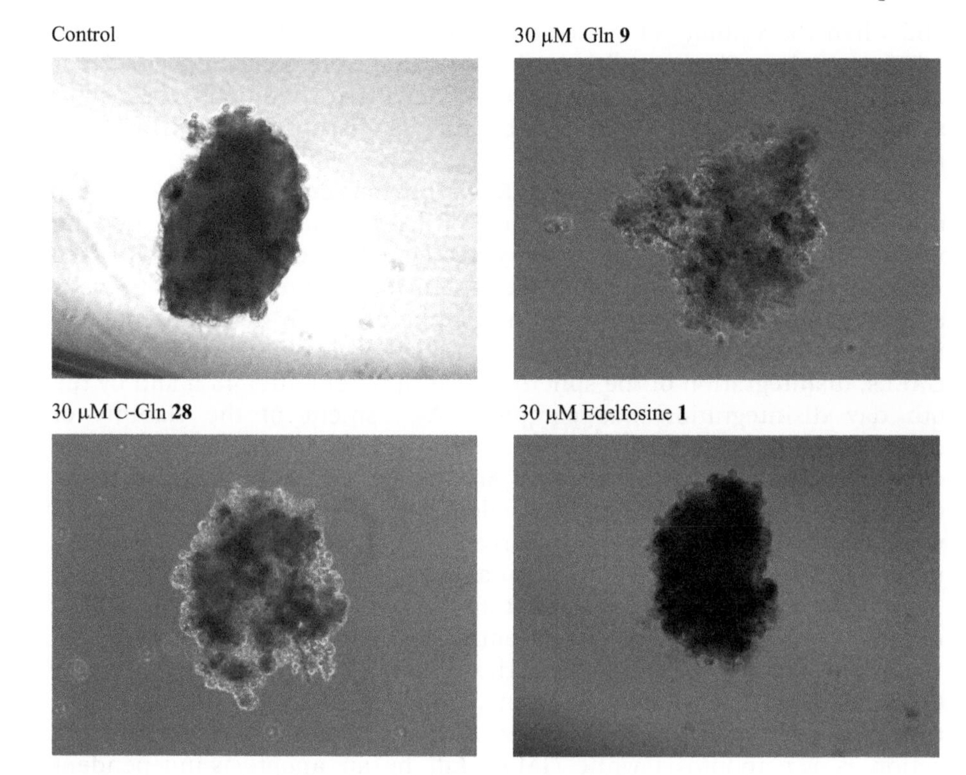

Figure 7.6 Effect of AEL on JIMT-1 stem cell integrity.

7.10 Conclusions and Perspectives

Cancer is a major health problem worldwide. The current and projected increase in the incidence of cancer, coupled with the high mortality and morbidity of the disease, makes it imperative to develop new effective cures. Although several classes of compounds are used clinically to treat the disease,[81] we still have high mortality rates for most cancers, especially if they have metastasized. Because these agents invariably act by inducing apoptosis to cause cell death, the ability of some cells in the tumor to resist apoptotic cell death, through a variety of mechanisms, renders the compounds ineffective. At the heart of the problem may be the cancer stem cells which have now been reported in virtually every tumor type,[73–78] and may be responsible for radio- and chemoresistance as well as the driving force in the metastatic process. There is therefore an urgent need to develop compounds that kill cells *via* mechanisms other than apoptosis to kill the cancer stem cells. We believe GAELs could ultimately be developed into clinically relevant anticancer drugs that will be effective against apoptosis-resistant cells that do not respond to chemotherapy or radiotherapy. As discussed above, some of the GAEL analogues we have developed are quite effective in not only preventing the growth of CSCs into tumorspheres but also cause the disintegration of the tumorspheres and ultimately kill the cells. We believe this is a function of their ability to kill cells *via* a non-apoptosis mechanism. Studies to elucidate the molecular details of the mechanism of cell death utilized by GAELs are clearly warranted as they could reveal novel molecules that could be targets to induce cell death in drug-resistant CSCs.

Structure–activity relationship studies have identified features of the GAEL molecule that can be used to develop more effective analogues than the existing compounds going forward. The synthesis of analogues with sugars other than glucose, mannose, and galactose, both pyranosides and furanosides, needs to be explored. Irrespective of the sugar, there seems to be a need for a cationic amine on the sugar for maximum activity. It is currently not known whether the amine is restricted to the 2-position of the sugar. It would be also worth exploring whether increasing the number of amino functions on the sugar will enhance the activity. As the α-anomeric sugars are on the whole more active than the β-anomers, the former would be the preferred configuration. The development of GAELs that will resist glycosidases *in vivo* is essential to move the development of the compounds forward from *in vitro* cell studies into animals. Although the *C*-glycosidic analogues of GAELs will meet the stability requirements, the arduous and involved nature of the synthesis makes this a last resort and the development of glycosidase-resistant analogues that can be made by facile synthesis should be pursued. The availability of such stable analogues will permit the compounds to be delivered either intravenously or orally. It may also be worthwhile to explore variations at the C-1 and C-2 positions of GAELs. While the methoxy at the C-2 position does not appear to be essential for activity, studies with ALPs have revealed unexpected results with respect to cell selectivity as a consequence of changes to the C-1 and C-2 moieties of the compounds.[82]

It is worth reiterating that very few compounds have been described that are capable of killing CSCs.[13] Thus the ability of GAELs to kill CSCs reveals their enormous potential. In view of the important role these cells have in cancer progression and metastases, further development of these compounds is clearly warranted to improve their activity, stability, and selectivity in order to incorporate them into treatment regimens in the fight to treat and cure cancer.

Acknowledgements

This work was supported by a grant from the Canadian Breast Cancer Foundation Prairies/NWT Region.

References

1. World Cancer Research Fund International; http://www.wcrf.org/cancer_statistics/world_cancer_statistics.php.
2. D. G. Kruh, *Oncogene*, 2003, **22**, 7262.

3. L. Gatti and F. Zunino, *Methods Mol. Med.*, 2005, **111**, 127.
4. D. B. Longley and P. G. Johnston, *J. Pathol.*, 2005, **205**, 275.
5. J. Plati, O. Bucur and R. Khosrvi-Far, *Integr. Biol.*, 2011, **3**, 279.
6. Y. Dai, T. S. Lawrence and L. Xu, *Am. J. Transl. Res.*, 2009, **1**, 1.
7. J. A. Magee, E. Piskounovva and S. J. Morrison, *Cancer Cell*, 2012, **21**, 283.
8. L. N. Abdullah and E. K.-H. Chow, *Clin. Transl. Med.*, 2013, **2**, 3.
9. L. Chang, P. H. Graham, J. Hao, J. Bucci, P. J. Cozzi, J. H. Kearsley and Y. Li, *Cancer Metastasis Rev.*, 2014, **33**, 469.
10. S. Fulda and S. Pervaiz, *Int. J. Biochem. Cell Biol.*, 2009, **42**, 31.
11. B. K. Garvalov and T. Acker, *J. Mol. Med.*, 2011, **89**, 95.
12. P. C. Hermann, S. L. Huber, T. Herrler, A. Aicher, J. W. Ellwart and M. Guba, *Cell Stem Cell*, 2007, **1**, 313.
13. R. Zobalova, M. Stantic, M. Stapelberg, K. Prokopova, J. Dong, J. Truksa *et al.*, in Drugs that Kill Cancer Stem-like Cells, ed. S. Shostak *Cancer Stem Cells Theories and Practice*, 2011, ISBN: 978-953-307-225-8.
14. P. Samadder, Y. Xu, F. Schweizer and G. Arthur, *Eur. J. Med. Chem.*, 2014, **78**, 225.
15. P. Samadder, R. Bittman, H.-S. Byun and G. Arthur, *Biochem. Cell Biol.*, 2009, **87**, 401.
16. L. Jahreiss, M. Renna, R. Bittman, G. Arthur and D. C. Rubinsztein, *Autophagy*, 2009, **5**, 835.
17. R. Bittman and G. Arthur in *Antitumor ether lipids: biological and biochemical effects in Liposomes: Rationale Design*, ed. A. S. Janoff, Marcel Dekker, New York, 1999, pp. 125–144.
18. W. E. Berdel, R. Andreesen and P. G. Munder, in *Phospholipids and Cellular Regulation*, ed. J. F. Kuo, CRC Press, Boca Raton, FL, 1985, vol. 2, pp. 41–73.
19. W. E. Berdel, *Br. J. Cancer*, 1991, **64**, 208.
20. W. E. Berdel, W. R. E. Bausert, U. Fink, J. Rastetter and P. G. Munder, *Anticancer Res.*, 1981, **1**, 345.
21. W. J. Houlihan, M. Lohmeyer, P. Workman and S. H. Cheon, *Med. Res. Rev.*, 1995, **15**, 157.
22. L. A. Smets, H. Van Rooij and G. S. Salmons, *Apoptosis*, 1999, **4**, 419.
23. W. J. van Blitterswijk and M. Verheij, *Biochim. Biophys. Acta*, 2013, **1831**, 663.
24. C. Gajate, A. Santos-Beneit, M. Modolell and F. Mollinedo, *Mol. Pharmacol.*, 1998, **53**, 602.
25. J. K. Jackson, H. M. Burt, A. M. Oktaba, W. Hunter, M. P. Scheid, F. Mouhajir, R. W. Lauemer, Y. Shen, H. Salari and V. Duronio, *Cancer Chemother. Pharmacol.*, 1998, **41**, 326.
26. F. Mollinedo, C. Gajate, S. Martin-Santamaria and F. Gago, *Curr. Med. Chem.*, 2004, **11**, 3163.
27. J. J. Gills and P. A. Dennis, *Curr. Oncol. Rep.*, 2009, **11**, 102.
28. H. Prinz, L. Six, K.-P. Ruess and M. Lieflander, *Liebigs Ann. Chem.*, 1985, 217.
29. N. Weber and H. Benning, *Chem. Phys. Lipids*, 1986, **41**, 93.

30. P. N. Guivisdalsky, R. Bittman, Z. Smith, M. L. Blank, F. Snyder, S. Howard and H. Salari, *J. Med. Chem.*, 1990, **33**, 2614.
31. A. Noseda, M. E. Berens, C. Piantadosi and E. J. Modest, *Lipids*, 1987, **22**, 878.
32. D. B. J. Hermann, E. Besenfelder, U. Bicker, W. Pahlke and E. Bohm, *Lipids*, 1987, **22**, 952.
33. X. Lu, K. Rengan, R. Bittman and G. Arthur, *Oncol. Rep.*, 1994, **1**, 933.
34. R. K. Erukulla, X. Zhou, P. Samadder, G. Arthur and R. Bittman, *J. Med. Chem.*, 1996, **39**, 1541.
35. K. C. Nicolaou, D. E. Lizos, D. W. Kim, D. Schlawe, R. G. de Noronha, D. A. Longbottom, M. Rodriquez, M. Bucci and G. Cirino, *J. Am. Chem. Soc.*, 2006, **128**, 4460.
36. P. Samadder, H.-S. Byun, R. Bittman and G. Arthur, *Anticancer Res.*, 2011, **31**, 3809.
37. A. M. Rogan, T. C. Hamilton, R. C. Young, R. W. Klecker Jr. and R. F. Ozols, *Science*, 1984, **224**, 994.
38. J. A. Green, D. T. Vistica, R. C. Young, T. C. Hamilton, A. M. Rogan and R. F. Ozols, *Cancer Res.*, 1984, **44**, 5472.
39. Y. Xu, M. Ogunsina, P. Samadder, G. Arthur and F. Schweizer, *ChemMedChem*, 2013, **8**, 511.
40. E. A. M. Fleer, C. Unger, D.-J. Kim and H. Eibl, *Lipids*, 1987, **22**, 856.
41. X. Lu, X. Zhou, D. Kardash and G. Arthur, *Biochem. Cell Biol.*, 1993, **71**, 122.
42. N. Weber and H. Benning, *Biochim. Biophys. Acta*, 1988, **959**, 91.
43. G. Arthur and P. Samadder, Unpublished observation 2013.
44. M. Ogunsina, H. Pan, P. Samadder, G. Arthur and F. Schweizer, *Molecules*, 2013, **18**, 15288.
45. G. Yang, R. W. Franck, R. Bittman, P. Samadder and G. Arthur, *Org. Lett.*, 2001, **3**, 197.
46. G. Yang, R. W. Franck, H.-S. Byun, R. Bittman, P. Samadder and G. Arthur, *Org. Lett.*, 1999, **1**, 2149.
47. A. Girault, J.-P. Haelters, M. Potier-Cartereau, A. Chantome, M. Pinault, S. Marionneau-Lambot, T. Oullier, G. Simon, H. Couthon-Gourves, P.-A. Jaffres, B. Corbel, P. Bougnoux, V. Joulin and C. Vandier, *Curr. Cancer Drug Targets*, 2011, **11**, 1111.
48. J. R. Marino-Albernas, R. Bittman, A. Peters and E. Mayhew, *J. Med. Chem.*, 1996, **39**, 3241.
49. H. S. Byun, R. Bittman, P. Samadder and G. Arthur, *ChemMedChem*, 2010, **5**, 967.
50. R. Andreesen, M. Modolell, H. U. Weltzein, H. Eibl, H. H. Common, G. W. Lohr and P. G. Munder, *Cancer Res.*, 1978, **38**, 3894.
51. L. Glasser, L. R. Somberg and W. R. Vogler, *Blood*, 1984, **64**, 1288.
52. R. Andreesen, M. Modolell and P. G. Munder, *Blood*, 1979, **54**, 519.
53. M. H. Runge, R. Andreesen, A. Pfleiderer and P. G. Munder, *J. Natl. Cancer Inst.*, 1980, **64**, 1301.

54. F. Mollinedo, J. de la Iglesi-Vincente, C. Gajate, A. E.-H. de Mendoza, J. A. Villa-Pulgrin, M. de Frais, G. Roue, J. Gil, D. Colomer, M. A. Campanero and M. J. Blanco-Prieto, *Clin. Cancer Res.*, 2010, **16**, 2046.
55. J. Storch and P. G. Munder, *Lipids*, 1987, **22**, 813.
56. G. Powis, M. J. Seewald, C. Gratas, D. Elder, J. Riebow and E. J. Modest, *Cancer Res.*, 1992, **52**, 2835.
57. I. Junghahn, J. Bergmann, P. Langen, I. Thun, C. Vollgraf and H. Brachwitz, *Anticancer Res.*, 1995, **15**, 449.
58. S. Egan, J. A. Wright, L. Jarolim, K. Yanagihara, R. H. Bassin and A. H. Greenberg, *Science*, 1987, **238**, 202.
59. P. Samadder, H.-S. Byun, R. Bittman and G. Arthur, *Anticancer Res.*, 1998, **18**, 465.
60. N. Weber and H. Benning, in *Topics in Lipid Research from Structural Elucidation to Biological Function*, ed. R. Klein and B. Schmititz, Royal Soceity of Chemistry, London, 1986, pp. 14–19.
61. I. Ahmad, J. J. Filep, J. C. Franklin, A. S. Janoff, G. R. Masters, J. Pattassery, A. Peters, J. J. Schupsky, Y. Zha and E. Mayhew, *Cancer Res.*, 1997, **57**, 1915.
62. J. V. Busto, E. Del Canto-Jañez, F. M. Goñi, F. Mollinedo and A. Alonso, *J. Chem. Biol.*, 2008, **1**, 89.
63. J. K. Jackson, H. Burt, W. Mok, H. Salari, E. R. Kumar, H.-S. Byun and R. Bittman, *Biochem. Cell Biol.*, 1994, **72**, 297.
64. D. S. Vallari, Z. L. Smith and F. Snyder, *Biochem. Biophys. Res. Commun.*, 1988, **156**, 1.
65. S. A. Susin, N. Zamzami, M. Castedo, T. Hirsch, P. Marchetti, A. Macho, E. Daugas, M. Geuskens and G. Kroemer, *J. Exp. Med.*, 1996, **184**, 1331.
66. N. Mizushima, A. Yamamoto, M. Hatano, Y. Kobayashi, Y. Kabeya, K. Suzuki, T. Tokuhisa, Y. Ohsumi and T. Yoshimori, *J. Cell Biol.*, 2001, **152**, 657.
67. Y. Takahashi, D. Coppola, N. Mattsushita, H. D. Cualing, M. Sun, Y. Sato, C. Liang, J. U. Jung, J. Q. Cheng, J. J. Mule, W. J. Pledger and H. G. Wang, *Nat. Cell Biol.*, 2007, **9**, 1142.
68. E.-L. Punnonen, K. Ryhanen and V. Marjomaki, *Eur. J. Cell Biol.*, 1998, **75**, 344.
69. L. A. Casiola-Rosen and A. L. Hubbard, *J. Biol. Chem.*, 1992, **267**, 8213.
70. W. A. Dunn, A. L. Hubbard and N. A. Aronson, *J. Biol. Chem.*, 1980, **255**, 5971.
71. P. C. Sullivan, A. L. Ferris and B. Storrie, *J. Cell. Physiol.*, 1987, **131**, 58.
72. H. Appelqvist, A.-C. Johansson, E. Linderoth, U. Johansson, B. Antonsson, R. Steinfeld, K. Kagedal and K. Ollinger, *Ann. Clin. Lab. Sci.*, 2012, **42**, 231.
73. M. Al-Hajj, M. S. Wicha, I. Weissman and M. F. Clarke, *Proc. Natl. Acad. Sci. U. S. A.*, 2003, **100**, 3983.
74. A. T. Collins, P. A. Berry, C. Hyde, M. Stower and N. J. Maitlan, *Cancer Res.*, 2005, **65**, 10946.

75. C. O'Brien, A. Pollet, S. Gallinger and J. E. Dick, *Nature*, 2007, **445**, 106.
76. C. Li, D. G. Heidt, P. Dalerba, C. F. Burant, L. Zhang, V. Adsav, M. Wicha, M. F. Clarke and D. M. Simeone, *Cancer Res.*, 2007, **67**, 1030.
77. Y. Pan and X. Huang, *Int. J. Clin. Exp. Med.*, 2008, **1**, 260.
78. N. A. Lobo, Y. Shimono, D. Qian and M. F. Clarke, *Annu. Rev. Cell Dev. Biol.*, 2007, **23**, 41.
79. R. Besancon, S. Valsesia-Wittman, A. Puisieux, C. C. de Fromental and V. Maguer-Satta, *Curr. Med. Chem.*, 2009, **16**, 394.
80. N. Y. Frank, T. Schatton and M. H. Frank, *J. Clin. Invest.*, 2010, **120**, 41.
81. L. R. Ferguson and A. E. Pearson, *Mutat. Res.*, 1996, **355**, 1.
82. H.-S. Byun, R. Bittman, P. Samadder and G. Arthur, *ChemMedChem*, 2010, **5**, 1045.
83. N. Weber and H. Benning, *J. Nutr.*, 1984, **114**, 247.

CHAPTER 8

Carbohydrates and Glycomimetics in Alzheimer's Disease Therapeutics and Diagnosis

CATARINA DIAS AND AMÉLIA P. RAUTER*

Centro de Química e Bioquímica, Faculdade de Ciências, Universidade de Lisboa, Campo Grande, 1749-016, Lisbon, Portugal
*Email: aprauter@fc.ul.pt

8.1 Introduction

The increasing average life expectancy in developed countries has led to an escalating concern associated with the emergence of age-related diseases. Amongst them, Alzheimer's disease (AD) is the most prevalent form of late-life dementia, affecting about 35 million individuals worldwide.[1–3] The disease is particularly noteworthy due to its devastating nature, unsuccessful treatment and high socio-economic impact. Alzheimer's disease is multi-faceted, characterized by a global decline in cognitive function and its pathogenesis is not yet fully understood. The main hallmark changes of AD brains include the deposition of senile plaques containing amyloid-β peptide aggregates, the deposition of neurofribrillary tangles (NFTs) containing hyperphosphorylated tau protein, the loss of synaptic function, inflammation and neuronal death.[1]

RSC Drug Discovery Series No. 43
Carbohydrates in Drug Design and Discovery
Edited by Jesús Jiménez-Barbero, F. Javier Cañada and Sonsoles Martín-Santamaría

Published by the Royal Society of Chemistry, www.rsc.org

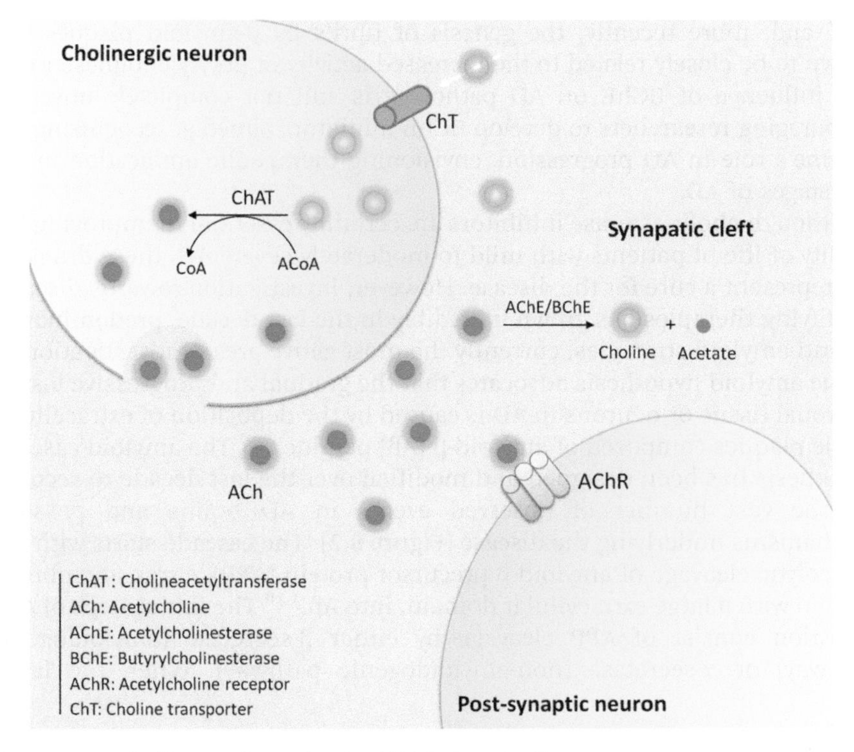

Figure 8.1 Cholinergic hypothesis of Alzheimer's disease.

The events involved in the cholinergic system are illustrated in Figure 8.1. The neurotransmitter acetylcholine (ACh) is produced in the presynaptic neuron by choline acetyltransferase and released in the synaptic cleft, where it binds to the ACh receptor on the postsynaptic membrane. Synaptic transmission is terminated by acetylcholinesterase (AChE), which hydrolyses ACh into acetate and choline. The profound loss of cholinergic function in the brain comprises an impressive decrease in the choline acetyltransferase level, choline uptake and ACh level in the neocortex and hippocampus.[4] It is well established that enhancing cholinergic transmission by blocking the activity of AChE slows down the AD-associated decline in behaviour and cognition. Indeed, currently prescribed drugs for the treatment of mild to moderate AD are AChE inhibitors, namely tacrine, donepezil and rivastigmine.[4] Nevertheless, these drugs have limited use because they only lead to a temporary symptomatic relief.[2,4]

Not only AChE, but also butyrylcholinesterase (BChE) is able to hydrolyze ACh. Although about 80% of the hydrolytic activity inside the healthy human brain is caused by AChE,[5] AChE levels decrease in late stages of AD, while those of BChE increase.[6] The ratio of BChE/AChE changes from 0.5 in the normal brain to 11 in the AD brain, and it is postulated that BChE may replace AChE in hydrolyzing brain acetylcholine at advanced stages of AD.[6] In fact, selective BChE inhibitors improve the cognitive performance of aged rats[7] and, more recently, the genesis of fibrils by β-amyloid plaques was shown to be closely related to the increased activity of butyrylcholinesterase.[8] The influence of BChE on AD pathology is still not completely unveiled, encouraging researchers to develop BChE inhibitors aimed at recognizing the enzyme's role in AD progression, envisioning therapeutic application in the late stages of AD.

Although cholinesterase inhibitors are certainly essential in improving the quality of life of patients with mild to moderately severe AD, these drugs do not represent a cure for the disease. However, investigation towards disease-modifying therapies has grown incredibly in the last decade, predominantly on anti-amyloid strategies, currently the most active area of investigation.[2,9]

The amyloid hypothesis advocates that the gradual and progressive loss of neuronal tissue or neurons in AD is caused by the deposition of extracellular senile plaques composed of amyloid-β (Aβ) peptide.[2,10] The amyloid cascade hypothesis has been designed and modified over the last decade to account for the vast number of observed events in AD brains and possible mechanisms underlying the disease (Figure 8.2). The cascade starts with the proteolytic cleavage of amyloid-β precursor protein (APP), a transmembrane protein with a large extracellular domain, into Aβ.[1,10] The initial steps of APP digestion consist of APP cleavage by either β-secretase (amyloidogenic pathway) or α-secretase (non-amyloidogenic pathway). When the latter

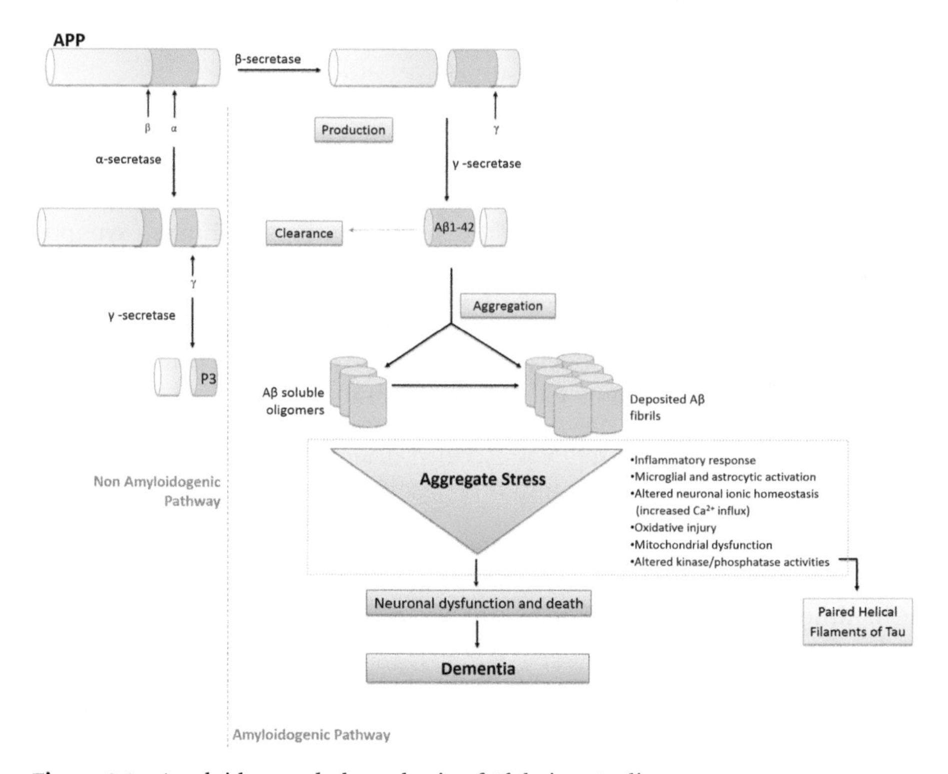

Figure 8.2 Amyloid cascade hypothesis of Alzheimer's disease.

product is further cleaved by γ-secretase the nontoxic P3 peptide is produced. Alternatively, cleavage of the β-secretase product by γ-secretase results in either the toxic Aβ1–42 or Aβ1–40. While Aβ1–40 is the most frequent form of Aβ, Aβ1–42 has a higher tendency to aggregate. Although the cause–effect phenomena associated with Aβ and the progression of the disease is not well understood, it is established that accumulation of Aβ1–42 is a key initiator of cellular damage in AD. Its aggregation results in soluble toxic oligomers and/or insoluble amyloid fibrils, whose exact molecular composition remains elusive due to their dynamic nature.[10] The smaller soluble forms of multimeric Aβ have been proposed as the most neurotoxic forms. However, it was recently proposed that Aβ oligomers possess distinct conformational polymorphisms and the unique combination of size, hydrophobicity, and conformation determine both its toxicity and that of the final oligomer aggregation state, thus contributing to AD pathology *via* different mechanisms.[11]

The term "aggregate stress" has been used to describe the events arising from Aβ aggregation and leading to cell apoptosis, including inflammatory response, microglial and astrocytic activation, altered neuronal ionic homeostasis (increased calcium influx), oxidative injury and mitochondrial dysfunction.[2,10] Kinase and phosphatase activities are also altered, leading to hyperphosphorylation of the tau protein, which results in the aggregation of tau into paired helical filaments and, ultimately, neurofibrillary tangles.[10] Although the precise temporal and mechanistic relationship between amyloid depositions and tau pathology is yet to be fully understood, scientists believe that the above-mentioned filaments of tau trigger a process of widespread neuronal/neuritic dysfunction and cell death, finally leading to neuronal loss and dementia.[10]

There are two therapeutic approaches targeting tau pathology: (1) inhibition of tau-phosphorylating kinases such as GSK-3, and (2) inhibition of tau aggregation and/or promotion of aggregate disassembly. Nevertheless, the few compounds targeting tau that reached clinical trials were disappointing regarding cognition.[9] Thus, since formation of tau NFTs is likely a cytological response to the gradual accumulation of Aβ, targeting amyloid seems to be the most sensible approach, and has gained strength as a modifying-disease strategy. For the treatment of amyloidosis, there are three main strategies: (1) reducing the production of Aβ (β- and γ-secretase inhibitors); (2) enhancing amyloid clearance; and (3) interfering with amyloid aggregation.[12]

The relevance of carbohydrates in drug discovery arises not only from their importance in life processes, but also because these stereochemically rich and polyfunctional molecules have the potential to be chemically manipulated in a multitude of ways. Moreover, as the brain requires a significant amount of glucose, there is a density of hexose transporters (GLUTs) at the blood–brain barrier (BBB), favouring carbohydrate-based drugs' access to the brain.[13] Thus, in addition to inherently bioactive carbohydrates, these molecules can also be linked to bioactive compounds, envisioning new derivatives and/or pro-drugs to improve targeting and pharmacological properties, such as water solubility and minimization of toxicity. Furthermore, carbohydrate-binding receptors expressed by microglia, resident immune cells of the central nervous system with the capacity to eliminate extracellular aggregates, are strongly involved in repair function during neurodegenerative or neuroinflammatory processes. Thus, carbohydrate-based drugs pursuing selective stimulation or inhibition of these receptors can contribute to a neural repair promoting microenvironment.[14]

Carbohydrate-based molecules have been extensively used over the last decade to meet the necessity of new neuroprotective compounds. Their structural diversity is overwhelming, including free and protected sugars, cyclitols, nucleosides, peptidomimetics and glycosaminoglycans. Nature has proven particularly resourceful in providing biologically active glycosides. Glycosylated terpenoids, flavonoids, iridoids, chromones and caffeoyl derivatives are amongst the vast number of secondary metabolites found in the literature addressing at least one of the above-mentioned strategies for AD treatment. Chemists have also been inspired in natural products to develop innovative structures with high pharmaceutical potential. Herein we will discuss the importance of carbohydrates and carbomimetic structures, addressing different aspects of neuroprotection currently under investigation, namely targeting amyloid and cholinergic hypotheses. The potential of carbohydrates in diagnosis will also be highlighted, focusing on both metabolic signatures of mitochondrial stress found in AD patients as well as imaging of amyloid aggregates with glyconanoparticles.

8.2 Carbohydrate-based Molecules and Glycomimetics in β-Amyloid Events

8.2.1 Terpene Glycosides

8.2.1.1 Ginsenosides

As previously stated, the plant kingdom has been highly important in drug discovery. The recognised nootropic and anti-aging effects of the well-known plant *Panax ginseng*, widely used in Chinese traditional medicine, have been attributed to ginsenosides, which are dammarane-type triterpene glycosides (see Figure 8.3). Although the neuroprotective activities of ginsenosides have been extensively studied and recently reviewed,[15–17] the investigation of their neuroprotective properties continues incredibly active, with a variety of recent studies concerning their potential for neurodegenerative diseases and their mechanism of action, in particular focusing on AD. Amongst the most recent findings lies the ability of ginsenoside Rg1 (Figure 8.3) to reduce Aβ production by inhibiting γ-secretase[18] and β-secretase, *via* stimulation of peroxisome proliferator-activated receptor-γ (PPARγ),[19,20] thus improving neuropathological and behavioural changes in mice overexpressing APP.[18] This ginsenoside also promotes non-amyloidogenic cleavage of APP by enhancing α-secretase activity,[21] while ginsenoside Rb1 is able to inhibit

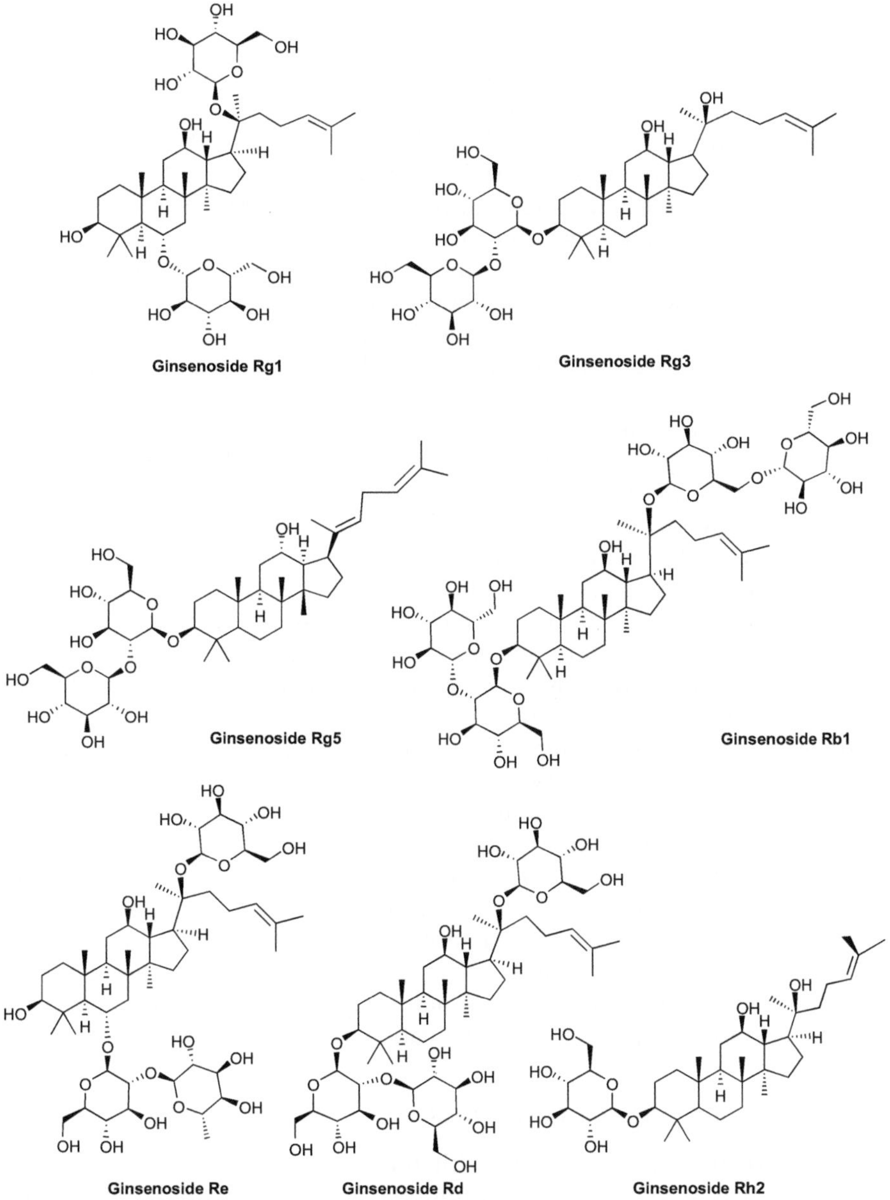

Ginsenoside Rg1 Ginsenoside Rg3

Ginsenoside Rg5 Ginsenoside Rb1

Ginsenoside Re Ginsenoside Rd Ginsenoside Rh2

Figure 8.3 Ginsenosides with anti-amyloid properties.

β-secretase *in vitro* and protects cells from toxicity induced by Aβ25–35,[17] the shortest peptide sequence retaining toxicity comparable with that of full-length Aβ1–42 and commonly used as an AD model. Along with Rb1, ginsenosides Rg1, Rd, Re, Rg3, Rg5 and Rh2 (Figure 8.3) are also reported to exert neuroprotective effects against Aβ-induced toxicity and inflammation.[16,17,22] For instance, Rg1 has a protective effect against Aβ1–40-induced neuronal injury and apoptosis in neuroblastoma cells, likely by inhibiting inflammatory cytokine production and by intervening in the apoptotic cascade.[23] Consistently, similar results were obtained in a comparable study using endothelial cells and Aβ25–35.[24] Rg1 also protects against Aβ-induced mitochondrial dysfunction, namely by increasing mitochondrial membrane potential, ATP levels and cytochrome *c* oxidase activity, rescuing neurons from oxidative stress.[25] Rg3 promotes Aβ clearance by microglial uptake, internalization and digestion and by increasing the gene expression of neprilysin, a rate-limiting enzyme in APP degradation into Aβ.[17] Treatment with this ginsenoside improved the LPS-induced cognitive impairments of rats, by inhibiting the expression of pro-inflammatory mediators.[26] This effect was also registered *in vitro* for Rg5, which additionally suppresses the production of reactive oxygen species (ROS) and pro-inflammatory cytokines in microglia.[27] Modulation of amyloid-induced neuroinflammatory responses in cultured microglia or stimulated brain by Rh2 was also investigated.[17] Finally, Rg1,[22,28,29] Rb1[30] and Rd[22,31] are all reportedly able to attenuate tau hyperphosphorylation, namely by regulating glycogen synthase kinase-3β (GSK-3β).[28,29,31]

In summary, advances in the research for ginsenosides' effects and mechanisms of action show that these naturally occurring glycosides intervene in a number of ladders of the amyloid cascade and ameliorate amyloid pathology in several animal models. Moreover, they act on the regulation of synaptic plasticity and neurotransmitter release, as described in Section 8.3.1.1 (below). More importantly, a single ginsenoside, namely Rg1, is able to act on multiple sites of action, and can be considered a candidate for development of a multi-target drug. These drugs are mostly pertinent in age-related neurodiseases, where multiple etiological and pathological targets are involved. Ginseng extracts have been studied in clinical trials, but the results are ambiguous and of difficult interpretation, due to the use of different extraction methods and different ginseng species.[16] Thus, given all the neuroprotective properties of these glycosides, allied with their lack of toxicity, randomised clinical trials to evaluate cognition improvement in AD patients are mandatory.

8.2.1.2 Other Terpene Glycosides

Other triterpene saponins along with monoterpene glycosides have also been described as neuroprotective, improving cognitive function in animal models. Asperosaponins A, B and C (Figure 8.4) are triterpene saponins isolated from *Dipsacus asper* that protect PC12 cells from Aβ25–35-induced cytotoxicity.[32] Also the triterpene glycoside hederacolchiside-E (Figure 8.4) protects against Aβ1–42-induced toxicity in SK-N-SH cells and enhances the cognitive function in mice with memory impairments.[33] In addition, a triterpene monoglycoside (Figure 8.4), identified in *Actaea racemosa*, reduces the production of Aβ1–42 with an IC_{50} of 0.1 μM in cultured cells overexpressing APP, by modulation of γ-secretase.[34]

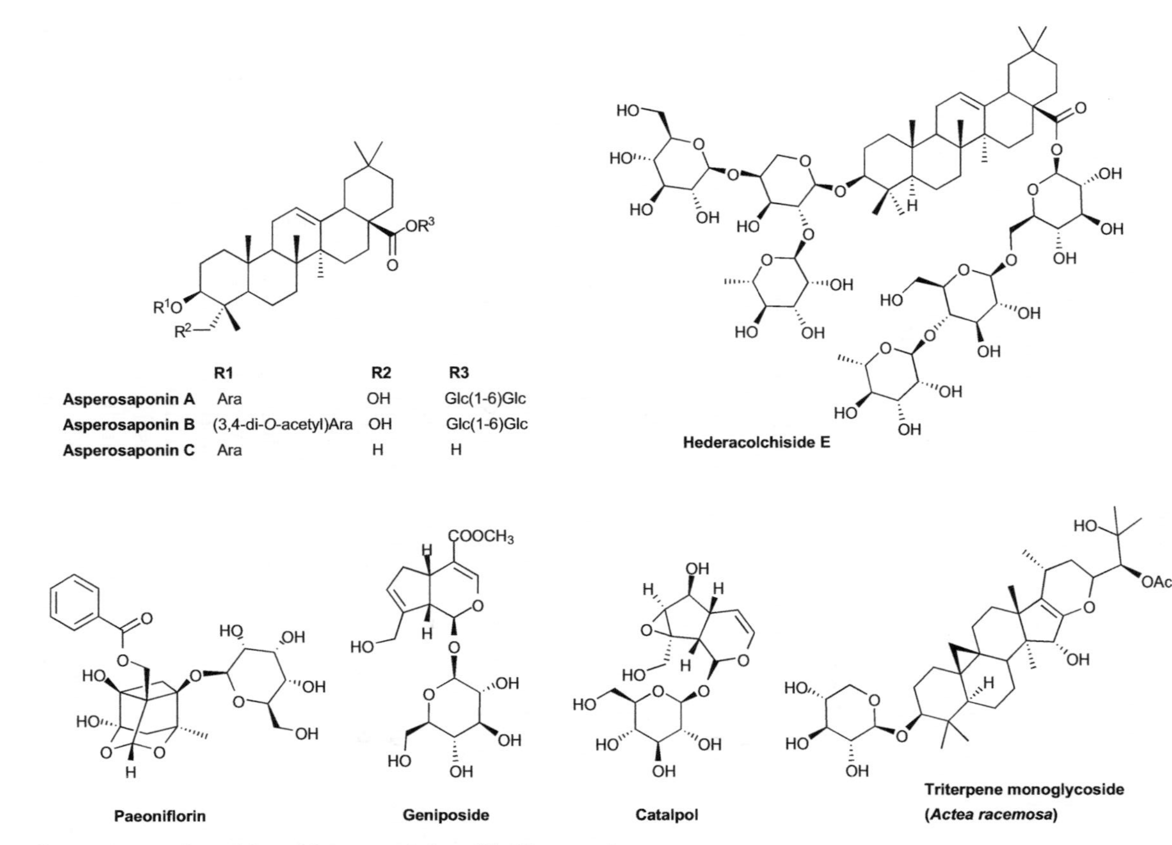

Figure 8.4 Other terpene glycosides able to modulate fibrillogenesis.

Paeoniflorin, a monoterpene glycoside isolated from *Radix paeoniae alba*, has been demonstrated to attenuate Aβ1–42-induced neurotoxicity by regulating calcium homeostasis and ameliorating oxidative stress in the hippocampus of rats,[35] and consequently reversed Aβ1–42-induced spatial learning and memory deficit.[36]

A number of *O*-glycosylated iridoids, monoterpenes embodying a cyclopentane ring fused to a six-membered oxygen heterocycle, have also been reported as neuroprotectors against amyloid-induced toxicity. For instance, geniposide (Figure 8.4) protects neuroblastoma cells from stress induced by formaldehyde, which promotes misfolding and aggregation of tau and Aβ proteins, through modulation of the apoptotic cascade.[37] Catalpol (Figure 8.4), an iridoid isolated from the roots of *Rehmannia glutinosa*, protects neuronal cells from Aβ1–42-triggered neurotoxicity to neurons, and also inhibits the glia-mediated inflammation process.[38]

8.2.2 Phenolic Glycosides

Natural occurring phenolic compounds are known for their beneficial health-promoting effects in chronic and degenerative diseases. The inhibition of amyloid fibril formation exerted by these antioxidants has been proposed to be associated with their ability to "break" β-sheets, as a consequence of structural constraints and specific aromatic interactions, with the phenolic rings being capable of disrupting peptide aromatic π–π stacking and the hydroxy groups forming competitive hydrogen bonding.[39] Given the biocompatible properties of carbohydrates, glycosylated phenolic compounds become highly significant in the quest for new drug candidates.

Acteoside (Figure 8.5), a phenylethanoid glycoside from *Orobanche minor*, strongly inhibits the aggregation of Aβ1–42, with an IC_{50} of 8.9 μM,[40] and protects against Aβ-induced cell injury by attenuation of ROS production, by modulation of apoptotic signal pathway through Bcl-2 family[41] and by upregulation of heme oxygenase-1.[42] Both the flavonol glucuronide hibifolin and the flavone glycoside linarin (Figure 8.5) protect cells against Aβ25–35-induced apoptosis.[43,44] While the former protects primary cortical neurons by several mechanisms, namely by abolishing Ca^{2+} mobilization, inhibiting caspase activation and suppressing DNA fragmentation,[43] the latter prevents neurotoxicity by acting on the apoptotic cascade through inhibition of GSK-3β and up-regulation of the anti-apoptotic protein Bcl-2.[44] 3-*O*-[β-D-Xylopyranosyl-(1→2)-β-D-galactopyranosyl]quercetin (Figure 8.5), isolated from *Panax notoginseng*, prevents Aβ-induced cell death by inhibiting Aβ aggregation, reducing brain damage in scopolamine-treated rats.[45] Interestingly, the aglycone itself did not show a comparable effect in amyloid aggregation, suggesting that the neuroprotective properties of this compound are strongly related to the presence of the sugar moiety.

A recent study showed how polyphenol glycosides and respective aglycons use different pathways to remodel and inactivate A-β oligomers.[46] Using the pairs phloretin/2′-*O*-(β-D-glucopyranosyl)phloretin, resveratrol/piceid,

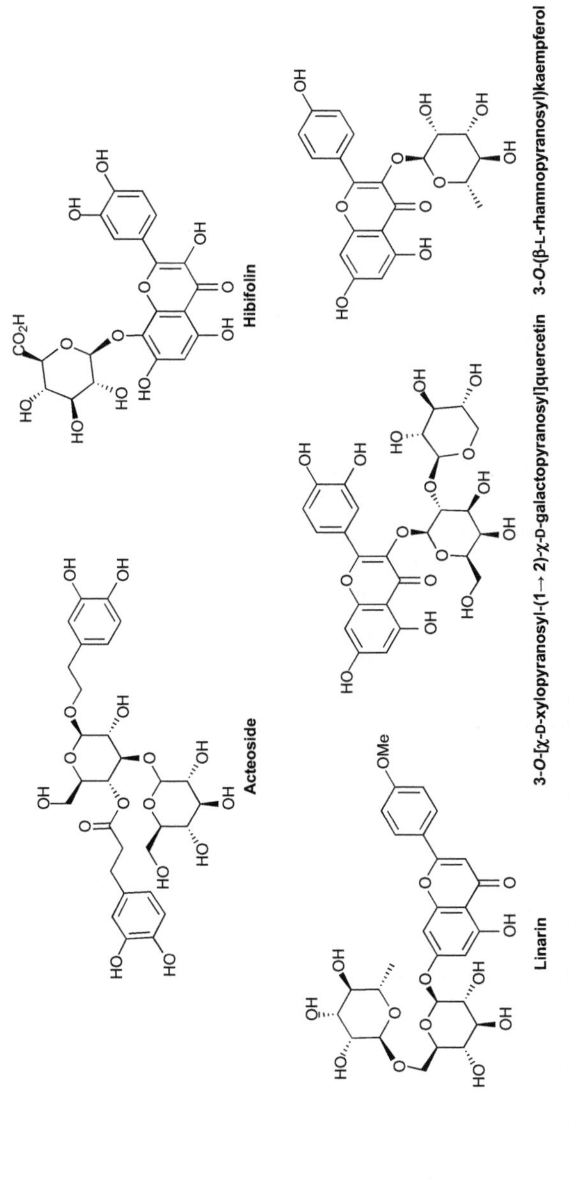

Figure 8.5 Phenolic glycosides intervening in amyloid events.

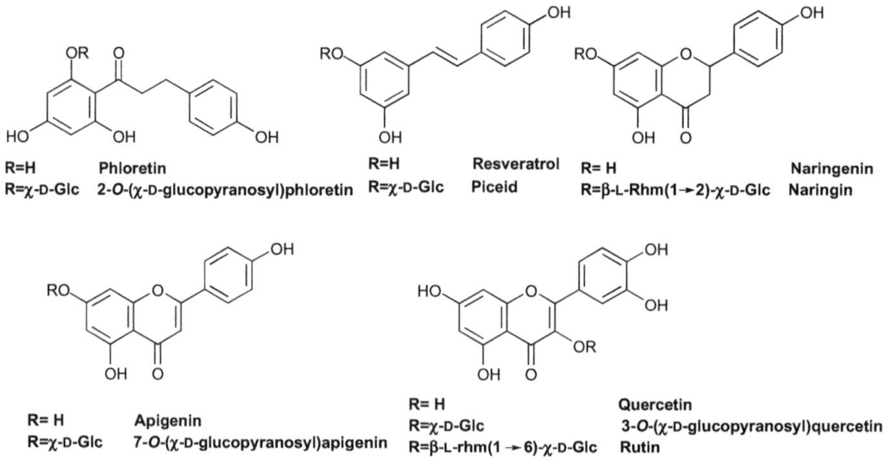

R=H Phloretin
R=χ-D-Glc 2-*O*-(χ-D-glucopyranosyl)phloretin

R=H Resveratrol
R=χ-D-Glc Piceid

R= H Naringenin
R=β-L-Rhm(1→2)-χ-D-Glc Naringin

R= H Apigenin
R=χ-D-Glc 7-*O*-(χ-D-glucopyranosyl)apigenin

R= H Quercetin
R=χ-D-Glc 3-*O*-(χ-D-glucopyranosyl)quercetin
R=β-L-rhm(1→6)-χ-D-Glc Rutin

Figure 8.6 Polyphenol glycoside/aglycon pairs used in the study conducted by Ladiwala *et al.*[46]

naringenin/naringin, apigenin/7-*O*-(β-D-glucopyranosyl)apigenin, quercetin/rutin and quercetin/3-*O*-(β-D-glucopyranosyl)quercetin (Figure 8.6), the authors showed that aglycons convert Aβ oligomers into large, off-pathway aggregates, while the respective glycosides rapidly dissociate Aβ oligomers into soluble, disaggregated peptide. Surprisingly, these polyphenol glycosides selectively target toxic Aβ conformers relative to non-toxic ones. Their disaggregation activity is believed to rely on the presence of both polyphenol and sugar in the same molecule, in which the aglycon initiates the remodelling process by disrupting intermolecular contacts between Aβ peptide through π-stacking interactions. The so-called CH–π-stacking interactions driven by association of the CH sugar bonds with π-electron cloud of the aromatic rings were suggested to cause interaction of sugars with the newly exposed aromatic rings of the Aβ residues, preventing their association. Interestingly, combinations of free polyphenols and sugars failed to disaggregate Aβ oligomers, highlighting the importance of polyphenol glycosides for the prevention of Aβ toxicity.[46] Coherently, a different study shows that 3-*O*-(α-L-rhamnosyl)kaempferol (Figure 8.5) protects cells against Aβ-induced toxicity, and morphological assays have proven that this flavonol glycoside inhibits fibrillogenesis of Aβ1–42, leading to the formation of smaller, soluble and non-toxic aggregates, and destabilizes pre-formed fibrils into non-toxic aggregates.[47]

Inspired in nature and taking advantage of synthetic tools, Raja and co-workers generated a clicked galactose–curcumin conjugate (see Figure 8.7) that is able to inhibit not only Aβ but also tau peptide aggregation at concentrations as low as 8 nM and 0.1 nM, respectively.[48] Curcumin is a polyphenol known for its multiple neuroprotective mechanisms, including inhibition of inflammation and disruption of A-β and tau peptide aggregation. However, its pharmaceutical use is restricted due to its poor water and plasma solubility and consequent low bioavailability. This non-toxic

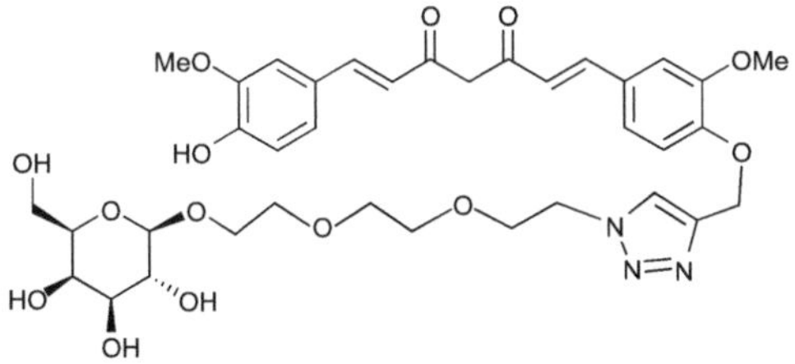

Figure 8.7 Sugar–curcumin conjugate.

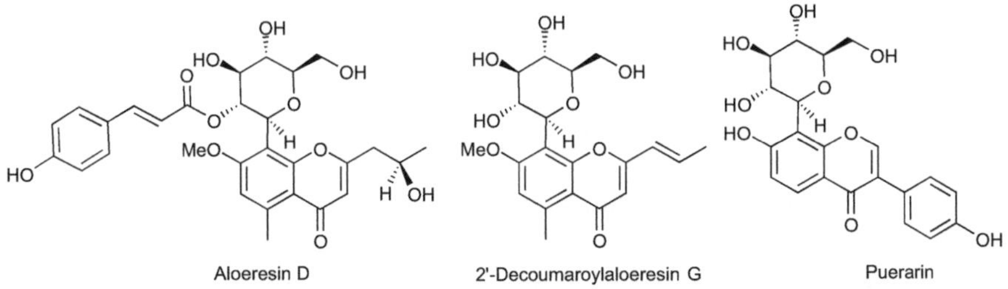

Figure 8.8 Examples of *C*-glycosyl compounds with anti-amyloid properties.

galactose–curcumin conjugate possessing a triazole-based linker is *ca.* 1000 times more soluble in water than curcumin, and exhibits enhanced ability to inhibit both A-β and tau aggregation.[48]

8.2.3 Other Glycosides and *C*-Glycosyl Compounds

Although terpene and phenolic glycosides have particular expression in the structural diversity of compounds that intervene in amyloid related events, other *O*- and *C*-glycosylated structures have been described with similar properties.

Amongst the *C*-glycosyl compounds, the chromones aloeresin D and C-2′-decoumaroyl-aloeresin G (Figure 8.8), both isolated from *Aloe vera*, inhibit β-secretase activity, with IC_{50} values of 39.0 and 20.5 μM, respectively.[49] The presence of a phenol group linked to the chromone scaffold at position 3 also resulted in a potential candidate for the treatment of AD, namely puerarin (Figure 8.8), a glucosyl isoflavone isolated from *Pueraria lobata*. It exhibits beneficial effects on various medicinal purposes due to its wide spectrum of pharmacological properties. These include neuroprotection, antioxidant, anti-inflammation and inhibition of alcohol intake, among others, as recently reviewed.[50] In particular, puerarin prevents Aβ-induced neurotoxicity by inhibiting neuronal apoptosis,[51,52] and attenuates Aβ25–35-induced lipid peroxidation and the overproduction of ROS, mainly by interrupting GSK-3β signalling, resulting in an increase of cell survival.[53,54] Administration of puerarin also reversed the Aβ1–42-induced impairment in spatial memory.[55]

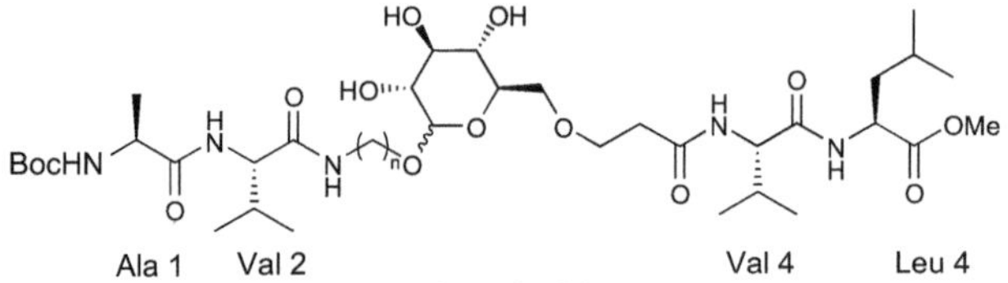

Figure 8.9 Peptidomimetics that inhibit amyloid-β aggregation.

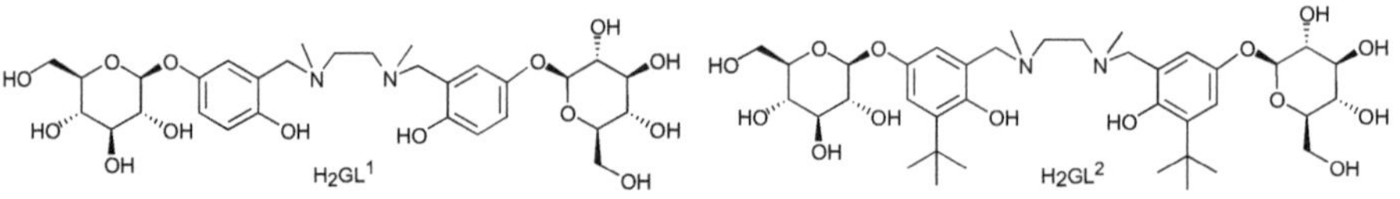

Figure 8.10 Carbohydrate-containing metal-ion chelators.

The glycosides embodying two short hydrophobic dipeptide units (Ala–Val and Val–Leu) (Figure 8.9) are peptidomimetics that intervene in A-β aggregation, in which the D-glucopyranosyloxy moiety is linked to the peptide *via* amidoalkyl (anomeric position) and amidoethyl (position 6) groups. These peptidomimetics are inhibitors of Aβ1–40 amyloid fibril formation, at the ratio of 0.1 : 1 (peptidomimetics/Aβ). While the central sugar moiety may perturb the regular interstrand hydrogen-bond network and destabilize β-sheet formation locally, the dipeptide units have the potential to interact with the hydrophobic regions of Aβ.[56] Hence, hydrophobic recognition and hydrophilic β-breakage strategies are combined in this innovative structure generated through glycochemistry approaches.

Other important characteristic of AD patients' brains with therapeutic implications is the high concentrations of Cu, Fe and Zn in amyloid plaques, and it is well established that interaction of Aβ with metal ions *in vitro* leads to Aβ aggregation, particularly with Zn^{2+}.[57] In this context, two glycosides incorporating metal-ion chelators, namely *N*,*N*′-bis{[5-(β-D-glucopyranosyloxy)-2-hydroxy]benzyl}-*N*,*N*′-dimethylethane-1,2-diamine (H_2GL1) and *N*,*N*′-bis{[3-*tert*-butyl-5-(β-D-glucopyranosyloxy)-2-hydroxy]benzyl}-*N*,*N*′-dimethylethane-1,2-diamine (H_2GL2) have been reported (Figure 8.10). Both compounds have moderate affinity for Cu^{2+} and Zn^{2+}, and were able to reduce Zn^{2+}-induced and Cu^{2+}-induced Aβ aggregation by approximately 50%,[57] thus acting as a starting point for the generation of new leads for metal-ion chelation therapy.

8.2.4 Sugars and Cyclitols

8.2.4.1 Free and Protected Sugars

As previously shown, a diversity of glycosides and *C*-glycosyl compounds intervene in amyloid-related events and the sugar moiety is highly involved

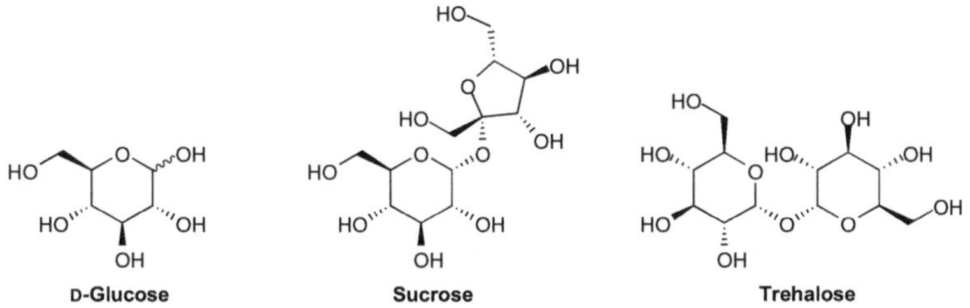

Figure 8.11 Free sugars able to retard amyloid fibril formation of 3HMut Wil protein.

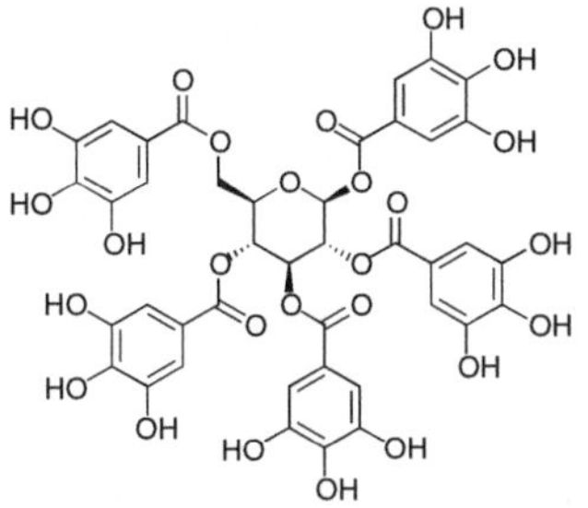

Figure 8.12 1,2,3,4,6-Penta-*O*-galloyl-β-D-glucopyranose.

in this interaction. However, free and protected sugars have also been reported to act on amyloid fibril formation.

The mechanism for retardation of amyloid fibril formation by free sugars was recently investigated using polypeptide chains in 3Hmut Wil protein as a model. The results showed that trehalose, sucrose and glucose (Figure 8.11) retarded the amyloid fibril formation of this protein, not by direct interaction, but by stabilizing the native state through preferential hydration.[58] Interestingly, it was shown that trehalose inhibits formation of Aβ1–40 fibrils and oligomers and reduces Aβ1–40-induced toxicity in SH-SY5Y cells, although these effects are not verified with Aβ1–42.[59]

The protected sugar 1,2,3,4,6-penta-*O*-galloyl-β-D-glucopyranose (PGG, Figure 8.12), a natural product isolated from *Paeonia suffruticosa* that incorporates acylated phenol groups as sugar substituents, inhibits Aβ fibril formation, destabilizes pre-formed Aβ fibrils *in vitro* and *in vivo* and protects neuronal cells (SK-N-SH) from Aβ-induced toxicity, restoring cell survival by 30%. Also, treatment with *Paeonia suffruticosa* extract improved long-term memory of transgenic mice overexpressing APP,[60] but the effect of PGG alone in the memory deficit of transgenic mice was not tested.

8.2.4.2 Cyclitols

Not only carbohydrates, but also carbohydrate mimetics have drawn the attention of researchers towards new compounds with anti-amyloidogenic properties. Amongst them, inositol stereoisomers (Figure 8.13) play an

myo-Inositol epi-inositol scyllo-inositol 1,4-di-O-methyl-scyllo-inositol

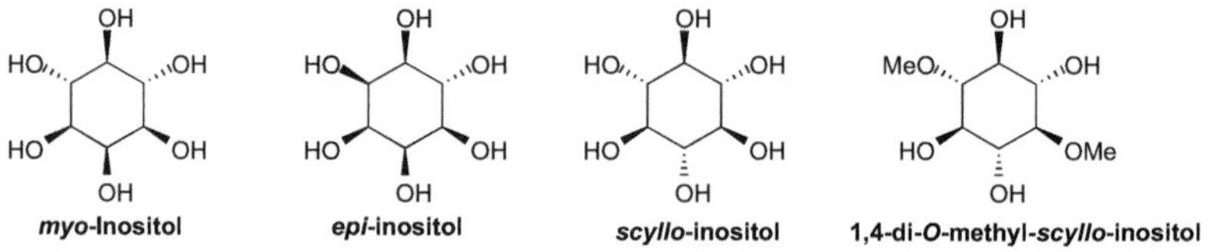

Figure 8.13 Inositol stereoisomers and the *scyllo*-inositol methylated derivative.

important role as Aβ aggregation inhibitors, leading to off-pathway high-molecular-weight oligomers in mice brain. Their formation resulted in improvement of the cognitive function, synaptic physiology changes and prevention of accelerated mortality of mice.[61] The most promising stereoisomer was, until recently, *scyllo*-inositol, which passes the blood–brain barrier using inositol transporters and is able to bind to Aβ, modulate its folding, inhibit its aggregation and dissociate pre-formed aggregates, with therapeutic effect on animals.[9] However, *scyllo*-inositol failed a phase II randomized clinical trial (RCT).[62] Diverse *scyllo*-inositol derivatives were recently generated with deoxy, fluoro, chloro and methoxy substitution. Amongst them, 1,4-di-*O*-methyl-*scyllo*-inositol (Figure 8.13) was the most effective one for the prevention of Aβ1–42 aggregation, although it was not more effective than *scyllo*-inositol.[63] Nonetheless, given the serious adverse effects registered by the administration of *scyllo*-inositol in high doses in the phase 2 RCT,[9,62] the search for new derivatives maintaining the same activity but possibly safer is highly significant.

8.2.5 Glycosaminoglycans

Glycosaminoglycans (GAGs) are basic building blocks of the extracellular matrix and are an important component of the cell membrane glycocalyx covering.[64] The association between GAGs and amyloid deposits and their possible involvement in the amyloidogenic pathway has been long studied and reviewed.[64–68] Several proteoglycans with GAG chains are known to co-localize with Aβ in senile plaques and to accumulate in the AD brain, such as dextran, heparan, keratin, deramatan and chondroitin sulfates.[67] The binding of proteoglycans and heparin to Aβ peptides is known to enhance both amyloid aggregation and fibril formation. It is also postulated that GAGs may protect aggregated Aβ from proteolytic degradation.[65] Moreover, proteoglycans or GAGs have been implicated in a number of other amyloid-related neurodegenerative diseases, namely prion diseases and Parkinson's disease.[67]

The eventual therapeutic implications of these facts and the possibility of a unique binding site mediating Aβ–GAG interaction encouraged scientists to search for a GAG-specific therapeutic intervention. Indeed, many efforts have been made to find compounds able to interfere with the interaction between endogenous GAGs and Aβ to prevent the formation of senile plaques or neurofibrillary tangles. There are two main approaches to the therapeutic potential of GAGs: one involves low molecular weight (LMW) GAGs and analogues inhibiting amyloid aggregation by competitive binding to Aβ, and a second one focuses on heparan sulfate (HS) biosynthesis inhibitors.

8.2.5.1 *Low Molecular Weight Glycosaminoglycans and Polysaccharides*

LMW enoxaparin and dalteparin, obtained from depolymerization of heparin and already marketed as anticoagulants, are two examples of LMW heparins that inhibited amyloidogenesis in an AA amyloid mouse model.[69] In particular, enoxaparin can cross the BBB and attenuates Aβ accumulation and deposition in the brain. Also, neuroparin (C3), a LMW glycosaminoglycan derived from heparin and composed of 4–10 oligosaccharides, was shown to have a positive effect in animal models with characteristic AD lesions, namely by reducing tau-2 immunoreactivity in the rat hippocampus, stimulated by injection of Aβ25–35. Indeed, C3, which is also known to cross the BBB, has reached human phase III clinical trials.[67] Also heparin itself can attenuate the neurotoxicity and proinflammatory effects of Aβ.[64]

Additionally, a novel acidic oligosaccharide rich in mannuronate blocks (MW ∼1300 Da), extracted from the brown algae *Echlonia Kurome Okam*, inhibited Aβ-induced toxicity and apoptosis in cells, by blocking Aβ aggregation.[70] A 30-day administration of this oligosaccharide to rats also attenuates memory impairment by scopolamine.[71]

Other example of oligomeric/polymeric structures studied in this context are those with a chitosan backbone conjugated with sialic acid analogues, such as ketodeoxynonulosonic acid (KDN) or tetrahydropyran-2-carboxylic acid (Figure 8.14).[72] These polysaccharides were developed previewing their potential to interact with cell surface glycoproteins and gangliosides. These structures, conjugated *via* EDC chemistry, are able to attenuate Aβ-induced toxicity in neuronal cells. The endocyclic oxygen and the multi –OH tail present in the sialic acid structure seemed to be essential for the neuroprotective effect.[72]

8.2.5.2 *Heparan Sulfate Biosynthesis Inhibition*

Owing to the important role of heparan sulfate (HS) biosynthesis in organ and cell function, its targeting should be carefully considered, because it may result in undesirable side effects. However, two 4-deoxyglucose derivatives,

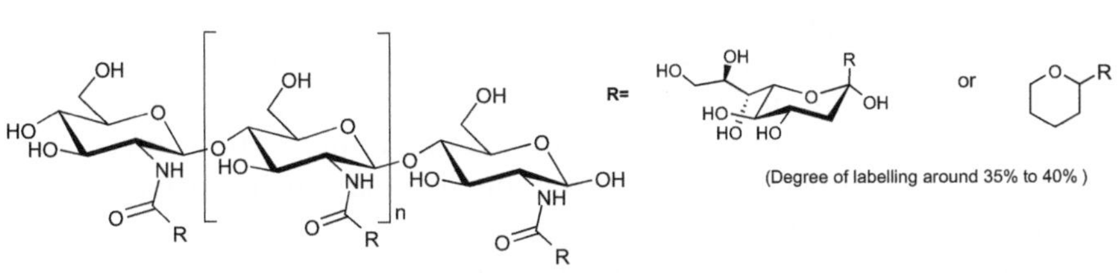

(Degree of labelling around 35% to 40%)

Figure 8.14 Polysaccharides with a chitosan backbone conjugated with KDN or tetrahydropyran-2-carboxylic acid.

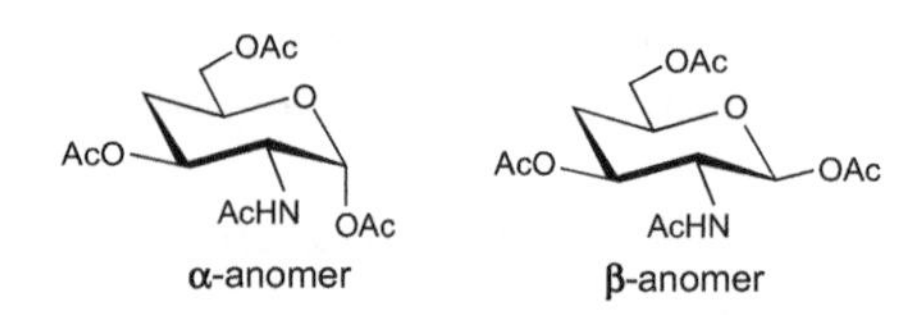

Figure 8.15 2-Acetamido-1,3,6-tri-*O*-acetyl-2,4-dideoxy-α- and -β-D-*xylo*-hexopyranose.

namely 2-acetamido-1,3,6-tri-*O*-acetyl-2,4-dideoxy-α- and -β-D-*xylo*-hexopyranoses (Figure 8.15) were reported to inhibit HS biosynthesis and, consequently, inhibited splenic AA amyloid deposition in mice (by 65–70%), at 0.1 mM concentration, with no toxicity associated.[73] Thus, despite of the possible side effects and the concerns it entails, inhibition of HS biosynthesis or modification of the HS structure could be a worthwhile research avenue for exploring new lead compounds with therapeutic applications in amyloidosis.

8.3 Carbohydrate-based Molecules Targeting the Cholinergic System

8.3.1 AChE Inhibitors

8.3.1.1 Saponins and Other Terpene Glycosides Including Ginsenosides

As previously discussed, nature has played an important role in drug discovery, and AChE inhibitors are not an exception. The significant role of terpene and terpenoid glycosides in the cholinergic hypothesis is highlighted in this subsection.

The extensively studied anti-amyloidogenic properties of ginsenosides have already been considered. However, the neuroprotective properties of these triterpene saponins can also be attributed to their effects in the cholinergic system. Rg1 and Rb1 (Figure 8.3) were found to increase the density of central muscarinic cholinergic receptors and the level of ACh in the central nervous system, which might result from choline acetyltransferase activity increase and acetylcholinesterase activity inhibition.[15,17] Moreover, both compounds increase neural plasticity and Rg1 also increases proliferation and differentiation of neural progenitor cells. This ginsenoside significantly improved memory deficits in aged rats, ovarioctomized rats and cerebral ischemia-reperfusion rats.[17]

Timosaponin AIII (Figure 8.16), a saponin from *Anemarrhena asphodeloides*, inhibited AChE *in vitro* (IC_{50} = 5.4 μM) and in scopolamine-treated mouse brain, with a comparable effect to that of tacrine.[74] These effects were reflected in an overall improvement of the scopolamine-induced learning and memory deficits in rats.[74]

The terpenoid glycosides cynatroside A and cynatroside B (Figure 8.16), from *Cynanchum atratum*, presented AChE inhibition with IC_{50} of 6.4 and 3.6 μM, respectively.[75] The latter compound inhibited AChE in a reversible

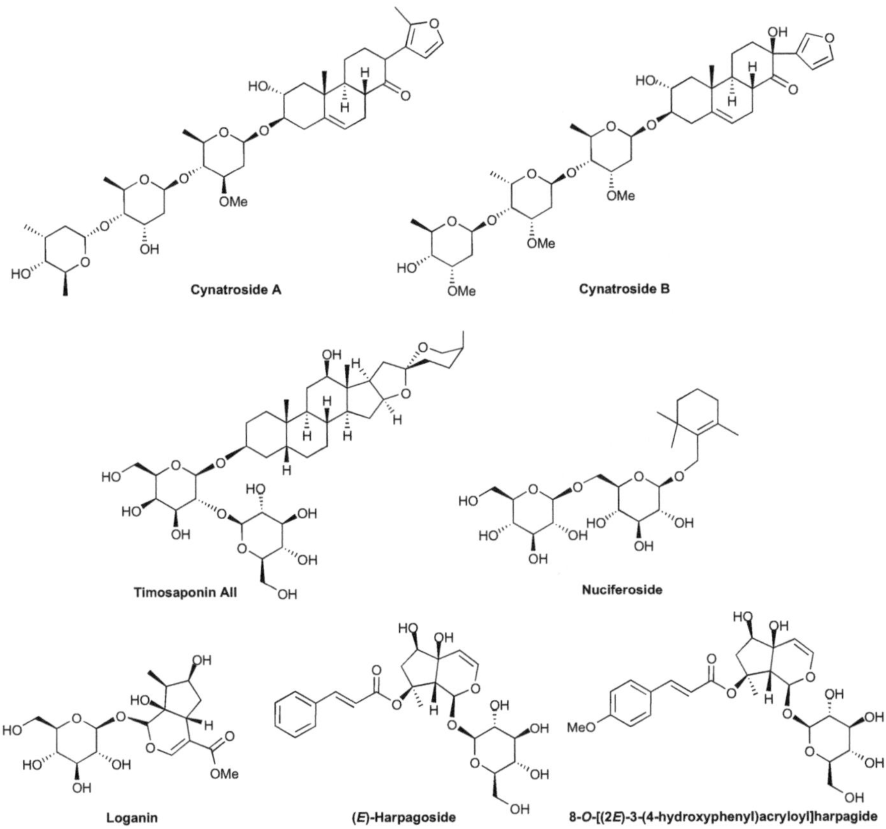

Figure 8.16 Terpene glycosides able to modulate cholinergic function.

and non-competitive manner, with a positive effect in both short-term and long-term memory in scopolamine-treated mice.[76]

A number of monoterpene glycosides were also described as AChE inhibitors. Paeoniflorin (Figure 8.4), whose anti-amyloidogenic properties have already been discussed, up-regulated the activity of choline acetyltransferase and down-regulated the activity of AChE in the hippocampus of Aβ1–42-treated rats.[36] Nuciferoside (Figure 8.16), a β-cyclogeraniol diglycoside identified in an extract from *Nelumbo nucifera*, showed significant non-competitive inhibition against AChE with an IC_{50} value of 3.20 μM.[77] In addition, the iridoid glycosides loganin, (*E*)-harpagoside, 8-*O*-[(*E*)-*p*-methoxycinnamoyl]harpagide and catalpol are all reported to inhibit AChE. Loganin, found in *Flos lonicerae*, *Fruit cornus* and *Strychonos nux vomica*, inhibits AChE activity in the hippocampus and frontal cortex of mice, resulting in a significant improvement of the scopolamine-induced memory impairments.[78] (*E*)-Harpagoside and 8-*O*-[(2*E*)-3-(4-hydroxyphenyl)acryloyl]harpagide (Figure 8.16), found in *Scrophularia buergeriana*, improve cholinergic function by decreasing the activity of AChE in the brain of senescent mice.[79] Both iridoid glycosides also have significant protective effects against glutamate-induced neurodegeneration in primary cultures of rat cortical neurons.[80] Catalpol (Figure 8.4) also registered anti-AChE activity in mice brain and increased choline acetyltransferase in positive neurons.[81]

8.3.1.2 *Flavonoid Glycosides*

Countless flavonoids can be found in the literature with anti-AChE activity. Amongst them, flavonoid glycosides assume a main role. In fact, screening 17 flavonoids by the TLC bioautographic assay pointed to the 7-glycosides tilianin[82] and linarin[82,83] (Figure 8.17) as the two most active compounds inhibiting AChE. This same study showed that the presence of a 4′-OMe group and a 7-*O*-glycosyl moiety were important for AChE inhibition.[82] Recently, icariin (Figure 8.17), also bearing the same structural features, improved the cognitive impairments in senescence-accelerated prone mice (SAMP) by significantly inhibiting AChE activity in the brain through monoamine level increase and oxidative damage inhibition.[84] This natural glycoside also decreased hippocampal levels of ACh, AChE and choline

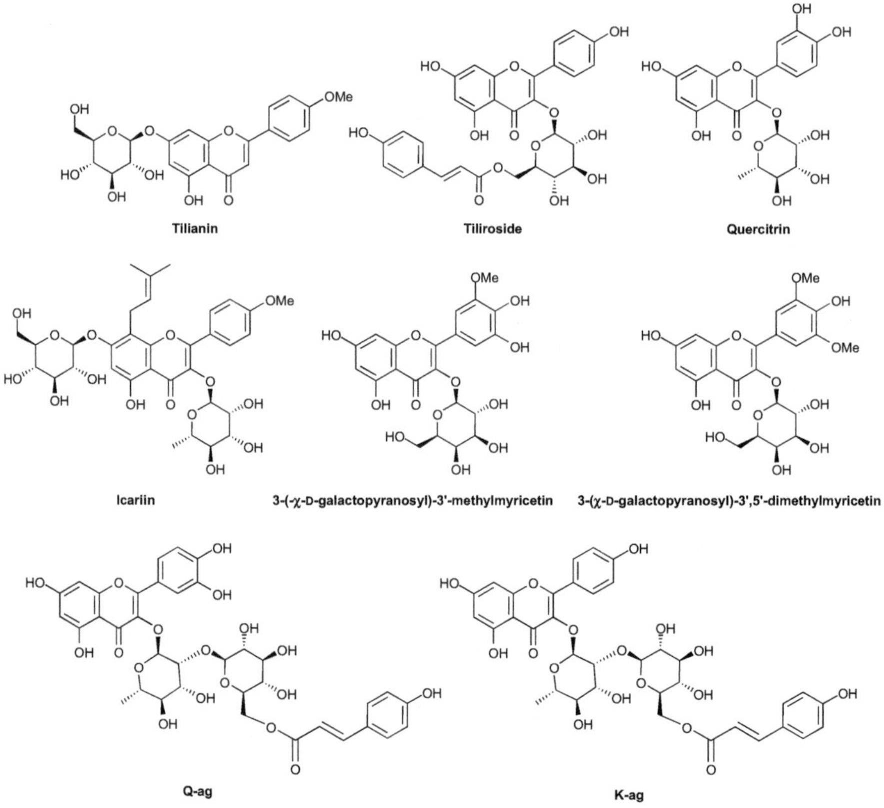

Figure 8.17 Flavonoid glycosides that inhibit AChE activity.

acetyltransferase in a rat model of chronic cerebral hypoperfusion.[85] In addition, the flavonol 3-glycosides tiliroside and quercitrin (Figure 8.17) from *Agrimonia pilosa*,[86] and the myricetin glycosides from *Cleistocalyx operlatus*,[87] also demonstrated AChE inhibitory activity. Tiliroside and quercitrin displayed IC_{50} values of 25.5 μM and 66.9 μM,[86] respectively, while 3-(β-D-galactopyranosyl)-3′-methylmyricetin and 3-(β-D-galactopyranosyl)-3′,5′-dimethylmyricetin presented IC_{50} values of 19.9 μM and 37.8 μM.[87] 3-({6-*O*-[(2*E*)-3-(4-Hydroxyphenyl)acryloyl]-β-D-glucopyranosyl}-(1→2)-α-L-rhamnopyranosyl)quercetin (Q-ag) and 3-({6-*O*-[(2*E*)-3-(4-hydroxyphenyl)acryloyl]-β-D-glucopyranosyl}-(1→2)-α-L-rhamnopyranosyl)kaempferol (K-ag) (Figure 8.17), from *Ginkgo biloba* standardized extract EGb-761®, increased extracellular levels of acetylcholine in rat brain by 151% compared to controls, at a dose of 10 mg kg^{-1}.[88]

8.3.1.3 Phenylethanoid Glycosides and Other Substituted Phenyl Glycosides

Several phenylethanoid glycosides and other substituted phenyl glycosides embodying phenolic group(s) were also reported as AChE inhibitors *in vitro*. Vanilloside (Figure 8.18), isolated from *Nelumbo nucifera*, inhibits AChE in a non-competitive manner, with an IC_{50} value of 4.55 μM.[77] The phenylpropanoid diglycoside rosavin and the rosavin analogues (*E*)-3-phenylprop-2-en-1-yl β-D-xylopyranosyl-(1→6)-β-D-glucopyranoside, (*E*)-3-(4-methoxyphenyl)prop-2-en-1-yl α-L-arabinopyranosyl-(1→6)-β-D-glucopyranoside and (*E*)-3-phenylprop-2-en-1-yl α-L-rhamnopyranosyl-(1→6)-β-D-glucopyranoside (Figure 8.18), isolated from *Rhodiola rosea L.*, displayed AChE inhibitory activity, with

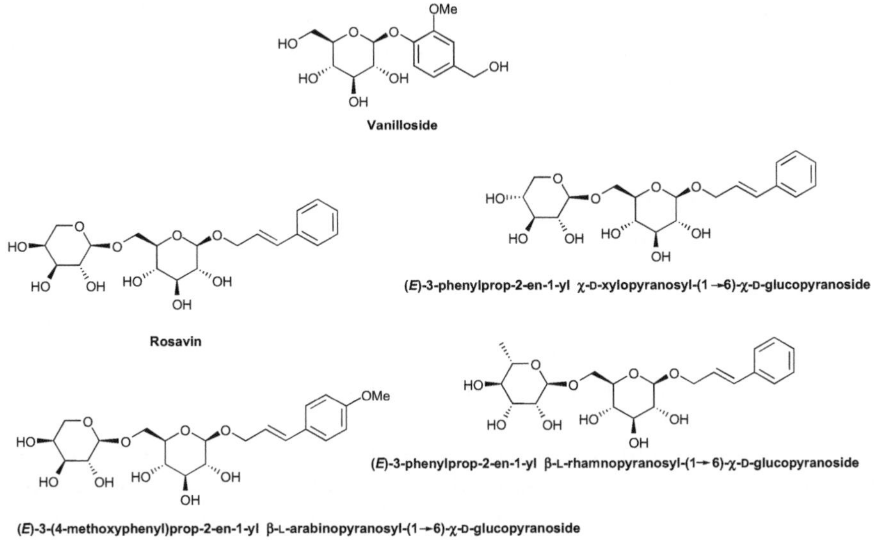

Vanilloside

Rosavin

(*E*)-3-phenylprop-2-en-1-yl χ-D-xylopyranosyl-(1→6)-χ-D-glucopyranoside

(*E*)-3-phenylprop-2-en-1-yl β-L-rhamnopyranosyl-(1→6)-χ-D-glucopyranoside

(*E*)-3-(4-methoxyphenyl)prop-2-en-1-yl β-L-arabinopyranosyl-(1→6)-χ-D-glucopyranoside

Figure 8.18 Vanilloside, rosavin and rosavin analogues.

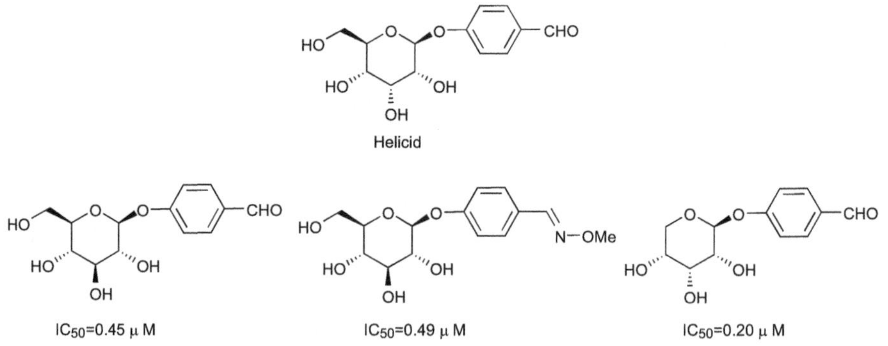

Figure 8.19 Helicid and helicid analogues with AChE inhibitory activity.

IC_{50} values of 1.72, 3.71, 4.23 and 2.05 μM, respectively.[89] Indeed, rosavin was the most potent AChE inhibitor amongst the natural compounds described so far. Inspired by these results, a small library of phenylpropanoid glycosides was generated, with analogues incorporating substituted phenyl groups with F, Cl and Br, and varying the methoxy and hydroxy substitution patterns. However, none of the synthesized derivatives was as active as the natural diglycosides shown in Figure 8.18.[89]

Derivatives of the natural antidepressant helicid were synthesized starting from 4-hydroxybenzaldehyde, followed by glycosylation, deprotection and condensation with amines.[90] These transformations afforded noteworthy AChE inhibitors with IC_{50} values under 10 μM, three of them under 0.55 μM (Figure 8.19). Interestingly, while helicid was not active up to 500 μM, its epimer at C-3 presented an IC_{50} of 0.45 μM. However, the most potent inhibitor, the 4-formylphenyl β-D-ribopyranoside, exhibits the same configuration of carbons 2, 3 and 4 as helicid and its hydroxymethyl group is replaced by a hydrogen atom, presenting an IC_{50} value of 0.20 μM on electric eel AChE, twice as active as galantamine.[90] These results highlight the close correlation of the bioactivity with the sugar structure.

8.3.2 Butyrylcholinesterase Inhibitors

8.3.2.1 Nucleosides

As already discussed, the recent association between BChE and maturation of senile plaques in AD brains has encouraged researchers to develop BChE inhibitors envisioning therapeutic application in late stages of AD. Recently, a series of N^7 and β-linked 2-acetamido-6-chloropurine nucleosides with a benzyl-protected sugar bicyclic moiety were developed by Rauter and co-workers, exhibiting a highly and selective BChE inhibition.[91] One of these N^7 nucleosides (Figure 8.20, A) displayed an IC_{50} value of 0.14 μM, of the same order of magnitude as that of rivastigmine.[91] Further studies on the impact of the sugar and purine structures in the activity led to a new and more effective structure, an N^7-linked 2-acetamido-6-chloro-α-D-mannosylpurine

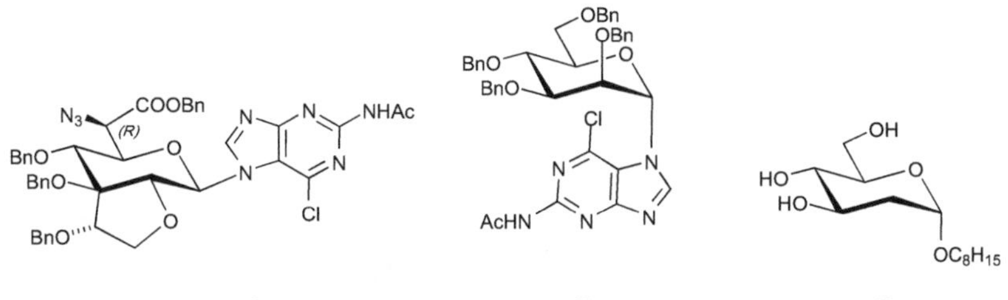

Figure 8.20 N^7 nucleosides (A and B) and octyl 2-deoxy-D-*arabino*-hexopyranoside (C) with selective BChE inhibitory activity.

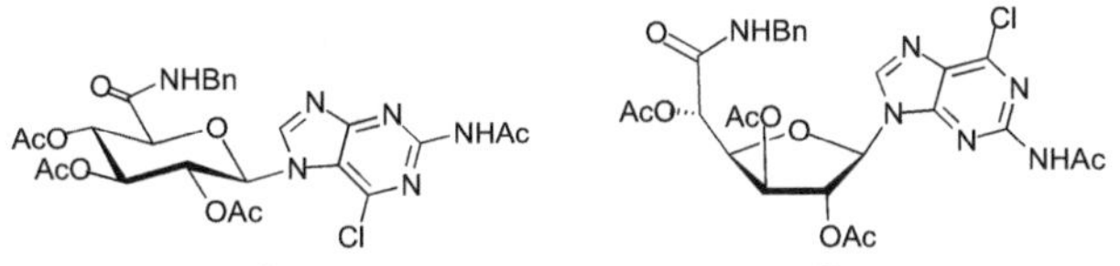

Figure 8.21 Nucleosides with selective AChE inhibitory activity.

(Figure 8.20, B). This nucleoside exhibited an inhibitory constant (K_i) of 50 nm and a selectivity factor of 340-fold for BChE over AChE.[92] Interestingly, the glucuronamide-based purine nucleosides bearing a six-membered ring protected with acetyl groups (Figure 8.21, A) were selective inhibitors of AChE. Also, those nucleosides linked to furanosyl moieties (Figure 8.21, B) showed the same trend.[93] The selectivity toward the BChE inhibition observed can be tentatively explained by the hindrance experienced by the inhibitor when accessing the active site of AChE, which has several bulky residues that hinder access to the enzyme pocket, while the catalytic site of BChE is easily accessed and large enough for the inhibitor to adopt its preferred conformation and form stable enzyme–inhibitor complexes, as observed for the selective BChE inhibitor octyl 2-deoxy-D-*arabino*-hexopyranoside (Figure 8.20, C).[94] This compound, which did not inhibit AChE, caused 100% inhibition of BChE at a concentration of 100 μM.[94]

8.3.2.2 Natural BChE Inhibitors

Also in the plant kingdom there are examples of selective BChE inhibitors, namely forsythoside B (Figure 8.22), a phenylethanoid glycoside isolated from *Verbascum xanthophoeniceum*,[95] and symcososide (Figure 8.22), a component of *Symplocos racemosa*.[96] Forsythoside B was able to inhibit 98% of BChE activity at a concentration of 100 μM, with only mild activity against AChE.[95] On the other hand, symcososide proved to be more selective towards BChE, displaying an IC_{50} of 21.1 μM for this enzyme, but no activity towards AChE.[96]

Although there is still plenty to learn about the role of BChE in the human brain and how selective BChE inhibitors can act on AD patients, a recent

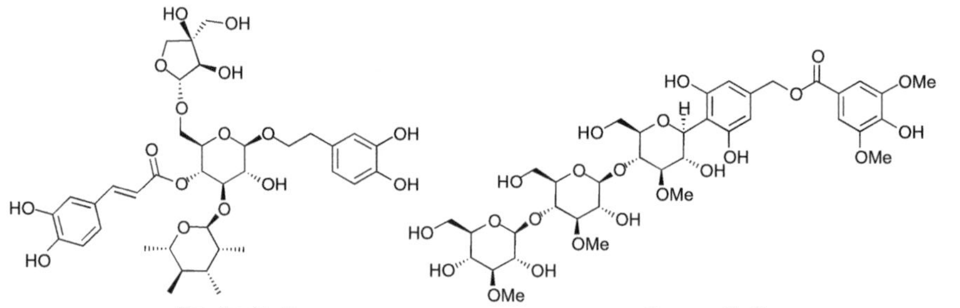

Forsythoside B

Symcososide B

Figure 8.22 BChE inhibitors of natural origin.

study showing that selective BChE inhibitors improved the cognitive performance of aged rats[6] supports that this is a valuable line of investigation.

8.4 Diagnosis

In clinical practice, diagnosis of "possible" or "probable" AD is currently obtained mainly by mental status testing, since a neuroimaging technique is not yet available to visually differentiate between the normal aging process and dementia. Definitive diagnosis is generally confirmed at post-mortem by the presence of intra- and extracellular protein aggregates in brain tissue.[97] Efforts have been made to find metabolic biomarkers for the diagnosis of AD. Metabolic pathway analysis of familial AD mouse models revealed significant alterations in the levels of metabolites involved in energy metabolism, including nucleotide metabolism, mitochondrial Krebs cycle, and carbohydrate and amino acid metabolic pathways.[98] Metabonomic profiling of transgenic AD mice also revealed that there are significant differences in both brain and plasma levels of different sugar metabolites, namely in D-fructose (brain) and in D-glucose, D-galactose and D-gluconic acid (plasma), when compared to healthy mice.[99] Carbohydrate metabolic pathway alteration in this model suggests the potential of carbohydrate structures for diagnosis. However, these biomarkers are yet to be investigated for their use in diagnosis.

Aβ *in vivo* imaging has been actively pursued over recent years, mainly regarding non-invasive techniques including positron emission tomography (PET), magnetic resonance imaging (MRI) and near-IR fluorescence imaging. Based on the fact that gangliosides and glycosaminoglycans bind to Aβ and may play significant roles in initiation of amyloid aggregation, Kouyoumdjian and co-workers developed superparamagnetic iron oxide glyconanoparticles whose magnetic cores are coated with sialic acid units (Figure 8.23) to detect Aβ.[100] Indeed, these non-toxic glyconanoparticles could selectively bind to Aβ. The superparamagnetic nature of the glyconanoparticles allowed detection of Aβ by MRI both *in vitro* and *ex vivo* in mouse brains. The authors suggest that other carbohydrates and carbohydrate-containing compounds, such as glycosaminoglycans and ganglioside GM1, can be utilized to further enhance

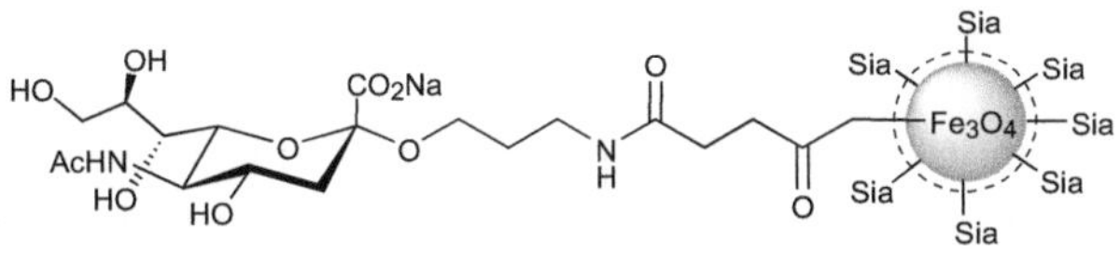

Figure 8.23 Glyconanoparticle coated with sialic acid units.

Aβ binding selectivity and specificity.[100] Nevertheless, these studies were conducted *ex vivo*, and diagnosis was performed post-mortem. For future imaging *in vivo*, it will be crucial to know whether these glyconanoparticles can access the plaques in the brain, passing the BBB.

8.5 Conclusions

Investigation towards effective therapeutic drugs for AD has proven challenging. Pursuing the understanding of the mechanisms underlying the disease was key to identifying new therapeutic possibilities. A diversity of glycostructures has shown potential for neuroprotection, as shown in Figure 8.24, summarizing the carbohydrate-based families of compounds acting on amyloidosis, cholinergic function and diagnosis. From natural to synthetic origin, from carbohydrates to carbomimetics, from small molecules to polysaccharides, many structures have been identified as potential future drugs for the prevention, treatment or diagnosis of this devastating disease. Amongst them, only the carbomimetic *scyllo*-inositol and the oligosaccharide neuroparin have reached clinical trials. While the former did not meet the endpoints, the results for neuroparin are yet to be known. The complex and multifactor aspects of AD may explain why drugs that address only one cause (one drug/one target strategy) afford

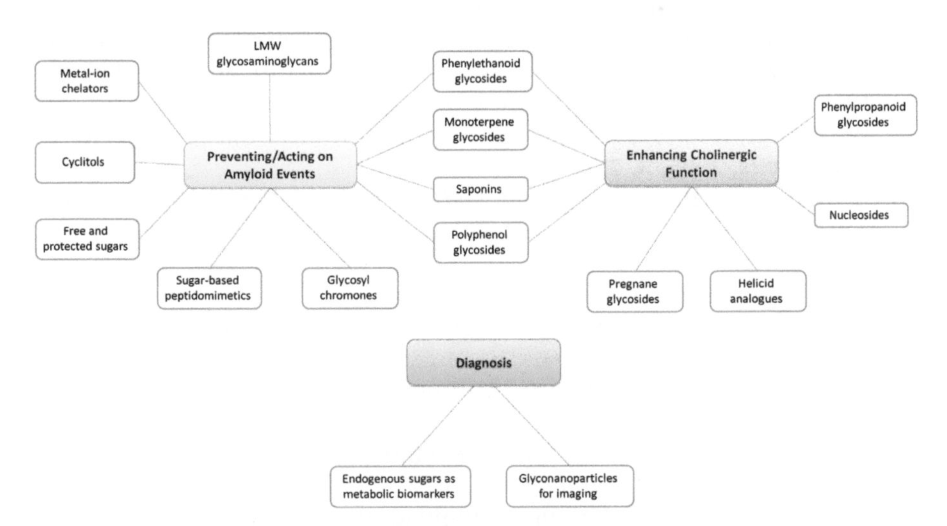

Figure 8.24 Carbohydrates in Alzheimer's disease therapeutics and diagnosis.

disappointing results in clinical trials. Many authors have discussed how a multi-target therapy based on a compound with poly-pharmacology is more likely to be efficient. Many of the compounds discussed in this review address more than one target, namely the ginsenosides Rg1 and Rb1, as well as the monoterpene glycoside paeoniflorin, which can act in both amyloid plaques and in the cholinergic system, with effects in the cognition of animals. Moreover, the use of carbohydrates linked to other bioactive molecules to favour their access to brain through the hexose transporters in the BBB, or to improve their pharmacological properties, *e.g.* a sugar–curcumin conjugate, highlights the importance of carbohydrate-containing drugs.

Also in diagnosis, the potential of carbohydrates is noticeable, either by the use of endogenous sugar metabolites as disease biomarkers or by the use of carbohydrate-based structures with high affinity to Aβ plaques. These findings strongly encourage further investigation in this area.

Acknowledgements

The authors thank Fundação para a Ciência e a Tecnologia (FCT) and Cipan for the PhD grant for C.D. (SFRH/BDE/51998/2012), and FCT for the financial support of the project Pest-OE/QUI/UI0612/2013. The European Commission is gratefully acknowledged for the approval of the INOVAFUNAGEING Commitment within the European Innovation Partnership for Active and Healthy Ageing - Action Group on Prevention of Functional Decline and Frailty (A3).

References

1. D. J. Selkoe, *Physiol. Rev.*, 2001, **81**(2), 741.
2. M. Citron, *Nat. Rev. Neurosci.*, 2004, **5**, 677.
3. R. W. Jones, in *Global Clinical Trials for Alzheimer's Disease: Design, Implementation, and Standardization*, ed. M. Bairu and M. Weiner, Elsevier Science, San Diego, CA, USA, 2013, Ch. 3, pp. 33–48.
4. W. H. Suh, K. S. Suslick and Y.-H. Suh, *Curr. Med. Chem.: Cent. Nerv. Syst. Agents*, 2005, **5**, 259.
5. E. Giacobini, in *Morphological and Biochemical Correlates of Neural Activity*, ed. M. M. Cohen and R. S. Snyder, Harper and Row, New York, 1964, pp. 15–38.
6. E. Giacobini, *Int. J. Geriatr. Psychiatry*, 2003, **18**, 1.
7. N. H. Greig, T. Utsuki, D. K. Ingram, Y. Wang, G. Pepeu, C. Scali, Q.-S. Yu, J. Mamczarz, H. W. Holloway, T. Giordano, D. Chen, K. Furukawa, K. Sambamurti, A. Brossi and D. K. Lahiri, *Proc. Natl. Acad. Sci. U. S. A.*, 2005, **102**, 17213.
8. S. Darvesh, M. K. Cash, G. A. Reid, E. Martin, A. Mitnitski and C. Geula, *J. Neuropathol. Exp. Neurol.*, 2012, **71**, 2.
9. F. Manglalasche, A. Solomon, B. Winblad, P. Mecocci and M. Kivipelto, *Lancet Neurol.*, 2010, **9**, 702.

10. E. Karran, M. Mercken and B. De Stropper, *Nat. Rev. Drug Discovery*, 2011, **10**, 698.
11. R. Kayed and C. A. Lasagna-Reeves, *J. Alzheimer's Dis.*, 2013, **33**, S67.
12. G. M. Hirschfield and P. N. Hawkins, *Int. J. Biochem. Cell Biol.*, 2003, **35**, 1608.
13. A. A. Qutub and C. A. Hunt, *Brain Res. Rev.*, 2005, **49**, 595.
14. B. Linnartz, L.-G. Bodea and H. Neumann, *Cell Tissue Res.*, 2012, **349**, 215.
15. Y. Cheng, L. Shen and J. Zhang, *Acta Pharmacol. Sin.*, 2005, **26**(2), 143.
16. K.-Y. Yoo and S.-Y. Park, *Molecules*, 2012, **17**, 3524.
17. H. J. Kim, P. Kim and C. Y. Shin, *J. Ginseng Res.*, 2013, **37**(1), 8.
18. F. Fang, X. Chen, T. Huang, L.-F. Lue, J. S. Luddy and S. S. Yan, *Biochim. Biophys. Acta*, 2012, **1822**, 286.
19. Q. Quan, J. Wang, X. Li and Y. Wang, *PLoS One*, 2013, **8**(3), e59155.
20. L. Chen, Z. Lin, Y. Zhu, N. Lin, J. Zhang, X. Pan and X. Chen, *Eur. J. Pharmacol.*, 2012, **675**, 15.
21. C. Shi, D. Zheng, L. Fang, F. Wu, W. H. Kwong and J. Xu, *Biochim. Biophys. Acta*, 2012, **1820**, 453.
22. J. Liu, X. Yan, L. Li, Y. Zhu, K. Qin, L. Zhou, D. Sun, X. Zhang, R. Ye and G. Zhao, *Neurochem. Res.*, 2012, **37**, 2738.
23. (a) W. Li, Y. Chu, L. Zhang, L. Yin and L. Li, *Brain Res. Bull.*, 2012, **88**, 501; (b) W. Li, Y. Chu, L. Zhang, L. Yin and L. Li, *Life Sci.*, 2012, **91**, 809.
24. J. Yan, Q. Liu, Y. Dou, Y. Hsieh, Y. Liu, R. Tao, D. Zhu and Y. Lou, *J. Ethnopharmacol.*, 2013, **147**, 456.
25. T. Huang, F. Fang, L. Chen, Y. Zhu, J. Zhang, X. Chen and S. S. Yan, *Curr. Alzheimer Res.*, 2012, **9**(3), 388.
26. B. Lee, B. Sur, J. Park, S.-H. Kim, S. Kwon, M. Yeom, I. Shim, H. Lee and D.-H. Hahm, *Biomol. Ther.*, 2013, **21**(5), 381.
27. Y. Y. Lee, J.-S. Park, J.-S. Jung, D.-H. Kim and H.-S. Kim, *Int. J. Mol. Sci.*, 2013, **14**, 9820.
28. X. Li, M. Li, Y. Li, Q. K. Quan and J. Wang, *Neural Regener. Res.*, 2012, **36**, 2860.
29. X. Song, J. Hu, S. Chu, Z. Zhang, S. Xu, Y. Yuan, N. Han, Y. Liu, F. Niu, X. He and N. Chen, *Eur. J. Pharmacol.*, 2013, **710**, 29.
30. H. Zhao, J. Di, W. Liu, H. Liu, H. Lai and Y. Lü, *Behav. Brain Res.*, 2013, **241**, 228.
31. L. Li, Z. Liu, J. Liu, X. Tai, X. Hu, X. Liu, Z. Wu, G. Zhang, M. Shi and G. Zhao, *Neurobiol. Dis.*, 2013, **54**, 320.
32. D. Ji, Y. Wu, B. Zhang, C.-F. Zhang and Z.-L. Yang, *Fitoterapia*, 2012, **83**, 843.
33. C.-K. Han, W. R. Choi and K.-B. Oh, *Plant Med.*, 2007, **73**, 665.
34. M. A. Findeis, F. Schroeder, T. D. McKee, D. Yager, P. C. Fraering, S. P. Creaser, W. F. Austin, J. Clardy, R. Wang, D. Selkoe and C. B. Eckman, *ACS Chem. Neurosci.*, 2012, **3**(11), 941.

35. S.-Z. Zhong, Q.-H. Ge, Q. Li, R. Qu and S.-P. Ma, *J. Neurol. Sci.*, 2009, **280**, 71.
36. Z. Lan, L. Chen, Q. Fu, W. Ji, S. Wang, Z. Liang, R. Qu, L. Kong and S. Ma, *Brain Res.*, 2013, **1498**, 9.
37. P. Sun, J. Chen, J. Li, M. Sun, W. Mo, K. Liu, Y. Meng, Y. Liu, F. Wang, R. He and Q. Hua, *BMC Complementary Altern. Med.*, 2013, **13**, 152.
38. B. Jiang, J. Du, J. Liu, Y.-M. Bao and L.-J. An, *Brain Res.*, 2008, **1188**, 139.
39. Y. Porat, A. Abramowitz and E. Gazit, *Chem. Biol. Drug Des.*, 2006, **67**, 27.
40. M. Kurisu, Y. Miyamae, K. Murakami, J. Han, H. Isoda, K. Irie and H. Shigemori, *Biosci. Biotechnol. Biochem.*, 2013, 77(6), 1329.
41. H.-Q. Wang, Y.-X. Xu, J. Yan, X.-Y. Zhao, X.-B. Sun, Y.-P. Zhang, J.-C. Guo and C.-K. Zhu, *Brain Res.*, 2009, **1283**, 139.
42. H.-Q. Wang, Y.-X. Xu and C.-Q. Zhu, *Neurotoxic. Res.*, 2012, **21**, 368.
43. J. T. T. Zhu, R. C. Y. Choi, H. Q. Xie, K. Y. Z. Zheng, A. J. Y. Guo, C. W. C. Bi, D. T. W. Lau, J. Li, T. T. X. Dong, B. W. C. Lau, J. J. Chen and K. W. K. Tsim, *Neurosci. Lett.*, 2009, **461**, 172.
44. H. Lou, P. Fan, R. G. Perez and H. Lou, *Bioorg. Med. Chem.*, 2011, **19**, 4021.
45. R. C. Choi, J. T. Zhu, K. W. Leung, G. K. Chu, H. Q. Xie, V. P. Chen, K. Y. Zheng, D. T. Lau, T. T. Dong, P. C. Chow, Y. F. Han, Z. T. Wang and K. W. Tsim, *J. Alzheimer's Dis.*, 2010, **19**(3), 795.
46. A. R. A. Ladiwala, M. Mora-Pale, J. C. Lin, S. S. Bale, Z. S. Fishman, J. S. Dordick and P. M. Tessier, *ChemBioChem*, 2011, **12**, 1749.
47. M. G. Sharoar, A. Thapa, M. Shahnawaz, V. S. Ramasamy, E.-R. Woo, S. Y. Shin and I.-S. Park, *J. Biomed. Sci.*, 2012, **19**, 104.
48. S. Dolai, W. Shi, C. Corbo, C. Sun, S. Averick, D. Obeysekera, M. Farid, A. Alonso, P. Banerjee and K. Raja, *Chem. Neurosci.*, 2011, **2**, 694.
49. L. Lv, Q.-Y. Yang, Y. Zhao, C.-S. Yao, Y. Sun, E.-J. Yang, K.-S. Song, I. Mook-Jung and W.-S. Fang, *Plant. Med.*, 2008, **74**, 540.
50. Y. X. Zhou, H. Zhang and C. Peng, *Phytother. Res.*, 2014, **28**(7), 961.
51. G. Xing, M. Dong, X. Li, Y. Zou, L. Fan, X. Wang, D. Cai, C. Li, L. Zhou, J. Liu and Y. Niu, *Brain Res. Bull.*, 2011, **85**, 212.
52. H.-Y. Zhang, Y.-H. Liu, H.-Q. Wang, J.-H. Xu and H.-T. Hu, *Cell Biol. Int.*, 2008, **32**, 1230.
53. F. Lin, B. Xie, F. Cai and G. Wu, *Drug Res.*, 2012, **62**(4), 187.
54. Y. Zou, B. Hong, L. Fan, L. Zhou, Y. Liu, Q. Wu, X. Zhang and M. Dong, *Free Radical Res.*, 2013, **47**(1), 55.
55. J. Li, G. Wang, J. Liu, L. Zhou, M. Dong, R. Wang, X. Li, X. Li, C. Lin and Y. Niu, *Eur. J. Pharmacol.*, 2010, **649**, 195.
56. B. Doregeret, L. Khemtémourian, I. Correia, J. Soulier, O. Lequin and S. Ongeri, *Eur. J. Med. Chem.*, 2011, **46**, 5959.
57. T. Storr, M. Merkel, G. X. Song-Zhao, L. E. Scott, D. E. Green, M. L. Bowen, K. H. Thompson, B. O. Patrick, H. J. Schugar and C. Orvig, *J. Am. Chem. Soc.*, 2007, **129**, 7453.

58. M. Abe, Y. Abe, T. Ohkuri, T. Mishim, A. Monji, S. Kanba and T. Ueda, *Protein Sci.*, 2013, **22**, 467.
59. R. Liu, H. Barkhordarian, S. Emadi, C. B. Park and M. R. Sierks, *Neurobiol. Dis.*, 2005, **20**, 74.
60. H. Fujiwara, M. Tabuchi, T. Yamaguchi, K. Iwasaki, K. Furukawa, K. Sekiguchi, Y. Ikarashi, Y. Kudo, M. Higuchi, T. C. Saido, S. Maeda, A. Takashima, M. Hara, N. Yaegashi, Y. Kase and H. Arai, *J. Neurochem.*, 2009, **109**, 1648.
61. J. McLaurin, M. E. Kierstead, M. E. Brown, C. A. Hawkes, M. H. L. Lambermon, A. L. Phinney, A. A. Darabie, J. E. Cousins, J. E. French, M. F. Lan, F. Chen, S. S. N. Wong, H. T. J. Mount, P. E. Fraser, D. Westaway and P. St. George-Hyslop, *Nat. Med.*, 2006, **12**, 801.
62. L. Hughes and S. Guthrie, in *Global Clinical Trials for Alzheimer's Disease: Design, Implementation, and Standardization*, ed. M. Bairu and M. Weiner, Elsevier Science, San Diego, CA, USA, 2013, ch. 10, pp. 159–177.
63. Y. Sun, G. Zhang, C. A. Hawkes, J. E. Shaw, J. McLaurin and M. Nitz, *Bioorg. Med. Chem.*, 2008, **16**, 7177.
64. B. Dudas and K. Semeniken, in *Handbook of Experimental Pharmacology Book Series – Heparin, a Century of Progress*, ed. R. Lever, B. Mulloy and C. P. Page, Springer, Berlin Heidelberg, 2012, vol. 207, pp. 325–343.
65. J. McLaurin, T. Franklin, X. Zhang, J. Deng and P. E. Fraser, *Eur. J. Biochem.*, 1999, **266**, 1101.
66. R. J. Castellani, D. A. DeWitt, G. Perry and M. A. Smith, *Med. Hypotheses Res.*, 2005, **2**, 393.
67. T. Ariga, T. Miyatake and R. K. Yu, *J. Neurosci. Res.*, 2010, **88**, 2303.
68. B. Dudas, M. Rose, U. Cornelli, A. Pavlovich and I. Hanin, *Neurodegener. Dis.*, 2008, **5**, 200.
69. H. Zhu, J. Yu and M. S. Kindy, *Mol. Med.*, 2001, **7**, 517.
70. J. Hu, M. Geng, J. Li, X. Xin, J. Wang, M. Tang, J. Zhang, X. Zhang and J. Ding, *J. Pharmacol. Sci.*, 2004, **95**, 248.
71. Y. Fan, J. Hu, J. Li, Z. Yang, X. Xin, J. Wang, J. Ding and M. Geng, *Neurosci. Lett.*, 2005, **374**, 222.
72. D. Dhavale and J. E. Henry, *Biochim. Biophys. Acta*, 2012, **1820**, 1475.
73. R. Kisilevsky, W. A. Szarek, J. B. Ancsin, E. Elimova, S. Marone, S. Bhat and A. Berkin, *Am. J. Pathol.*, 2004, **164**, 2127.
74. B. Lee, K. Jung and D.-H. Kim, *Pharmacol., Biochem. Behav.*, 2009, **93**, 121.
75. K. Y. Lee, S. H. Sung and Y. C. Kim, *Helv. Chim. Acta*, 2003, **86**, 474.
76. K. Y. Lee, J. S. Yoon, E. S. Kim, S. Y. Kang and Y. C. Kim, *Plant. Med.*, 2005, **71**, 7.
77. H. A. Jung, Y. J. Jung, S. K. Hyun, B. S. Min, D. W. Kim, J. H. Jung and J. S. Choi, *Biol. Pharm. Bull.*, 2010, **33**(2), 267.
78. S.-H. Kwon, H.-C. Kim, S.-Y. Lee and C.-G. Jang, *Eur. J. Pharmacol.*, 2009, **619**, 44.
79. E. J. Jeong, K. Y. Lee, S. H. Kim, S. H. Sung and Y. C. Kim, *Eur. J. Pharmacol.*, 2008, **588**, 78.

80. S. R. Kim, K. Y. Lee, K. A. Koo, S. H. Sung, N.-G. Lee, J. Kim and Y. C. Kim, *J. Nat. Prod.*, 2002, **65**, 1696.
81. X. Zhang, C. Jin, Y. Li, S. Guan, F. Han and S. Zhang, *Food Chem. Toxicol.*, 2013, **58**, 50.
82. P. Fan, A.-E. Hay, A. Marston and K. Hostettmann, *Pharm. Biol.*, 2008, **46**(9), 596.
83. P. P. Oinonen, J. K. Jokel, A. I. Hatakka and P. M. Vuorela, *Fitoterapia*, 2006, 77, 429.
84. X.-L. He, W.-Q. Zhou, M.-G. Bi and G.-H. Du, *Brain Res.*, 2010, **1334**, 73.
85. R.-X. Xu, Q. Wu, Y. Luo, Q.-H. Gong, L.-M. Yu, X.-N. Huang, A.-S. Sun and J.-S. Shi, *Clin. Exp. Pharmacol. Physiol.*, 2009, **36**, 810.
86. M. Jung and M. Park, *Molecules*, 2007, **12**, 2130.
87. B. S. Min, T. D. Cuong, J.-S. Lee, B.-S. Shin, M. H. Woo and T. M. Hung, *Arch. Pharm. Res.*, 2010, **33**(10), 1665.
88. J. Kehr, S. Yoshitake, S. Ijiri, E. Koch, M. Nöldner and T. Yoshitake, *Int. Psychogeriatr.*, 2012, **24**(Sup. 1), S25.
89. X.-D. Li, S.-T. Kang, G.-Y. Li, X. Li and J.-H. Wang, *Molecules*, 2011, **16**, 3580.
90. H. Wen, C. Lin, L. Que, H. Ge, L. Ma, R. Cao, Y. Wan, W. Peng, Z. Wang and H. Song, *Eur. J. Med. Chem.*, 2008, **43**, 166.
91. F. Marcelo, F. V. M. Silva, M. Goulart, J. Justino, P. Sinay, Y. Blériot and A. P. Rauter, *Bioorg. Med. Chem.*, 2009, **17**, 5106.
92. S. Schwarz, R. Csuk and A. P. Rauter, *Org. Biomol. Chem.*, 2014, **12**, 2446.
93. N. M. Xavier, S. Schwarz, P. D. Vaz, R. Csuk and A. P. Rauter, *Eur. J. Org. Chem.*, 2014, 2770.
94. A. Martins, M. S. Santos, C. Dias, P. Serra, V. Cachatra, J. Pais, J. Caio, V. H. Teixeira, M. Machuqueiro, M. S. Silva, A. Pelerito, J. Justino, M. Goulart, F. V. Silva and A. P. Rauter, *Eur. J. Org. Chem.*, 2013, 1448.
95. M. Georgiev, K. Alipieva, I. Orhan, R. Abrashev, P. Denev and M. Angelova, *Food Chem.*, 2011, **128**, 100.
96. V. Ahmad, M. Zubair, M. Abbasi, F. Kousar, M. Rasheed, N. Rasool, J. Hussain, S. Nawaz and M. Choudhary, *Polish J. Chem.*, 2006, **80**(3), 403.
97. R. Barone, L. Sturiale, A. Palmigiano, M. Zappia and D. Garozzo, *J. Proteomics*, 2012, **75**, 5123.
98. E. Trushina, E. Nemetlu, S. Zhang, T. Christensen, J. Camp, J. Mesa, A. Siddiqui, Y. Tamura, H. Sesaki, T. W. Wengenack, P. P. Dzeja and J. F. Poduslo, *PLoS One*, 2012, 7(2), e32737.
99. Z.-P. Hu, E. R. Browne, T. Liu, T. E. Angel, P. C. Ho and E. C. Y. Chan, *J. Proteome Res.*, 2012, **11**, 5903.
100. H. Kouyoumdjian, D. C. Zhu, M. H. El-Dakdouki, K. Lorenz and J. Chen, *ACS Chem. Neurosci.*, 2013, **4**, 575.

CHAPTER 9

Galactofuranose Biosynthesis: Discovery, Mechanisms and Therapeutic Relevance

GUILLAUME EPPE, SANDY EL BKASSINY AND STÉPHANE P. VINCENT*

University of Namur, Département de Chimie, Laboratoire de Chimie Bio-Organique, rue de Bruxelles 61, B-5000 Namur, Belgium
*Email: stephane.vincent@unamur.be

9.1 Introduction

Galactofuranose, the atypical and thermodynamically disfavored form of D-galactose, has in reality a very old history in chemistry and biochemistry. Indeed, synthetic methods to produce galactofuranosides selectively date back to the beginning of the 20th century.[1] Moreover, in 1937, Haworth *et al.* found that the extracellular polysaccharide "galactocarolose" produced by *Penicillium charlesii* contains galactose in a furanose configuration.[2] The search for biosynthetic precursors rapidly began and, in 1970, it was found that when galactocarolose biosynthesis was inhibited, UDP-galactofuranose **1** (UDP-Gal*f*) accumulated.[3] UDP-Gal*f* was thus identified as a likely biosynthetic precursor of galactofuranosides. One year later, while studying the T1 antigen of *Salmonella typhimurium*, Nikaido and Sarvas speculated that UDP-galactopyranose **2** (UDP-Gal*p*) was the precursor of UDP-Gal*f*.[4] However, the first gene that, unambiguously, could be attributed to the conversion of

RSC Drug Discovery Series No. 43
Carbohydrates in Drug Design and Discovery
Edited by Jesús Jiménez-Barbero, F. Javier Cañada and Sonsoles Martín-Santamaría

Published by the Royal Society of Chemistry, www.rsc.org

UDP-Gal*p* **2** into its furanose form was only identified and cloned in 1996[5] and the first galactofuranosyltransferase was discovered at the very beginning of the 21st century.[6] As we will detail below, the Gal*f*-processing enzymes happen to proceed through unique or rarely occurring mechanisms. Meanwhile, galactofuranose moieties have also been identified in the cell wall of major pathogens. Therefore, interest into the biosynthesis of galactofuranose and the search for glycomimetics of galactofuranose have grown rapidly in recent years. The purpose of this chapter is to give an overview of the fundamental aspects of the galactofuranose biosynthesis, from the biological occurrence to the search for inhibitors. Several review articles partially covering some aspects of this topic have been published in the literature and will be cited in each section.

9.2 Occurrence and Therapeutic Relevance of β-Galactofuranosides

In mammals, D-galactose is only found in the ubiquitous pyranose form, for example as a disaccharide with glucose (lactose), in complex glycoproteins (ABO blood group antigens) and glycolipids. On the other hand, D-galactofuranose (Gal*f*), while totally absent in mammals, is frequently found in bacteria, protozoa and fungi,[7] as well as in other microorganisms including lichens, some marine organisms and nematodes.[8] The β-anomer of D-galactose (β-Gal*f*) has drawn a lot of attention, because of its presence in severe human pathogens like *Mycobacterium tuberculosis*,[9] *Leishmania major*[10] and *Trypanosoma cruzi*.[11] This topic has been well reviewed recently,[7b,7c,12] and therefore the following pages will only include the most representative and pathogenic microorganisms.

9.2.1 In Protozoa

Two parasites, *Leishmania major* and *Trypanosoma cruzi*, possess well-characterized β-D-Gal*f* conjugates. *L. major* is the causative agent of leishmaniasis, affecting 12 million people worldwide. *T. cruzi* is the causative agents of Chagas disease, affecting 8–10 million people.[13] In *T. cruzi*, two classes of compounds contain D-Gal*f*: the glycoinositolphospholipids (GIPLs) and the mucins. In mucins, two cores have been identified, Gal*p*-(β1→4)-GlcNAc and Gal*f*-(β1→4)-GlcNAc **3** (Figure 9.1A). The less infective strains contain Gal*f*, which could be related to the fact that Gal*f* units are antigenic epitopes, thus triggering immunogenic responses.[14] In *Leishmania*, Gal*f* units are found in lipophosphoglycan conjugates **4** (LPG, Figure 9.1B) and GIPLs as part of the external surface.[10] The metabolism of those two glycolipid complexes (LPG and GIPLs) is critical in *L. major* and *T. cruzi*, since these conjugates are mediating not only the recognition and attachment of host macrophages but also the binding to the intestine of the vector insect.[10,15]

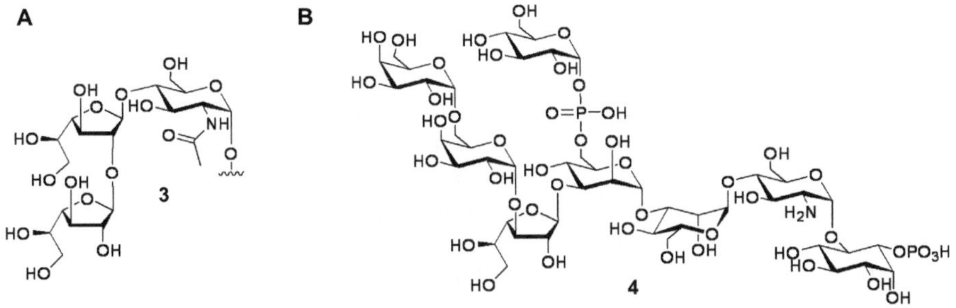

Figure 9.1 (A) Structure of mucins from *Trypanozoma cruzi*. (B) Lipophosphoglycan core from *Leishmania*.

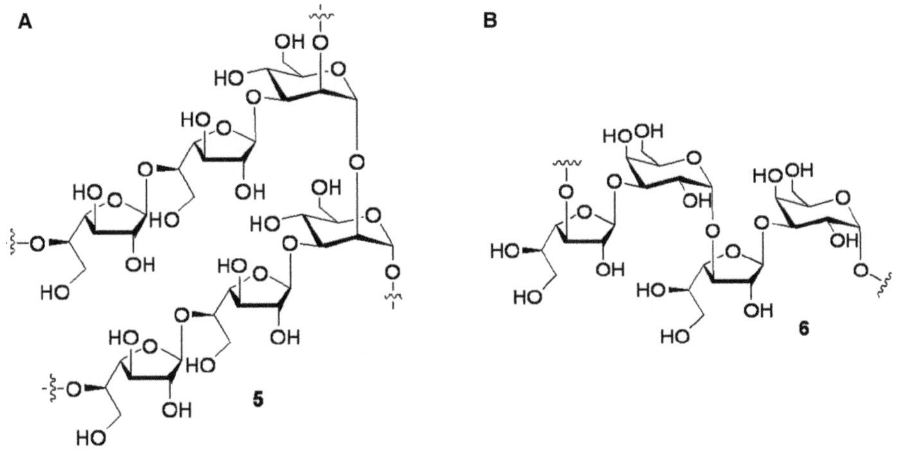

Figure 9.2 (A) Galactomannan pattern as found in *Aspergillus fumigatus*. (B) Galactan moiety of the O-antigen of *Klebsiella pneumoniae*.

9.2.2 In Fungi

Fungi are seen as a growing threat for immunodeficient patients, causing harmful infections such as candidoses and aspergillosis. D-Gal*f* is part of the major antigen circulating in patients with invasive aspergillosis.[16] This disease is caused by the most-studied β-D-Gal*f*-containing fungus: *Aspergillus fumigatus*. At least four different molecules that include β-D-Gal*f* units in their structure are found in *A. fumigatus*. A galactomannan secreted by *Aspergillus* is principally constituted by a core chain **5** composed of (1→2)- and (1→6)-linked α-D-mannopyranosides, with side chains of around 4–10 units of (1→5)-linked β-D-galactofuranoside residues (Figure 9.2A).[17] Again, D-Gal*f* units were shown to be immunodominant in those galactomannans.[18]

9.2.3 In Bacteria

D-Gal*f* has been found in Gram-positive and Gram-negative bacteria.[7b,12,19] Some of those bacteria are highly pathogenic and the glycoconjugates containing β-D-Gal*f* are involved in immunological reactions and/or in the virulence of the infectious agent. Good examples of such glycoconjugates are the lipopolysaccharide (LPS)[20] O-antigen from *Escherichia coli* K-12[21] or from the enteroinvasive *E. coli* O16,[22] as well as the galactan-I O-antigen **6** from *Klebsiella pneumoniae* (Figure 9.2B).[23] *K. pneumoniae* is an opportunistic pathogen threatening the lungs of hospitalized and immunocompromised patients. This bacterium is often very resistant to antibiotics and causes lung infections and urinary tracts infections, as well as bacteremia.

9.2.4 In Mycobacteria

The genus *Mycobacterium* encompasses major pathogens such as *Mycobacterium leprae* (causing leprosy), *Mycobacterium avium* (an opportunistic pathogen commonly infecting HIV-positive persons) as well as *Mycobacterium tuberculosis* (the causative agent of tuberculosis).

Mycobacteria are characterized by a very thick, hydrophobic cell-wall displaying a very limited permeability (Figure 9.3).[9,24] β-D-Gal*f* plays a unique

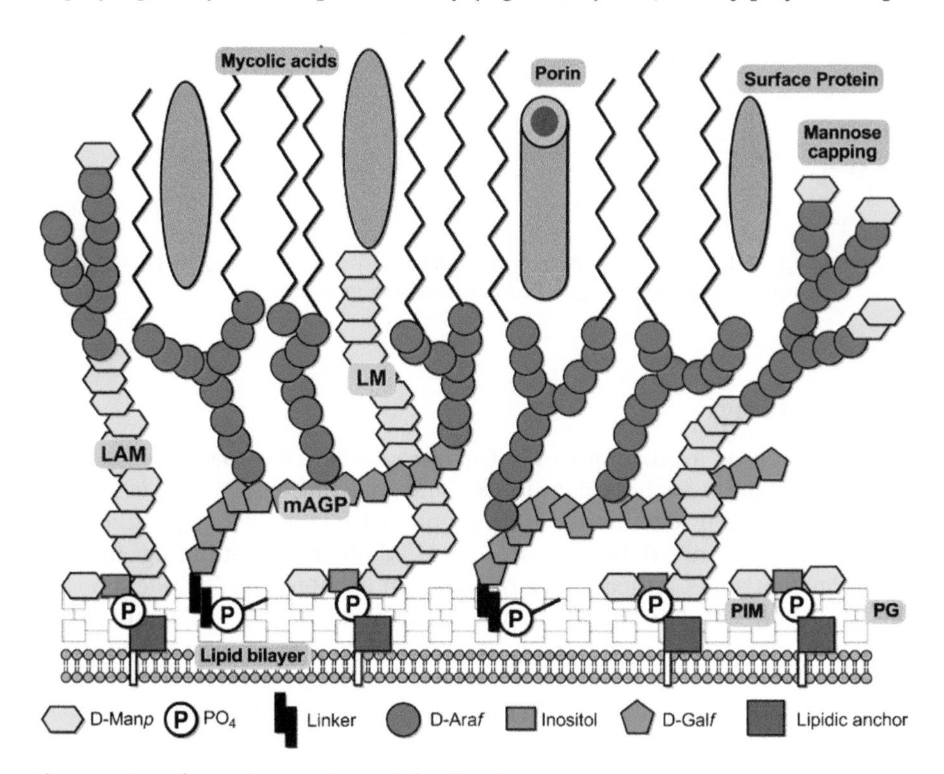

Figure 9.3 Schematic mycobacterial cell-wall structure.

role as a key building block of the mycobacterial cell wall. The bottom layer of this defensive structure is the peptidoglycan (PG): a rigid network of alternating *N*-acetylglucosamine (GlcNAc) and a glycolic *N*-acylated muramic acid (MurNGly). The muramic acids bear small peptides (three to five amino acids) that crosslink in between each other.[25] This rigid structure holds several glycolipids consisting in the main subunits of the cell wall: mycolyl-arabinogalactan-peptidoglycan complexes (mAGP), lipoarabinomannans (LAM), lipomannans (LM) and phospho-*myo*-inositol-mannosides (PIMs) (Figure 9.3).

The mAGP complex 7 consists of four main components attached one to another following this sequence: a specific disaccharide linking the PG, the galactan, the arabinan and the mycolic acids (Figure 9.4). The PG is covalently linked *via* a phosphate to α-L-Rha*p*-(1→3)-α-D-Glc*p*NAc disaccharide,[26] which in turn is linked to the galactan. The galactan is a linear chain of 30–35 alternating β-(1→5)- and β-(1→6)-Gal*f* linkages. Every galactan chain bears three arabinans and the linkage is positioned at the O-5 of the Gal*f* unit. The arabinans have an average length of 35 linear (1→5)-α-D-Ara*f* units along with some (1→2)- and (1→3)-α-Ara*f* ramifications. Approximately two-thirds of those arabinans are terminated by a cluster of four mycolic acids that are very long lipids of heterogeneous structures.[24b,27]

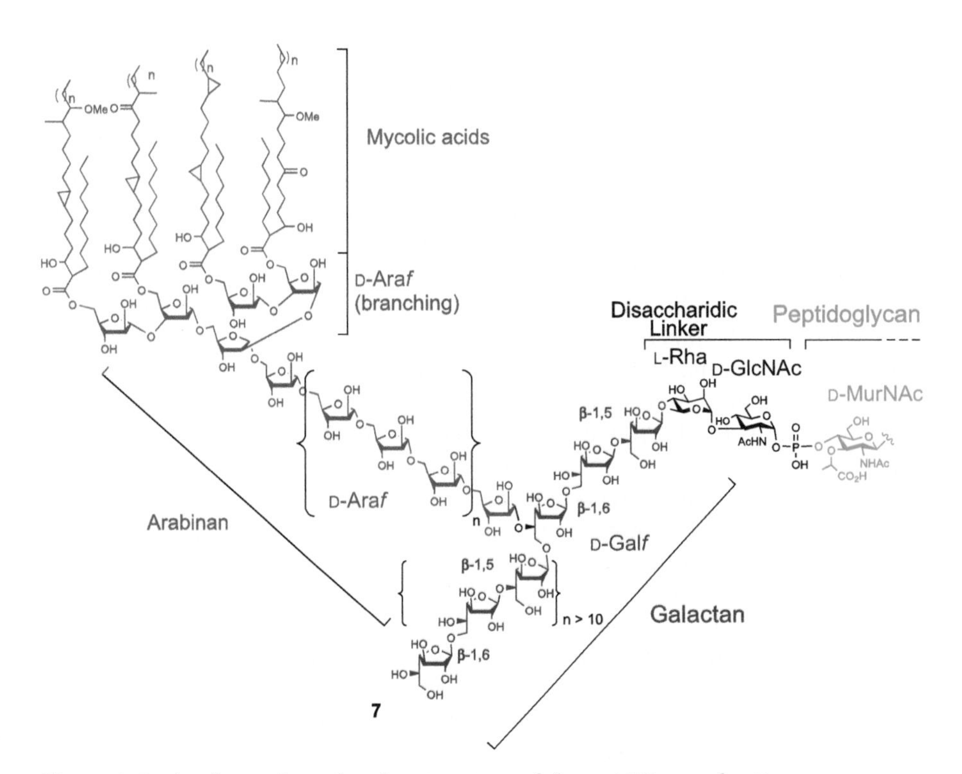

Figure 9.4 A schematic molecular structure of the mAGP complex 7.

Mycolic acids are a family of around 500 molecularly related structures that quantitatively represent the major components of the mycobacterial cell wall.[24a] Their functions are numerous. They give mycobacteria very unique features: resistance to injury, low permeability to hydrophilic molecules (glucose, glycerol, antibiotics such as β-lactams, *etc.*), resistance to dehydration and, most of all, the ability to persist and even develop inside their own natural "predator" —the macrophage—thus hiding from it.[28]

9.2.5 Therapeutic Relevance of the Galactofuranose Biosynthesis

As components of the cell wall of pathogenic species, the Gal*f* residues may play a role in the virulence of the infectious agent, or, in certain cases, the absence of this sugar can compromise the viability of the cell. In the first case, targeting the Gal*f* biosynthetic pathway with drug-like molecules may lead to antivirulent agents. In the second case, novel antibiotics may be discovered. However, the exact role of Gal*f* residues is not always clear, and this important question is still under debate for some microorganisms. For instance, *Aspergillus fumigatus*, a mutant lacking UDP-galactopyranose mutase (UGM), remained as pathogenic and as resistant to cell-wall inhibitors and phagocytes as the wild-type parental strain,[29] whereas it was shown that Gal*f* participates in the virulence of this fungus.[30] In nematodes such as *C. elegans*, the situation is very different since Carlow *et al.* provided data indicating that Gal*f* likely has a pivotal role in the maintenance of surface integrity, supporting investigation of UGM as a drug target in parasitic species.[31] In the parasite *Leishmania major*, generation of a mutant devoid of Gal*f* resulted in attenuated virulence, which highlights an important role of Gal*f* for eukaryotic pathogens.[32]

As detailed above, Gal*f* is an important component of the cell wall of mycobacteria, including *M. tuberculosis*, the causative agent of tuberculosis (TB). Each year, *M. tuberculosis* is responsible for 8 million human infections and 2 million deaths. Today's therapies are more and more obsolete with the emergence of new multi-drug resistant strains (MDR-TB) and extensively resistant strains (XDR-TB) which have become a major cause of mortality.[33] While MDR-TB requires a longer treatment including "second line" drugs which have more side effects and are more expensive, XDR-TB is almost unresponsive to any known drugs, and both XDR-TB and MDR-TB are particularly deadly for immunocompromised (*e.g.* HIV positive) patients.[33b]

In this context, where new drugs are needed, the biosynthesis of the cell wall was validated as a target in a growing number of studies.[7c,24c,33b,34] By hooking the ramifications of mycolic acids on the core peptidoglycan, D-Gal*f* is a key structural component in the biosynthesis of this cell wall: galactan biosynthesis is thus essential for the growth of mycobacteria.[35] This key role added to the absence of Gal*f* in humans makes the inhibition of its biosynthesis attractive in the fight against TB. As explained in the following

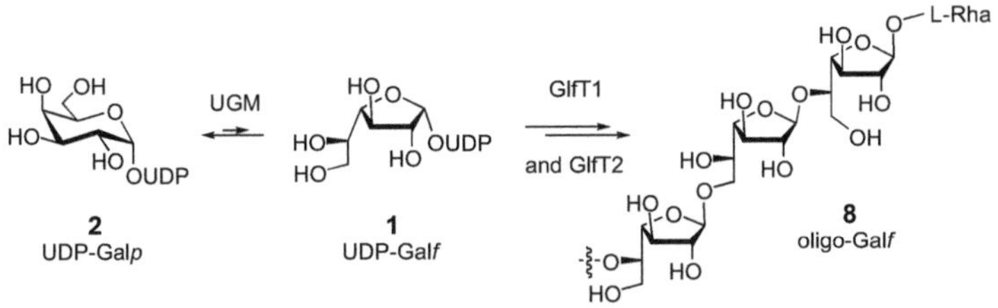

Scheme 9.1 The two main steps in the mycobacterial galactan biosynthesis.

pages, the galactan biosynthesis proceeds through some unusual biochemical steps involving unique enzymes.

We should also mention that, beside the isomerase UGM and galactofuranosyltransferases, a third class of Gal*f*-processing enzymes has also been discovered. Indeed, some galactofuranosidases have also been isolated and characterized, and some inhibitors have been produced.[36] However, it is not yet clear whether these enzymes are therapeutic targets.

9.2.6 Biosynthesis of the Mycobacterial Galactan

The mycobacterial galactan is biosynthesized in two main steps (Scheme 9.1). The UDP-galactopyranose mutase (UGM) catalyzes a ring contraction of UDP-Gal*p* **2** into its furanose form UDP-Gal*f* **1**. Once produced, UDP-Gal*f* **1** becomes the D-Gal*f* donor in an oligomerization process catalyzed by two galactofuranosyltransferases (GlfT1 and GlfT2). These two steps give rise to a galactan conjugate **8**, precursor of the mAGP complex 7 (Figure 9.4).

In the following sections, we will focus on the efforts realized to understand the mechanisms and the development of inhibitors of UGM (Section 9.3) and galactofuranosyltransferases (Section 9.4).

9.3 UDP-Galactopyranose Mutase (UGM)

In 1996, the gene encoding for UGM was identified and cloned for the first time in *E. coli*.[5] The enzyme turned out to be a flavo-enzyme non-covalently bound to a flavine adenine dinucleotide (FAD) cofactor (Figure 9.5). Depending on the oxidation state of its cofactor, UGM can be in the oxidized (FAD, **9**) or reduced state ($FADH_2$, **11**). Once deprotonated ($FADH^-$, **10**), the latter is the active state and the kinetic parameters of UGM_{Ec}[†] were determined to be $K_m = 22$ μM (for UDP-Gal*f*) and $k_{cat} = 27\ s^{-1}$ for the reaction performed from the furanoside **1** to the pyranoside **2**.[37]

At that time, it was the first enzyme—from any organism—shown to catalyze the conversion of a pyranose into its furanose form. The authors

[†]The different UGMs will be differentiated as a function of their origin: from *E. coli* (UGM_{Ec}), from *K. pneumoniae* (UGM_{Kp}), from *M. tuberculosis* (UGM_{Mt}), from *T. cruzi* (UGM_{Tc}) *etc...*

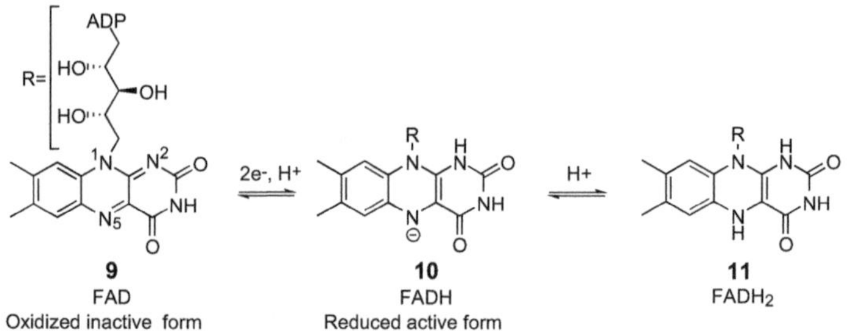

R=
ADP
HO
OH
HO
2e-, H+
H+

9
FAD
Oxidized inactive form

10
FADH
Reduced active form

11
$FADH_2$

Figure 9.5 The main redox states of FAD.

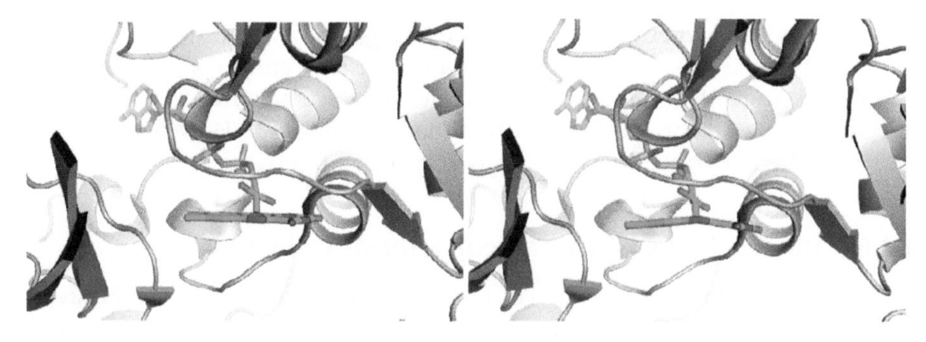

Figure 9.6 FAD is planar when oxidized (*left*; PDB: 2B17) while $FADH^-$ (reduced) is bent (*right*; PDB: 1WAM).

demonstrated that the equilibrium is strongly shifted toward the pyranose form (Scheme 9.1), thus explaining why UGM kinetics are usually studied in the biosynthetic backward direction, from the furanose to the pyranose. A few months later, new UGMs were identified in *K. pneumoniae* and *M. tuberculosis*.[38] Today, UGMs are known to be present in *Aspergillus* spp.,[30,39] in protozoa[40] and in worms[8,31] (*Caenorhabditis elegans*).[19] Moreover, Lowary *et al.* discovered and characterized a UGM from *C. jenuni* using UDP-*N*-acetylgalactosamine as a substrate,[41] and a mutase converting UDP-arabinopyranose into its furanose form was characterized in plants.[42] In this chapter, we will focus on the most studied UGMs, *i.e.* from *M. tuberculosis*, *K. pneumoniae*, *A. fumigatus* and *E. coli*.

9.3.1 UGM 3D Structures

In 2001, the first crystal structure of the UGM from *E. coli* (UGM_{Ec}) revealed a mixed α/β dimeric protein having a crescent shape.[43] The FAD cofactor is buried, non-covalently bound to the enzyme in a cleft. The putative binding site, located in this cleft, displayed good candidates for substrate complexation: Trp156 with the uracil; Arg247 and Arg276 with the phosphates; Tyr151, Tyr181, Tyr311, Tyr346, Glu298, Asp348 and His59 with the galactose. Electrochemical analysis demonstrated that the reduced form of UGM is anionic ($FADH^-$; Figure 9.6, right).[44]

In 2005, the first crystal structure of UGM in the active (reduced) state was published by Naismith *et al.* The crystal structure of *M. tuberculosis* UGM (UGM_{Mt}) in the oxidized state and *K. pneumoniae* (UGM_{Kp}) in oxidized and reduced states pointed out differences with UGM_{Ec}.[45] The residues at the interfaces of the dimers are not conserved as well as some residues surrounding the cofactor. In the oxidized state, the isoalloxazine ring is aromatic, thus planar (Figure 9.6, left). In contrast, the reduced $FADH^-$ has a bending *Re* face, with N-1 carrying the charge. Importantly, for $FADH^-$ (unlike FAD), N-5 has its lone pair oriented toward the putative binding site.[45,46]

Directed mutagenesis on UGM_{Kp} clearly showed the crucial role of two arginines (Arg174 and Arg280).[46] While mutations often resulted in a decrease of the catalytic activity, the mutation of any of these two arginines results in a total loss of activity. Since Arg280 is close to the two phosphates, its importance is easy to understand: it stabilizes the two negative charges by coulombic interactions. However, Arg174 is located on a mobile loop,[47] which would be able to move and place this residue toward the inner side of the active site.

Already in Naismith's early study of UGM_{Ec}, four tyrosines (conserved among all UGMs) were highlighted in UGM_{Kp}; the loss of one of them (Tyr155, Tyr185, Tyr 314 or Tyr 348) causes an increase in K_m from 43 μM to a range of 390–820 μM without affecting the k_{cat}.[46] The four tyrosines could be implicated in the substrate's binding either through aromatic–carbohydrate interactions or through H-bonding. Two negatively charged residues (Glu301 and Asp351) appeared to be essential, thus suggesting positively charged intermediates. The key role of Trp160 was then confirmed, as it was suspected to bind uridine by π-stacking.

A few years later, the crystal structures of *A. fumigatus* UGM_{Af},[48] *D. radiodurans* UGM_{Dr}[49] and *T. cruzi* UGM_{Tc}[50] were determined; more recently, 3D structures of UGM in complex with UDP,[48b,49a] UDP-galactopyranose,[48b,49b,51] the *C*-glycosidic analogue of UDP-galactopyranose[49b] and UDP-glucose[52] were obtained. The latter results gave valuable information regarding the binding modes of the UDP and the galactopyranose moieties. In addition, binding models suggested that the FAD could play the role of catalytic base for the proton transfer between the positions O-4 and O-5 of the galactose moiety. A density functional theory (DFT) study strongly supported this unusual role for a redox cofactor.[53]

However, the binding mode of the galactose residue in its furanose form is still unknown. This complementary structural information would be extremely useful, since all UGMs have a much stronger binding affinity for the furanose form compared to the pyranose one.

9.3.2 UGM Mechanism

9.3.2.1 Gander's Early Hypothesis

Twenty years before UGM was discovered, a mechanism of enzymatic pyranose-to-furanose conversion was hypothesized by Gander *et al.*, based

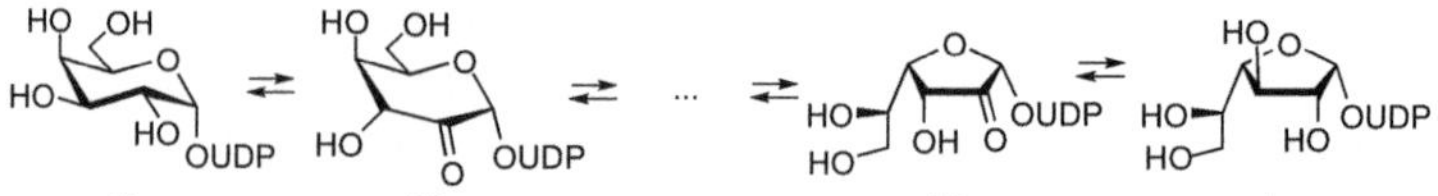

Scheme 9.2 First mechanistic hypothesis involving 2-keto intermediates.

on the analysis of reductive titrations on crude extracts of *P. charlesii*. Interestingly, this mechanism involved an oxidation of the galactose at the C-2 position of UDP-galactopyranose (Scheme 9.2).[54] Gander was even speculating that the presence of a ketone would force the pyranose to adopt a boat conformation that would facilitate the attack of the 4-OH group. This was a very modern way to describe an enzyme mechanism, more than 30 years before the publication of the first 3D structures of UGM in complex with UDP-Gal*p* analogues.

However, this hypothesis was ruled out by the use of fluorinated analogues of UGM substrates. Indeed, several teams investigated the mechanism and the UGM binding properties with fluorosugars.

9.3.2.2 *Fluorinated Carbohydrates as Binding and Mechanistic Probes*

In 2001, Liu *et al.* demonstrated that C-2 and C-3 fluorofuranose analogues **14** and **15** (Figure 9.7)[55] are substrates of UGM_{Ec}, despite the fact that they cannot be oxidized at positions C-2 and C-3, thus ruling out Gander's hypothesis.[55]

The fact that the 2-fluoro analogue **14** has a much lower k_{cat} but an almost unmodified K_m shows that the C-2 hydroxyl of UDP-Gal*f* does not strongly interact with the protein and that an oxycarbenium intermediate is likely involved in the mechanism. The comparison of the K_m values of **14** and **15** with **1** suggests either strong attractive interactions between the 3-hydroxyl of UDP-Gal*f* **1** and the enzyme, or that the 3-fluoro group in **15** caused significant repulsive interactions.[55] With the non-reduced UGM_{Kp}, saturated transfer difference NMR (STD-NMR) studies involving **14** and **15** showed high K_d values for both substrates (up to 800 μM).[56] Given the tolerance for fluorination at the 2- and 3-positions, the synthesis of tetrafluorinated UDP-galactose analogues **17** and **21** were later synthesized.[57] Surprisingly, competition assays and STD-NMR techniques clearly showed that **17** had a low micromolar affinity to UGM. Compared to the apparent loss of affinity resulting from the fluorination at the 3-position (analogue **15**), these results showed that tetrafluorination can still have a beneficial effect on binding when monofluorination at the same position provokes a dramatic increase of K_m. For carbohydrate processing enzymes, this was the first unambiguous case of affinity enhancement through local sugar perfluorination.

Other monofluorinated UDP-galactose derivatives **16** and **18–20** were also synthesized and compared to UDP-deoxygalactose analogues or

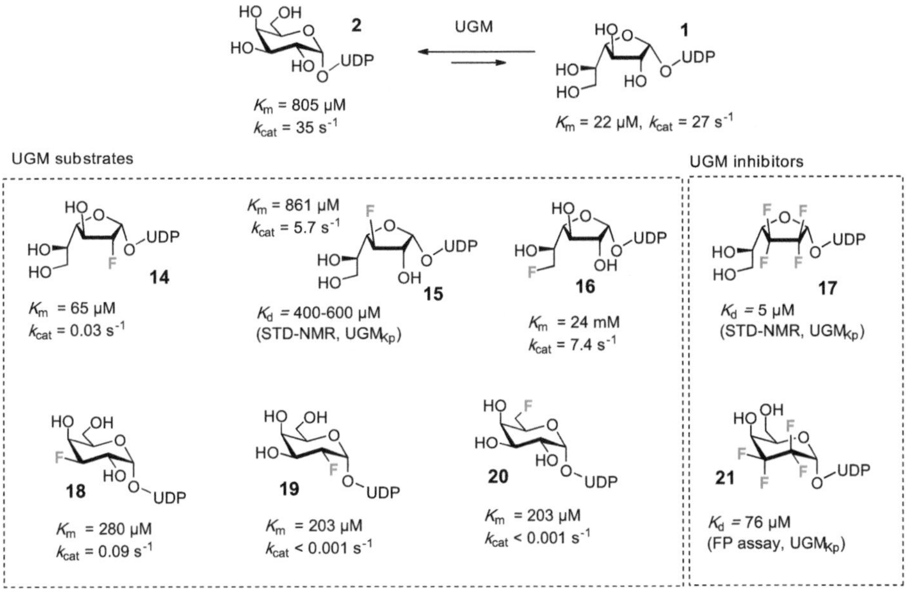

Figure 9.7 Fluorinated substrate analogues designed as UGM mechanistic and binding probes.

UDP-β-L-Ara*f*.[58] Noteworthy, Field *et al.* performed a systematic study on a large collection of UDP-Gal analogues to probe the effect of modifications at the C-2, C-3 and C-4 positions of the pyranose substrate on the catalytic turnover.[58c] They demonstrated that none of those hydroxyl groups are essential for binding or turnover, while UDP-α-D-talopyranose, the C-2 epimer of the natural substrate, could not be isomerized by UGM_{Kp}.[58c]

9.3.2.3 1,4-Anhydrogalactose as a Putative Intermediate

Blanchard's team elegantly demonstrated through positional isotope exchange experiments (PIX) that the UGM mechanism involves the cleavage of the anomeric bond and the release of UDP.[59] They followed by ^{13}C NMR the reaction of UDP-galactopyranose **2** radiolabelled at the anomeric carbon and oxygen (^{13}C and ^{18}O) in the presence of UGM_{Kp}. From these experiments, the authors postulated a high-energy intermediate with the structure of a 1,4-anhydrogalactose **24** (Scheme 9.3). Indeed, such an intermediate may be the direct precursor of a pyranose **22/23** or a furanose **25/26**, depending on which C–O bond is cleaved. The nucleophilic displacement may occur either by an S_N2-like mechanism (Scheme 9.3, path a) or by an S_N1 cationic mechanism with anchimeric assistance (Scheme 9.3, path b).

However, this hypothesis does not imply any redox process nor a key role for the FAD, although the redox state of the cofactor has a dramatic impact on the reaction rate.[59] Our group started this investigation by synthesizing

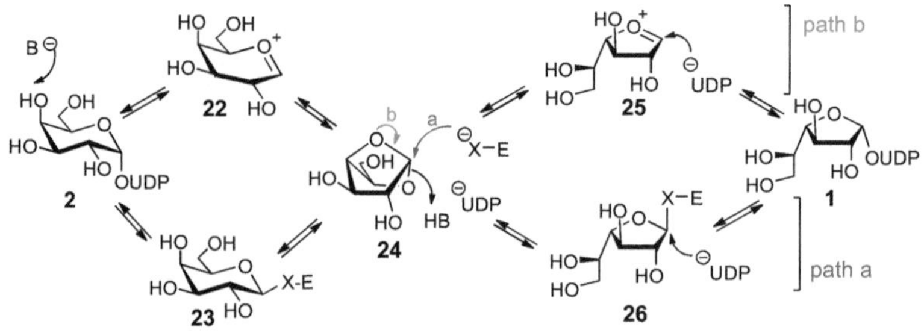

Scheme 9.3 Blanchard's UGM mechanism involving a central intermediate **24** (X–E = a putative catalytic nucleophile, as proposed by Blanchard in this study).[59]

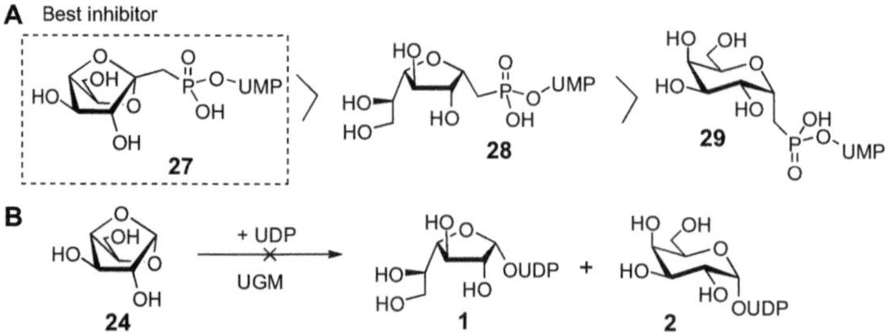

Figure 9.8 (A) Conformational probes assayed against UGM_{Ec} under reducing conditions. (B) The postulated intermediate **24** does not react with UDP and UGM_{Ec}.

mechanistic probes (Figure 9.8A) locked in key conformations: the bicyclic ^{1,4}B boat galactose **27** and a furanose-locked phosphonate **28**.[60] Later on these molecules were compared to the pyranose-locked phosphonate **29**.[61] It could be shown that the boat-locked molecule has the strongest affinity for UGM_{Ec}, followed by the furanoside **28**, then the pyranoside **29** (Figure 9.8A). Such a result suggests that the galactose residue adopts this conformation either as a high-energy intermediate or at a transition state. However, the mechanism involving the 1,4-anhydrogalactose **24** as a low-energy intermediate was ruled out in 2006.[62] Indeed, molecule **24** was synthesized and the incubation with UDP and the enzyme did not produce UDP-galactose **1** or **2** (Figure 9.8B), which implies that **24** cannot be postulated as a low-energy intermediate. However, the inhibition study involving the boat-locked analogue **27** strongly suggests that the conformation of one of the high-energy intermediates or one of the transition states is close to a ^{1,4}B boat.[60]

In addition, this inhibition study also demonstrated that there is a significant difference in the binding affinities of substrate analogues as a function of the redox state of UGM.[60]

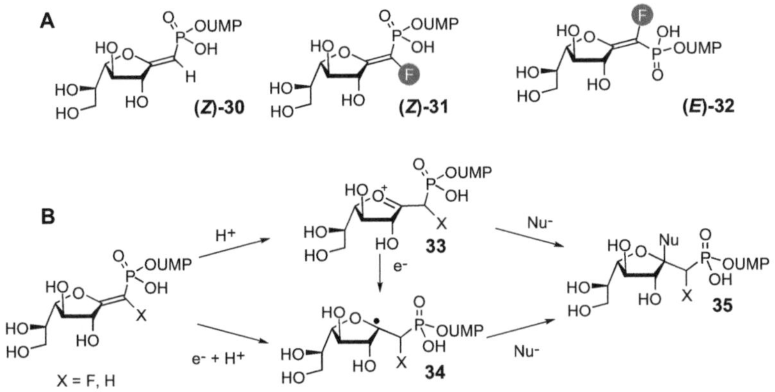

Figure 9.9 (A) *exo*-Glycals as inactivators of UGM. (B) The inactivation mechanism by *exo*-glycals does not involve single electron transfers (SETs).

9.3.2.4 exo-Glycals as Mechanistic Probes

From 2004, another generation of inhibitors, the three *exo*-glycals [(*Z*)-**30**, (*Z*)-**31** and (*E*)-**32**] have been designed as potential UGM inhibitors (Figure 9.9).[63] Unexpectedly, these molecules displayed a time-dependent inactivation profile against the non-reduced UGM_{Ec} rather than a competitive inhibition behavior. Interestingly, the above-mentioned fluorinated analogues **14** and **15** (Figure 9.7) were also found to behave as time-dependent inactivators of UGM_{Ec}.[55] These experiments suggested that the mechanism of this isomerization involved a covalent intermediate.[55,63] To investigate further the inactivation mechanism, the fluorinated enol ether derivatives **31** and **32** were synthesized.[63b] The latter molecules were also found time-dependent inactivators, but with much slower kinetics than their non-fluorinated analogue **30.** This result supports a two-electron process involving cationic intermediates such as **33** as a key step of the UGM inactivation rather than single-electron transfers (that could have been promoted by the FAD cofactor). All attempts failed to identify by mass spectrometry a covalent adduct **35** (Figure 9.9) of the inactivator bound to the enzyme. Therefore, the nature of the active site's nucleophile has not been identified yet.

9.3.2.5 An Unexpected Covalent Catalysis

A major step in the UGM mechanistic investigation was achieved in 2004 when Kiessling and co-workers evidenced a new catalytic role for FAD. Indeed, this team characterized a covalent adduct of the FAD cofactor and a radiolabeled UDP-Gal*p* **42** by mass spectrometry after having reduced the postulated iminium intermediate **41** with sodium cyanoborohydride (Figure 9.10B).[64] These results strongly suggested that the three FAD–galactose covalent adducts **38–40** are intermediates of the reaction. These adducts can be formed either through a direct attack of the nucleophilic $FADH^-$ cofactor onto the oxycarbenium **22** and **25** (Figure 9.10A, path A),

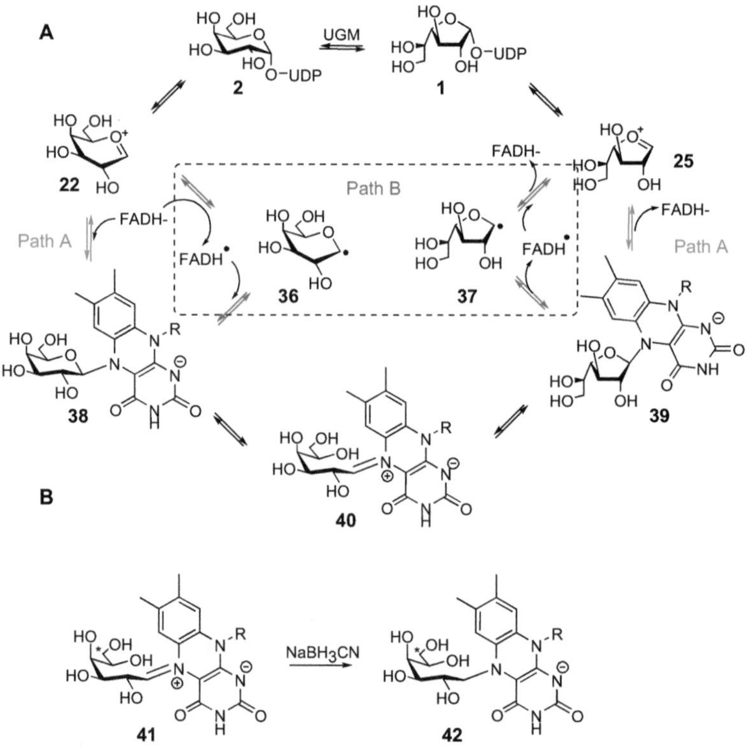

Figure 9.10 (A) Covalent nucleophilic catalysis by the N-5 of the FAD. (B) Reduction of the radiolabeled iminium intermediate **41**.

or thanks to a preliminary SET followed by radical coupling (path B). While SETs are very common in the chemistry of flavoproteins, a nucleophilic covalent catalysis of the FAD on an anomeric carbon has never been evidenced. Kiessling *et al.* could also characterize the iminium adduct by UV spectroscopy. In the presence of dithionite as a reducing agent and a high concentration of UDP-Gal*p*, the UV spectra of the enzyme changed: the absorbance increased at 500 nm and an isobestic point appeared at 475 nm. These key experiments were further improved[51,52] and applied to other UGMs,[65] including eukaryotic enzymes, and always lead to the conclusion that the FAD played the role of the catalytic nucleophile. This mechanism (*i.e.* the involvement of intermediates **38–40**, Figure 9.10) is nowadays the most accepted one.

However, some aspects of the UGM mechanism are still under debate: for instance, the implication of radical intermediates as well as the S_N1/S_N2 character of the substitution reactions.

9.3.2.6 *Single Electron Transfers?*

In 2003, Liu *et al.* reconstituted UGM_{Ec} with 1-deaza and 5-deaza analogues of FAD.[66] The analysis of the reaction kinetics tended to highlight a radical mechanism and clearly evidenced a crucial role for the N-5 of the flavin. Moreover, it also confirmed the importance of the cofactor. In the same series of experiments, the photo-reduction of FAD demonstrated a 26-fold enhancement rate of specific activity. UGM is active with FAD or 1-deaza-FAD but not with 5-deaza-FAD. Both FAD and 1-deaza-FAD are known to be able to participate in one- or two-electron processes, but 5-deaza-FAD is limited solely to two-electron processes.[66] A fluorescence spectroscopic study of the UGM indicated a pK_a of 6.7 for the cofactor, thus showing that the FAD is in its anionic semiquinone form under physiological pH.[64] Thanks to potentiometric analysis, Naismith *et al.* showed that the neutral semiquinone ($FADH^{\bullet}$) is stabilized in the presence of substrate and the fully reduced flavin is the anionic $FADH^-$ rather than the neutral $FADH_2$.[44] The anionic $FADH^-$ has the potential to act as a rapid one-electron donor/acceptor without being slowed by a coupled proton transfer and is therefore an ideal crypto-redox cofactor. Furthermore, based on their crystallographic data, Naismith and co-workers argued that a mechanism based on a covalent adduct would require strong conformational changes while a redox process would necessitate very few.[45] From computational simulations of UGM_{Kp} in complex with UDP-galactose, they reported a distance of 4.8 Å between the N-5 of the FAD and the anomeric carbon of the galactose and concluded it is too far for a covalent attack.[45] Further STD NMR experiments demonstrated that this distance is closer to 3.0 Å, and thus compatible with both mechanisms through electron transfer and covalent catalysis.[47] At present, the question of involvement of radical species in the UGM mechanism is still under debate. No experiment has ruled out or demonstrated this hypothesis.

9.3.2.7 S_N1- or S_N2-like Mechanism?

Different biochemical investigations have been set up to define whether S_N1- or S_N2-type displacements occur in his transformation, in other words whether cationic intermediates such as **22** and **25** are involved (Figure 9.9). For instance, a linear free energy relationship (LFER) was used to monitor the kinetic variations as a function of the substitution pattern of the FAD.[67] It was thus concluded that these results support a concerted S_N2-type displacement where N-5 of the reduced FAD is the nucleophile and would rule out the cationic oxycarbenium transition states. However, this conclusion is in clear disagreement with the interpretation of the results obtained with fluorinated analogues **14** and **15** that support substantial cationic character at the transition state(s).[55] Recent DFT calculations suggest that the two substitutions at the anomeric position would follow an S_N2 mechanism.[53]

9.3.2.8 Comparison of Prokaryotic and Eukaryotic UGMs

Later, Sobrado *et al.* studied the kinetic and chemical mechanisms of eukaryotic UGMs from *Trypanosoma cruzi* by using steady state kinetics, fluorescence anisotropy and trapping the iminium intermediate.[65a]

The main difference with prokaryotic UGMs is that this flavoenzyme employs NADPH as a redox cofactor. It was found that the presence of UDP-Gal*p* in the binding site is necessary to protect the enzyme from oxidation. The kinetic data showed the formation of an iminium species which, after addition of $NaCNBH_3$, provides a flavin–galactitol adduct **42** (Figure 9.10), characterized by mass spectrometry, a result in line with Kiessling's mechanism. No formation of a flavin semiquinone or oxycarbenium intermediate has been detected. This biochemical investigation globally shows that the mechanism discovered for bacterial UGMs applies for UGM_{Tc} and, overall, the results suggest S_N2 processes.

9.3.3 UGM Inhibitors

9.3.3.1 UGM Inhibition Assays

Owing to the therapeutic relevance of UGMs, inhibition assays were quickly developed after their discovery. The first assay is still employed nowadays and consists in monitoring the conversion of UDP-Gal*f* **1** into its pyranose form **2** by HPLC.[68] For evaluation of the inhibitor's potency, this assay is rather tedious and usually only leads to the measurement of inhibition percentages, but it remains the most widely used activity assay.[55,60,69] Moreover, it is the only competition assay against UGM's natural substrate **1** available to date. In addition, it is performed under reducing conditions, which is of paramount importance because UGM must be reduced to be kinetically competent. In 2003, a microtiter plate assay was developed, based on the selective release of tritiated formaldehyde from radiolabeled UDP-galactofuranose **1** thanks to a periodate oxidation that cannot occur with UDP-galactopyranose **2**.[70] Because of this technique, uridine-based inhibitors could be identified from a library. In 2004, a fluorescence polarization assay was developed as a high-throughput technology to identify UGM inhibitors.[71] In this assay, potential inhibitors and a fluorescent UGM ligand (UDP-fluorescein **43**, Figure 9.11) compete for the occupancy of the non-reduced UGM binding pocket.

It was also shown, in one study, that the intrinsic fluorescence of the FAD cofactor bound to UGM could be exploited for the evaluation of binding constants. Fluorescence titrations of UGM were thus performed and allowed the measurements of K_d values.[69b]

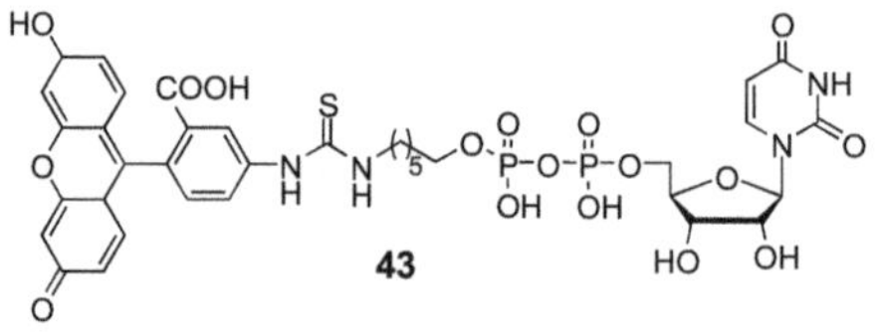

Figure 9.11 UDP-fluorescein **43** used for the fluorescence polarization assay.

9.3.3.2 UGM Inhibitors

In order to study the mechanism of UGM and to probe the possible interactions within the active site, various inhibitors have been designed. Owing to their similarity to the substrate or to their analogy with the hypothetical transition states involved in the mechanism of UGM, they can constitute new tools to better understand the modes of interactions between the enzyme and the substrate. As described in the preceding section, some molecules have been developed as mechanistic and conformational probes: fluorinated carbohydrates (Figure 9.7), *exo*-glycals (Figure 9.9) and phosphonates (Figure 9.8). In the following section we describe the most important classes of inhibitors that can be classified into the following categories.

9.3.3.3 Cationic UDP-Gal*f* Analogues

Very early on, Fleet *et al.* synthesized the iminocyclitols **44** and **45** (Figure 9.12) and found that they were moderate *in vitro* inhibitors of UGM_{Ec}, but also of the mycobacterial galactan biosynthesis.[69a]

A series of ammonium, selenonium and sulfonium ions **46–48** (Figure 9.12) were prepared by Pinto *et al.* as mimics of the oxycarbenium intermediate **25** (Figure 9.10).[69c] In a similar approach, the conformationally locked bicyclic iminocyclitol **49** was also developed.[69d] Molecules **46–49** did not exhibit a strong inhibition profile under their assay conditions. Moreover, Martin and co-workers developed the synthesis of iminosugars **50** featuring both an endocyclic secondary amine and the UMP subunit.[72] Moderate inhibition activities were also obtained against UGM_{Ec}. These results probably show that, in contrast to other glycosyl processing enzymes such as glycosidases, mimicking an oxycarbenium species does not constitute an efficient strategy to obtain glycomimetics displaying strong affinities

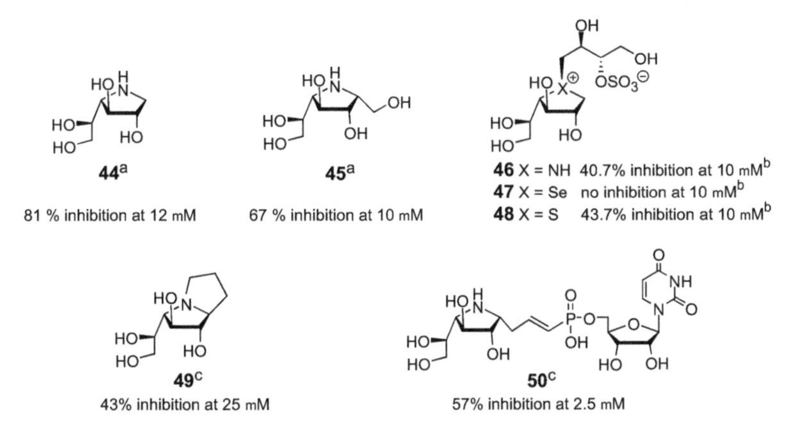

Figure 9.12 Cationic furanosides as UGM inhibitors. All the assays were performed under reducing conditions. [a][UDP-Galf] = 70 μM; UGM_{Ec}. [b][UDP-Galf] = 63 μM; UGM_{Kp}. [c][UDP-Galf] = 200 μM; UGM_{Ec}.

for UGM. These results are in line with the mechanistic studies suggesting S_N2-like processes for the UGM catalyzed isomerization.

9.3.3.4 *Acyclic Inhibitors*

To determine which high-energy intermediates are stabilized by UGM, different acyclic sugars **51–57** (Figure 9.13) were designed to mimic the central intermediate **40** (see UGM mechanism, Figure 9.10). For instance, Itoh *et al.* generated the UDP-D-galactitol **51**, which exhibited 54% inhibition of UGM_{Ec} at 2.5 mM (which corresponded to a $K_d = 46$ μM by a fluorescence assay).[69b] This promising result prompted our group to synthesize similar compounds. First, Pan *et al.* prepared molecules **52** and **53** featuring not only the D-galactitol chain but also an amino group at the anomeric position to mimic the positive charge of intermediate **40**.[73] Unfortunately, these molecules display weak inhibition due to the absence of the UMP moiety. More recently, Ansiaux *et al.* designed a novel series of UDP-galactitol analogues **54–57** bearing different electrophilic substituents between the galactose and the UDP moiety (Figure 9.13).[74] The inhibition study against UGM_{Ec} could clearly show that slight changes in the relative orientation of the UDP and the galactitol moieties resulted in dramatic variations of binding properties. Compared to known inhibitors, the epoxide derivative **54** displayed a tight, reversible, inhibition profile. Moreover, a time-dependent inactivation study showed that none of the electrophilic structures **54–57** could react with UGM, or its FAD cofactor. Compared to the inhibition levels obtained with iminocyclitols **44–50** (Figure 9.12), the latter results showed that glycomimetics of acyclic intermediate **40** displayed significantly better affinities than analogues of the oxycarbenium **25**.

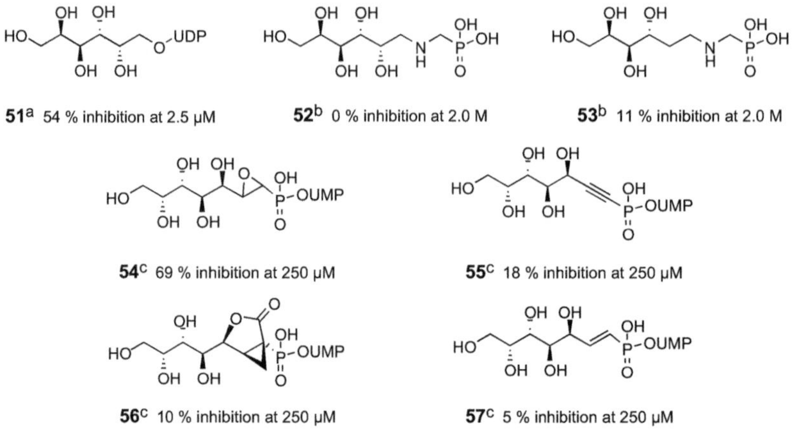

51[a] 54 % inhibition at 2.5 μM

52[b] 0 % inhibition at 2.0 M

53[b] 11 % inhibition at 2.0 M

54[c] 69 % inhibition at 250 μM

55[c] 18 % inhibition at 250 μM

56[c] 10 % inhibition at 250 μM

57[c] 5 % inhibition at 250 μM

Figure 9.13 Acyclic substrate analogues as mimics of the acyclic flavin-galactose intermediate; all the assays were performed under reducing conditions. [a][UDP-Galf] = 50 μM; [UGM_{Kp}] = 8.2 nM. [b][UDP-Galf] = 150 μM; [UGM_{Ec}] = 15 nM. [c][UDP-Galf] = 105 μM; [UGM_{Ec}] = 22 nM.

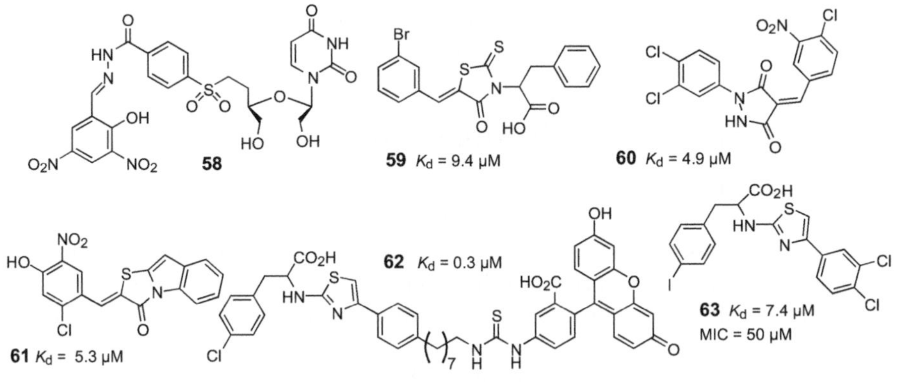

Figure 9.14 Library of aromatic compounds as UGM inhibitors.

9.3.3.5 *Aromatic and Heterocyclic Compounds*

From the first microtiter-plate assay developed against UGM, McNeil *et al.* managed to identify the uridine derivative **58** (Figure 9.14).[70] This molecule displayed a low-micromolar IC_{50} but was not active against *M. tuberculosis*. Kiessling *et al.* took advantage of the above-mentioned fluorescence polarization assay to screen a library of drug-like compounds in a high-throughput manner.[71] Eleven heterocyclic ligands were identified from a 16 000 chemical entities library.[71,75] This allowed the discovery of the thiazolidinone **59** that displayed a K_d value of 9.4 μM against UGM_{Kp}. From this hit, a structure–activity relationship was realized by screening new libraries and by synthesizing analogues of **59** by solid-phase synthesis.[76] Two examples (molecules **60** and **61**) that have been assayed against UGM_{Mt} are represented in Figure 9.14. Later on, Dykhuizen and co-workers prepared compound **62**, which appeared to be an effective inhibitor with an IC_{50} of 0.3 μM; interestingly, this potent inhibitor is a hybrid molecule featuring both the fluorescein of fluorescent probe **43** and inhibitor **60**.[77] Moreover, 2-aminothiazole **63** was assayed against *Mycobacterium smegmatis* and was found to prevent the bacterial growth with a MIC (minimum inhibitory concentration) of 50 μM, thus validating UGM as a pharmacological target for the treatment of tuberculosis.

9.3.4 Intermediate Conclusion

The UGM mechanism and structure(s) have drawn a lot of attention and, despite the different hypotheses and divergent conclusions, some key features of this biocatalytic transformation can be reasonably accepted. The UGM is active only when the FAD is reduced and it is the N-5 of this FAD that would act as a nucleophile either through an S_N1 or S_N2 mechanism. Although never ruled out, the hypothesis of single electron transfers seems less likely according to the experimental data gathered to date. Recently, other flavoenzymes have been found to covalently use reduced flavin in order to catalyze non-redox transformations.[78]

9.4 Galactofuranosyltransferases

Glycosyltransferases catalyze glycosylation reactions. Mechanistically, this reaction corresponds to the nucleophilic substitution of an *acceptor* on the electrophilic anomeric position of an activated sugar, named the *donor*. In mycobacteria, the galactofuranosyltranferases (Gl*f* Ts) use UDP-Gal*f* as a D-Gal*f* donor and these units are stereoselectively added one to another (Figure 9.15).

After the isomerization catalyzed by UGM, the newly produced UDP-Gal*f* is transported from the cytosol into the Golgi, where the glycosylation takes place.[32] While such a transporter has not been yet identified in mycobacteria, the first Golgi UDP-Gal*f* transporter was identified in *A. fumigatus*.[30] Several galactofuranosyltransferases have been cloned, for instance from *E. coli*,[79] *Aspergillus nidulans*[80] and *Trypanosoma rangeli*,[81] but the most studied ones are the two Gl*f* Ts from *M. tuberculosis* (Rv3782 or Gl*f* T1[82] and Rv3808c or Gl*f* T2[6]).[24b] The mycobacterial galactan biosynthesis is

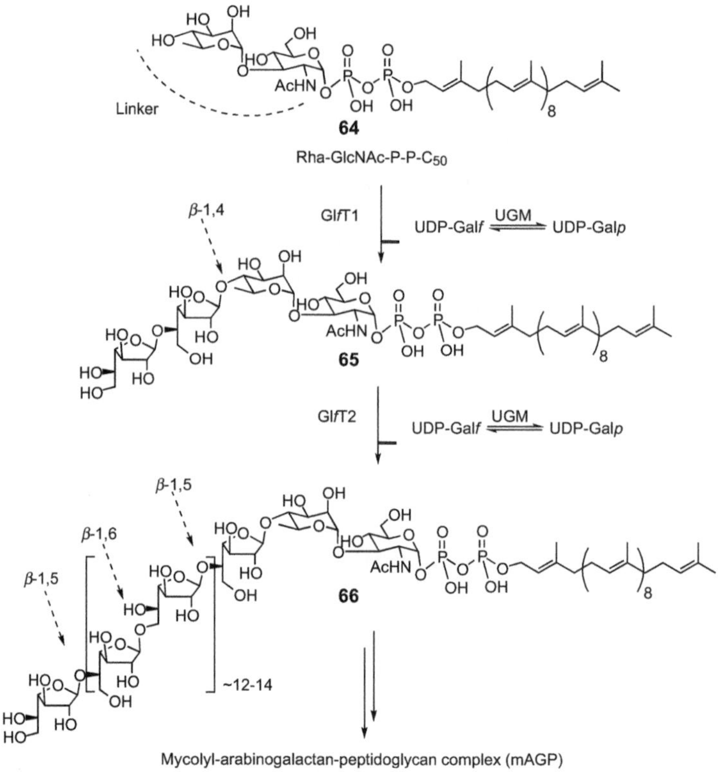

Figure 9.15 Two Gl*f* Ts (Gl*f* T1 and Gl*f* T2) assemble the mycobacterial galactan.

represented in Figure 9.15. The acceptor **64** contains an mAGP linker connected to a decaprenyl phosphate. Two Gal*f* units are first added to the rhamnose unit by Gl*f* T1 and Gl*f* T2 catalyzes the transfer of over 20 Gal*f* residues through a unique controlled polymerization sequence. The galactan is then translocated outside the cytosol, where it is transferred to the peptidoglycan core.[83] Then, arabinosyltransferases connect three arabinans on each galactan and the whole new molecule is esterified with mycolic acids.

9.4.1 Gl*f* T Mechanisms

Both enzymes use UDP-Gal*f* as donor and belong to the GT-2 family in the CAZy glycosyltransferase classification (www.cazy.org). However, they have a low degree of sequence homology. The first one, Gl*f* T1, is particularly efficient in catalyzing the addition of the two first Gal*f* units on the α-L-Rha*p*-(1→3)-Glc*p*NAc-phosphate **64** lipidic acceptor. Then, Gl*f* T2 carries on the polymerization until the galactan reaches an average of 27 units.

Gl*f* T2 is a genuine processive polymerase: it sequentially transfers monomers to the growing chain without interruption.[84] This was elegantly demonstrated by competitive distraction assays: preincubated Gl*f* T2 with unlabeled substrate, followed by addition of an isotopically labeled one, gave mostly unlabeled galactan. The ratio between isotopically labeled and unlabeled biosynthesized galactan was determined by MALDI-TOF spectrometry.

Since both enzymes can transfer the monomer in β-(1→5) and β-(1→6) alternating linkages, they are both bifunctional.[82b] It must be noted that such a double function for a glycosyltransferase is very rare. The "one enzyme, one sugar linkage" motto from Hagopian and Eylar was an early dogma in glycobiology.[85] Other known examples of oligosaccharides built up on a disaccharidic repeating pattern are hyaluronic acid, chondroitin sulfate, heparan sulfate[86] and some extracellular capsules in pathogenic bacteria.[87] Those bifunctional enzymes have drawn a lot of attention in order to address the mechanism and the number of active sites used. In the case of heparan sulfate, the enzyme is thought to have two active sites due to the identification of two transferase domains in its sequence. Very often, however, the number of active sites is unknown.

Lowary *et al.* used STD NMR experiments to realize an epitope mapping and demonstrated that two trisaccharidic acceptors with different terminal linkages bind competitively to Gl*f* T2, thus strongly suggesting a single active site.[88] Pushing the investigation further, Kiessling and co-workers confirmed that only one active site is necessary for bifunctional catalysis.[89]

9.4.2 Gl*f* T Acceptor Substrates

For *in vitro* inhibition assays of Gl*f* T1 and Gl*f* T2, a synthetic acceptor mimic is needed since the natural one is not available. More than 30 different synthetic acceptors have been prepared,[7c] some representative examples

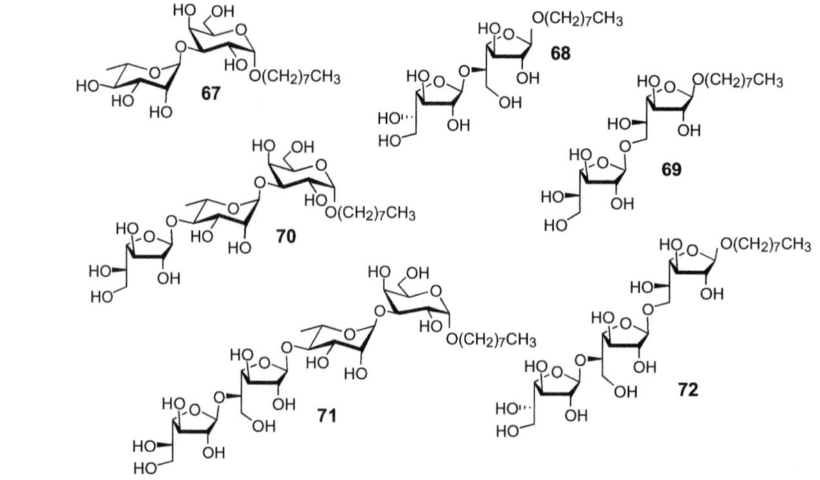

Figure 9.16 Some representative acceptors synthesized as substrates of Gl*f* T1 and Gl*f* T2.

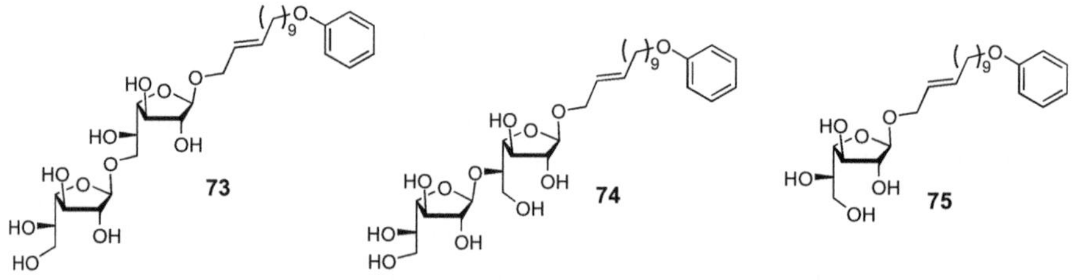

Figure 9.17 Initiator substrates of Gl*f* T2.

being depicted in Figure 9.16.[6a,82b,84,88,90] In the case of Gl*f* T2 (the most studied Gl*f* T by far), the best acceptors bear at least three Gal*f* units.

In their investigation of the Gl*f* T2 mechanism, Kiessling's group synthesized a series of initiator substrates with a ruthenium-catalyzed cross-metathesis as the penultimate transformation.[90b] It was found that the degree of polymerization highly depended on the lipid part and the saccharide. For instance, with molecules **73–75** (Figure 9.17), all displaying a terminal phenyl ring, 27, 35 and 46 Gal*f* residues could be transferred, respectively. Surprisingly, this investigations indicate that lipids displaying a single galactofuranose residue **75** can act as substrates for Gl*f* T2, suggesting a relaxed specificity for the enzyme.

9.4.3 Gl*f* T Assays with Donor Analogues

Just after the cloning of Gl*f* T2, Besra *et al.* developed an inhibition assay based on the use of an *O*-alkyl β-D-Gal*f*-(1→6)-β-D-Gal*f* neoglycolipid acceptor **69**.[6a,82c] The assays were performed in the presence of Gl*f* T2, resulting in

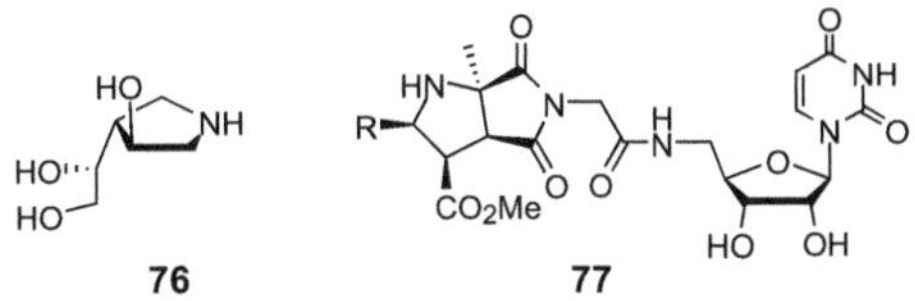

Figure 9.18 Gl*f* T2 inhibitors.

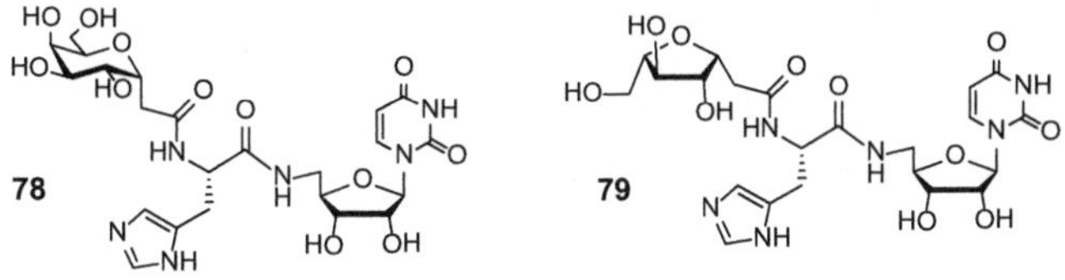

Figure 9.19 Nucleoside–sugar mimetics as Gl*f* T2 inhibitors.

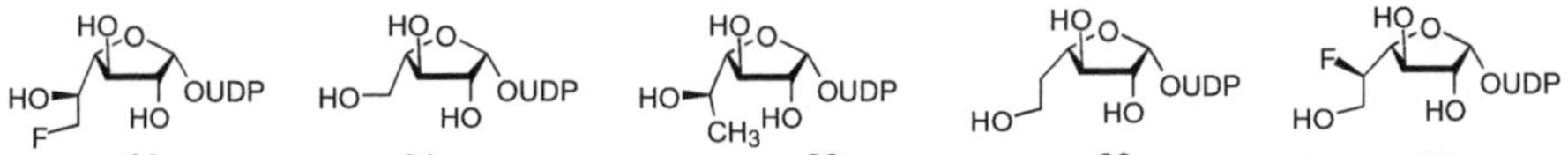

Figure 9.20 Chain terminator UDP-Gal*f* analogues.

[^{14}C]Gal*f* incorporation from UDP-[^{14}C]Gal. In collaboration with Thomas *et al.*, iminocyclitols such as **76** were identified as weak inhibitors of Gl*f* T2 (Figure 9.18).[91] Later on, a series of uridine analogues were produced and assayed against Gl*f* T2. Some molecules such as **77** displayed moderate inhibition properties.

From 2006,[90e] Lowary and co-workers implemented a continuous coupled spectrophotometric assay in 384-well plates to screen Gl*f* T2 inhibitors.[92] This assay measures the formation of UDP liberated from UDP-Gal*f* upon Gl*f* T2-mediated transfer of Gal*f* to a synthetic acceptor. Lactate dehydrogenase (LDH), coupled with pyruvate kinase (PK), oxidizes NADH as UDP is released. In collaboration with Zhou's team, analogues **78** and **79** could be identified as moderate inhibitors (Figure 9.19).[93] The IC_{50} value obtained for **78** (332 μM) suggests an affinity in the same range as the donor substrate UDP-Gal*f* **1** ($K_m = 380$ μM).[92]

In 2010, Daniellou and Mikušová published a general study in which they described the synthesis of four derivatives **80–83** of UDP-Gal*f* modified at C-5 and C-6 using a chemoenzymatic route (Figure 9.20).[94] The assay consisted in following the incorporation of Gal*f* residues from a reaction mixture containing the mycobacterial Gl*f* Ts and the radiolabeled acceptor glycolipid **64** (Figure 9.15), as well as the natural donor substrate UDP-Gal*f* **1** and/or its synthetic analogues. In these cell-free assays, compounds **80–83** prevented the formation of mycobacterial galactan, *via* the production of short "dead-end" intermediates resulting from their incorporation into the growing

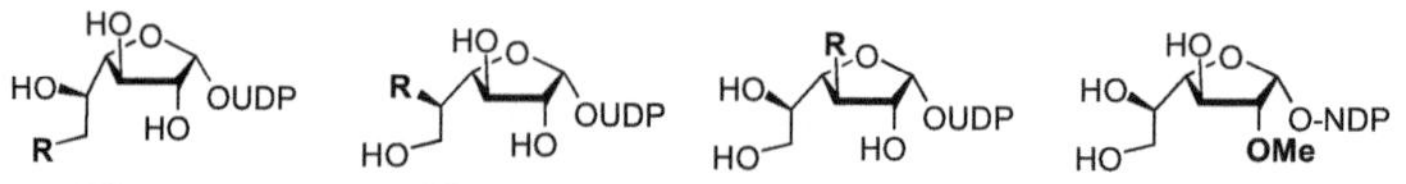

Figure 9.21 Methylated and deoxygenated UDP-Gal*f* analogues.

oligosaccharide chain. It appeared that Gl*f*T1 incorporates two residues from UDP-β-L-Araf **81**, and only one residue from the three other compounds **80**, **82** and **83**. It can thus be expected that, particularly in the case of the 5-deoxy analogue **83**, chain termination should already occur at the first Gl*f*T1 reaction.

In 2012, Kiessling *et al.* reported a novel synthesis of 6-fluoro-6-deoxy-Gal*f*-UDP **80** and 5-fluoro-5-deoxy-Gal*f*-UDP **84** (Figure 9.20), as chain termination agents and probes of the sequence specificity of Gl*f*T2.[90c] A glycolipid containing six Gal*f* units was used as an acceptor and the polymerization was followed by MALDI-TOF mass spectrometry. In all experiments, chain termination occurred just prior to the generation of a mistaken linkage, suggesting that the pattern of alternating β-(1,5) and β-(1,6) glycosidic bonds is essential to galactan biosynthesis.

In 2012, Lowary's group determined the kinetic parameters (K_m, k_{cat}) or the inhibition constants K_i of deoxygenated or methylated analogues **85–89** of UDP-Gal*f* on the C-3, C-5 and C-6 positions against Gl*f*T2 (Figure 9.21).[95] In general, the results revealed critical enzyme–substrate interactions and indicated that hydrogen bonding orients the donor to an optimal position in the binding pocket. The results obtained with all methylated analogues showed that the binding pocket is sterically crowded. The UDP-3-deoxy-Gal*f* derivative **87** had its relative activity decreased to 1%. Noteworthy, it is the best inhibitor known to date against Gl*f*T2 [$K_i = 1.2\ (\pm 0.2) \times 10^2$ μM]. This result showed that the C-3 hydroxyl group orients the donor for catalysis. The study also demonstrated that the C-6 hydroxyl group of UDP-Gal*f* is not essential for substrate activity. It could also be shown that substrate analogues deoxygenated at the 5- and 6-positions **82** and **83** (as well as with the arabino derivative **81**) lead to "dead-end" products stopping the polymerization process, thus demonstrating the strict dual activity of Gl*f*T2.

9.4.4 A Processive Galactan Polymerization Mechanism

As mentioned above, Gl*f*T2 is an unusual glycosyltransferase catalyzing a polymerization, which leads to key questions on the number and structures of active sites as well as on the way the enzyme controls the length of the polymer. Indeed, the mechanism responsible for the chain termination in processive enzymes, such as polymerase producing nucleic acids or proteins, usually relies on templates. For instance, for the protein biosynthesis, the polymerization is stopped and the peptide is released when the ribosome reaches a specific codon. On the other hand, polysaccharide synthesis occurs without the aid of a template: it is governed only by the specificity and selectivity of glycosyltransferases. The final biophysical and biochemical properties of the polysaccharides fully depend on the regioselectivity, stereoselectivity and substrate specificity of glycosyltransferases. Because they are not synthesized through template polymerization and despite the fact that polysaccharides are the most abundant organic compounds in nature, until recently it was still unknown how their length is controlled.

In 2009, Kiessling *et al.* showed that isolated recombinant Gl*f*T2 can catalyze the synthesis of polymers with degrees of polymerization that are matching with values observed in mycobacteria, suggesting that length control by Gl*f*T2 is intrinsic.[90d] Investigations using synthetic substrates revealed that Gl*f*T2 is processive. These data indicated that Gl*f*T2 may control the galactan length by using a substrate tether, distal from the catalytic site.

A model for the Gl*f*T2 chain-termination (*i.e.* length control) mechanism was developed and hypothetically extended to other carbohydrate-based polymerases.[90d] The basis would be the existence of two close binding pockets, one being the active site and the second dedicated only to bind the decaprenyl moiety of the acceptor. Although only short oligomers were obtained when the acceptor has a small lipid, if the lipid was longer, long galactans could be obtained.[90d] The authors thus postulated that the lipid part is used as a tether. Additionally, in such a tethering mechanism, unreacted acceptors compete for the lipidic subsite. Therefore higher concentrations of unelongated acceptor will increase the rate of complete dissociation, and thereby promote polymer termination. Indeed, an increase in the population of short polymers was observed at higher acceptor concentrations. Accordingly, longer polymers were obtained at low acceptor concentration.

Later on, to address whether Gl*f*T2 utilizes a processive or distributive mechanism, Kiessling *et al.* developed a mass spectrometry assay.[84] These investigations suggested that Gl*f*T2 possesses subsites for Gal*f* residue binding and that substrates that can fill these subsites to undergo efficient processive polymerization.

Interestingly, the obtention of the crystallographic structure of Gl*f*T2, described in the next section, brought novel elements to the GL*f*T2 polymerization mechanism.

9.4.5 Gl*f*T2 Structure

Two crystal structures were determined for Gl*f*T2: unliganded and in complex with UDP.[96] Protomers of a 90×50×50 Å size are assembled in C_4-symmetric homotetramers of 100×100×75 Å. The center is hollow with an open conical shape of diameter inferior to 10 Å on the N-terminal side and superior to 40 Å on the opposing side. A manganese bivalent cation is bound to the "DXD" motif: Asp256 and Asp258.

For inverting glycosyltransferases, a carboxylate group activates an hydroxyl from the acceptor.[97] As postulated by earlier reports,[88,95] Asp372 is appropriately positioned in the active site and could play the role of general base. Indeed, site directed mutagenesis resulted in no activity for the replacement by a serine. By mutating other amino acids, the authors highlighted a potential binding site for the acceptor where His296, Glu300 and Tyr344 play a key role. This site is close to Asp372, where the acceptor is activated, which is consistent with the previous reports of the same group.[95] This hydrophobic block of residues could fit both β-(1→5) and β-(1→6) linkages, but the difference in length would mechanically put the acceptor in two different positions. The longer β-(1→6) linkage disposes the terminal galactofuranose deeper in the active site, thus resulting in the activation of the 5-OH. For the β-(1→5) linkage the terminal galactofuranose is less exposed, leaving the 6-OH accessible. Each of the two positions would lead to a different activation: the terminal 5-OH or the terminal 6-OH, then producing an alternated β-(1→5);β-(1→6) oligosaccharide. This hypothesis has an intriguing consequence: it positions the growing chain into the central cavity of the homotetrameric complex.

The C face of the tetramer includes a large amount of exposed positively charged and hydrophobic residues. This face could be linked to the phospholipidic membrane, the decaprenyl-phospholipid is probably buried in the membrane and the growing polysaccharide is in the hole inside the tetramer. Compared to the model proposed by Kiessling and co-workers,[90d] the authors propose that the decaprenyl moiety of the acceptor would rather diffuse along the lipidic bilayer. The inherent steric constraint due to the size of the central hollow chamber could limit the length of the polymer to an average of 30 units.

9.5 General Conclusions

From the pioneering studies on the galactofuranose biosynthetic pathways in the 20th century, who could have imagined that this carbohydrate would lead the scientific community to the discovery of mechanistically unique enzymes and potential therapeutics against the most prevalent infectious diseases, in less than 15 years? However, many aspects of this biosynthetic pathway remain obscure, both from the enzymology and the biological functions of the Gal*f*-processing proteins. For instance, little is known on the biological functions of galactofuranosidases[36a] as well as Gal*f*-binding lectins.

Indeed, beyond the biosynthesis of naturally occurring Gal*f*-containing glycoconjugates, the synthesis of artificial galactofuranosides has seen several interesting applications at the interface of chemistry and biomedicine. For instance, oligo-Gal*f* glycoconjugates have been developed for *in vitro* aspergillosis diagnosis.[98] In an approach more related to medicinal chemistry, simple alkyl-[99] or sulfonamide-galactofuranosides[100] have been assayed as antimycobacterial agents against *Mycobacterium smegmatis*.

However, the mode of action of these antibacterial agents has not been determined and it is not clear whether the biological activities observed in both cases are dependent on the carbohydrate structure. More recently, Ferrières and Robert-Gangneux conducted an investigation on the *in vitro* antileishmanial activity of a series of alkyl glycosides against *Leishmania donovani*.[101] They could show that octyl β-D-galactofuranoside showed the most promising effects by inhibiting promastigote growth at an IC_{50} of 8.96 μM. The mechanism of action was investigated by EPR and NMR, and structural alterations were analyzed by transmission electron microscopy. Interestingly, the furanoside was active while the corresponding galactopyranose had no effect on parasite membrane fluidity or growth. It was then demonstrated that octyl β-D-galactofuranoside induced apoptosis in >90% of promastigotes.

These promising results along with the beautiful results obtained in the field of Gal*f* mechanistic enzymology clearly indicate that galactofuranose will continue to inspire the scientific community in the next decades.

Acknowledgements

Our own activities in the field of galactofuranose biosynthetic pathway were funded by an ARC grant from the Académie Louvain (A.R.C. 08/13-012) and by the FNRS (PDR T.0170.13).

References

1. J. W. Green and E. Pacsu, *J. Am. Chem. Soc.*, 1938, **60**, 2056–2057.
2. W. N. Haworth, H. Raistrick and M. Stacey, *Biochem. J.*, 1937, **31**, 640–644.
3. A. G. Trejo, G. J. Chittenden, J. G. Buchanan and J. Baddiley, *Biochem. J.*, 1970, **117**, 637–639.
4. M. Sarvas and H. Nikaido, *J. Bacteriol.*, 1971, **105**, 1063–1072.
5. P. M. Nassau, S. L. Martin, R. E. Brown, A. Weston, D. Monsey, M. R. McNeil and K. Duncan, *J. Bacteriol.*, 1996, **178**, 1047–1052.
6. (a) L. Kremer, L. G. Dover, C. Morehouse, P. Hitchin, M. Everett, H. R. Morris, A. Dell, P. J. Brennan, M. R. McNeil, C. Flaherty, K. Duncan and G. S. Besra, *J. Biol. Chem.*, 2001, **276**, 26430–26440; (b) K. Mikušová, T. Yagi, R. Stern, M. R. McNeil, G. S. Besra, D. C. Crick and P. J. Brennan, *J. Biol. Chem.*, 2000, **275**, 33890–33897.
7. (a) R. M. De Lederkremer and W. Colli, *Glycobiology*, 1995, **5**, 547–552; (b) P. Peltier, R. Euzen, R. Daniellou, C. Nugier-Chauvin and V. Ferrières, *Carbohydr. Res.*, 2008, **343**, 1897–1923; (c) M. R. Richards and T. L. Lowary, *ChemBioChem*, 2009, **10**, 1920–1938; (d) L. L. Pedersen and S. J. Turco, *Cell. Mol. Life Sci.*, 2003, **60**, 259–266.
8. D. A. Wesener, J. F. May, E. M. Huffman and L. L. Kiessling, *Biochemistry*, 2013, **52**, 4391–4398.

9. D. C. Crick, S. Mahapatra and P. J. Brennan, *Glycobiology*, 2001, **11**, 107R–118R.
10. M. J. McConville and M. A. J. Ferguson, *Biochem. J.*, 1993, **294**, 305–324.
11. R. M. de Lederkremer and R. Agusti, *Glycobiology of Trypanosoma cruzi, 62*, in *Advances in Carbohydrate Chemistry and Biochemistry*, 2009, pp. 311–366.
12. C. Marino, C. Gallo-rodriguez and R. M. Lederkremer, in *Galactofuranosyl-containing Glycans: Occurrence, Synthesis and Biochemistry*, ed. H. M. Mora-Montes, Nova Science Publishers, 2012, pp. 207-268.
13. A. Rassi and J. A. Marin-Neto, *Lancet*, 2010, **375**, 1388–1402.
14. M. V. De Arruda, W. Colli and B. Zingales, *Eur. J. Biochem.*, 1989, **182**, 413–421.
15. D. L. Sacks, G. Modi, E. Rowton, G. Späth, L. Epstein, S. J. Turco and S. M. Beverley, *Proc. Natl. Acad. Sci. U.S.A.*, 2000, **97**, 406–411.
16. J. P. Latgé, *Med. Mycol.*, 2009, **47**, S104–S109.
17. J. P. Latgé, H. Kobayashi, J. P. Debeaupuis, M. Diaquin, J. Sarfati, J. M. Wieruszeski, E. Parra, J. P. Bouchara and B. Fournet, *Infect. Immun.*, 1994, **62**, 5424–5433.
18. J. E. Bennett, A. K. Bhattacharjee and C. P. J. Glaudemans, *Mol. Immunol.*, 1985, **22**, 251–254.
19. I. Chlubnova, L. Legentil, R. Dureau, A. Pennec, M. Almendros, R. Daniellou, C. Nugier-Chauvin and V. Ferrières, *Carbohydr. Res.*, 2012, **356**, 44–61.
20. M. R. Leone, G. Lackner, A. Silipo, R. Lanzetta, A. Molinaro and C. Hertweck, *Angew. Chem., Int. Ed.*, 2010, **49**, 7476–7480.
21. C. Whitfield, *Trends Microbiol.*, 1995, **3**, 178–185.
22. M. Linnerborg, A. Weintraub and G. Widmalm, *Eur. J. Biochem.*, 1999, **266**, 460–466.
23. B. R. Clarke, D. Bronner, W. J. Keenleyside, W. B. Severn, J. C. Richards and C. Whitfield, *J. Bacteriol.*, 1995, **177**, 5411–5418.
24. (a) P. J. Brennan, *Tuberculosis*, 2003, **83**, 91–97; (b) S. Berg, D. Kaur, M. Jackson and P. J. Brennan, *Glycobiology*, 2007, **17**, 35–56R; (c) F. E. Umesiri, A. K. Sanki, J. Boucau, D. R. Ronning and S. J. Sucheck, *Med. Res. Rev.*, 2010, **30**, 290–326; (d) P. H. Tam and T. L. Lowary, *Curr. Opin. Chem. Biol.*, 2009, **13**, 618–625.
25. A. K. Mishra, N. N. Driessen, B. J. Appelmelk and G. S. Besra, *FEMS Microbiol. Rev.*, 2011, **35**, 1126–1157.
26. M. Daffe, P. J. Brennan and M. McNeil, *J. Biol. Chem.*, 1990, **265**, 6734–6743.
27. M. McNeil, M. Daffe and P. J. Brennan, *J. Biol. Chem.*, 1991, **266**, 13217–13223.
28. C. E. Barry III and K. Mdluli, *Trends Microbiol.*, 1996, **4**, 275–281.
29. C. Lamarre, R. Beau, V. Balloy, T. Fontaine, J. Wong Sak Hoi, S. Guadagnini, N. Berkova, M. Chignard, A. Beauvais and J. P. Latgé, *Cell Microbiol.*, 2009, **11**, 1612–1623.

30. P. S. Schmalhorst, S. Krappmann, W. Vervecken, M. Rohde, M. Müller, G. H. Braus, R. Contreras, A. Braun, H. Bakker and F. H. Routier, *Eukaryotic Cell*, 2008, 7, 1268–1277.
31. J. F. Novelli, K. Chaudhary, J. Canovas, J. S. Benner, C. L. Madinger, P. Kelly, J. Hodgkin and C. K. S. Carlow, *Dev. Biol.*, 2009, **335**, 340–355.
32. B. Kleczka, A. C. Lamerz, G. Van Zandbergen, A. Wenzel, R. Gerardy-Schahn, M. Wiese and F. H. Routier, *J. Biol. Chem.*, 2007, **282**, 10498–10505.
33. (a) E. Marris, *Nature*, 2006, **443**, 131; (b) A. Koul, E. Arnoult, N. Lounis, J. Guillemont and K. Andries, *Nature*, 2011, **469**, 483–490.
34. (a) L. Deng, K. Mikusová, K. G. Robuck, M. Scherman, P. J. Brennan and M. R. McNeil, *Antimicrob. Agents Chemother.*, 1995, **39**, 694–701; (b) K. Mikušová, R. A. Slayden, G. S. Besra and P. J. Brennan, *Antimicrob. Agents Chemother.*, 1995, **39**, 2484–2489.
35. (a) F. E. I. Pan, M. Jackson, Y. Ma, M. McNeil and M. M. C. Neil, *J. Bacteriol.*, 2001, **183**, 3991–3998; (b) E. C. Dykhuizen, J. F. May, A. Tongpenyai and L. L. Kiessling, *J. Am. Chem. Soc.*, 2008, **130**, 6706–6707.
36. (a) C. Marino, K. Mariño, L. Miletti, M. J. M. Alves, W. Colli and R. M. De Lederkremer, *Glycobiology*, 1998, **8**, 901–904; (b) E. Repetto, C. Marino, M. L. Uhrig and O. Varela, *Bioorg. Med. Chem.*, 2009, **17**, 2703–2711.
37. Q. Zhang and H. w. Liu, *J. Am. Chem. Soc.*, 2000, **122**, 9065–9070.
38. (a) R. Köplin, J. R. R. Brisson, C. Whitfield and C. Whitield, *J. Biol. Chem.*, 1997, **272**, 4121–4128; (b) A. Weston, R. J. Stern, R. E. Lee, P. M. Nassau, D. Monsey, S. L. Martin, M. S. Scherman, G. S. Besra, K. Duncan and M. R. McNeil, *Tubercle Lung Dis.*, 1997, **78**, 123–131.
39. (a) A. M. El-Ganiny, D. A. R. Sanders and S. G. W. Kaminskyj, *Fungal. Genet. Biol.*, 2008, **45**, 1533–1542; (b) R. A. Damveld, A. Franken, M. Arentshorst, P. J. Punt, F. M. Klis, C. A. M. J. J. Van Den Hondel and A. F. J. Ram, *Genetics*, 2008, **178**, 873–881.
40. (a) S. M. Beverley, K. L. Owens, M. Showalter, C. L. Griffith, T. L. Doering, V. C. Jones and M. R. McNeil, *Eukaryotic Cell*, 2005, **4**, 1147–1154; (b) H. Bakker, B. Kleczka, R. Gerardy-Schahn and F. H. Routier, *Biol. Chem.*, 2005, **386**, 657–661; (c) M. Oppenheimer, A. L. Valenciano and P. Sobrado, *Biochem. Biophys. Res. Commun.*, 2011, **407**, 552–556.
41. (a) M. B. Poulin, H. Nothaft, I. Hug, M. F. Feldman, C. M. Szymanski and T. L. Lowary, *J. Biol. Chem.*, 2010, **285**, 493–501; (b) M. B. Poulin, Y. Shi, C. Protsko, S. A. Dalrymple, D. A. R. Sanders, B. M. Pinto and T. L. Lowary, *ChemBioChem*, 2014, **15**, 47–56.
42. T. Konishi, T. Takeda, Y. Miyazaki, M. Ohnishi-Kameyama, T. Hayashi, M. a. O'Neill and T. Ishii, *Glycobiology*, 2007, **17**, 345–354.
43. D. A. R. Sanders, A. G. Staines, S. A. McMahon, M. R. McNeil, C. Whitfield and J. H. Naismith, *Nat. Struct. Mol. Biol.*, 2001, **8**, 858–863.

44. S. W. B. Fullerton, S. Daff, D. A. R. Sanders, W. J. Ingledew, C. Whitfield, S. K. Chapman and J. H. Naismith, *Biochemistry*, 2003, **42**, 2104–2109.
45. K. Beis, V. Srikannathasan, H. Liu, S. W. B. Fullerton, V. A. Bamford, D. A. R. Sanders, C. Whitfield, M. R. McNeil and J. H. Naismith, *J. Mol. Biol.*, 2005, **348**, 971–982.
46. J. M. Chad, K. P. Sarathy, T. D. Gruber, E. Addala, L. L. Kiessling and D. A. R. Sanders, *Biochemistry*, 2007, **46**, 6723–6732.
47. Y. Yuan, X. Wen, D. A. R. Sanders and B. M. Pinto, *Biochemistry*, 2005, **44**, 14080–14089.
48. (a) R. Dhatwalia, H. Singh, M. Oppenheimer, D. B. Karr, J. C. Nix, P. Sobrado and J. J. Tanner, *J. Biol. Chem*, 2012, **287**, 9041–9051; (b) K. E. van Straaten, F. H. Routier and D. A. R. Sanders, *J. Mol. Biol.*, 2012, **287**, 10780–10790; (c) R. Dhatwalia, H. Singh, L. M. Solano, M. Oppenheimer, R. M. Robinson, J. F. Ellerbrock, P. Sobrado and J. J. Tanner, *J. Am. Chem. Soc.*, 2012, **134**, 18132–18138.
49. (a) S. K. Partha, K. E. van Straaten and D. A. R. Sanders, *J. Mol. Biol.*, 2009, **394**, 864–877; (b) S. K. Partha, A. Sadeghi-Khomami, K. Slowski, T. Kotake, N. R. Thomas, D. L. Jakeman and D. A. R. Sanders, *J. Mol. Biol.*, 2010, **403**, 578–590.
50. R. Dhatwalia, H. Singh, M. Oppenheimer, P. Sobrado and J. J. Tanner, *Biochemistry*, 2012, **51**, 4968–4979.
51. T. D. Gruber, W. M. Westler, L. L. Kiessling and K. T. Forest, *Biochemisty*, 2009, **48**, 9171–9173.
52. T. D. Gruber, M. J. Borrok, W. M. Westler, K. T. Forest and L. L. Kiessling, *J. Mol. Biol.*, 2009, **391**, 327–340.
53. W. Huang and J. W. Gauld, *J. Phys. Chem. B*, 2012, **116**, 14040–14050.
54. W. S. Fobes and J. E. Gander, *Biochem. Biophys. Res. Commun.*, 1972, **49**, 76–83.
55. Q. Zhang and H. w. Liu, *J. Am. Chem. Soc.*, 2001, **123**, 6756–6766.
56. Y. Yuan, D. W. Bleile, X. Wen, D. a. R. Sanders, K. Itoh, H.-w. Liu and B. M. Pinto, *J. Am. Chem. Soc.*, 2008, **130**, 3157–3168.
57. I. N'Go, S. Golten, A. Ardá, J. Cañada, J. Jiménez-Barbero, B. Linclau and S. P. Vincent, *Chem. Eur. J.*, 2014, **20**, 106–112.
58. (a) A. Burton, P. Wyatt and G. J. Boons, *J. Chem. Soc., Perkin Trans. 1*, 1997, 2375–2382; (b) J. N. Barlow and J. S. Blanchard, *Carbohydr. Res.*, 2000, **328**, 473–480; (c) J. C. Errey, M. C. Mann, S. A. Fairhurst, L. Hill, M. R. McNeil, J. H. Naismith, J. M. Percy, C. Whitfield and R. A. Field, *Org. Biomol. Chem.*, 2009, 7, 1009–1016; (d) G. Eppe, P. Peltier, R. Daniellou, C. Nugier-Chauvin, V. Ferrières, S. P. Vincent and V. Ferrieres, *Bioorg. Med. Chem. Lett.*, 2009, **19**, 814–816; (e) Q. Zhang and H.-w. Liu, *Bioorg. Med. Chem. Lett.*, 2001, **11**, 145–149.
59. J. N. Barlow, M. E. Girvin and J. S. Blanchard, *J. Am. Chem. Soc.*, 1999, **121**, 6968–6969.
60. A. Caravano, D. Mengin-Lecreulx, J.-M. Brondello, S. P. Vincent and P. Sinaÿ, *Chem. Eur. J.*, 2003, **9**, 5888–5898.
61. A. Caravano and S. P. Vincent, *Eur. J. Org. Chem.*, 2009, 1771–1780.

62. A. Caravano, P. Sinaÿ and S. P. Vincent, *Bioorg. Med. Chem. Lett.*, 2006, **16**, 1123–1125.
63. (a) A. Caravano, S. P. Vincent and P. Sinaÿ, *Chem. Commun.*, 2004, **33**, 1216–1217; (b) A. Caravano, H. Dohi, P. Sinaÿ and S. P. Vincent, *Chem. Eur. J.*, 2006, **12**, 3114–3123.
64. M. Soltero-Higgin, E. E. Carlson, T. D. Gruber and L. L. Kiessling, *Nat. Struct. Mol. Biol.*, 2004, **11**, 539–543.
65. (a) M. Oppenheimer, A. L. Valenciano, K. Kizjakina, J. Qi and P. Sobrado, *PLoS One*, 2012, 7, e32918; (b) I. O. Fonseca, K. Kizjakina and P. Sobrado, *Arch. Biochem. Biophys.*, 2013, **538**, 103–110.
66. Z. Huang, Q. Zhang and H. w. Liu, *Bioorg. Chem.*, 2003, **31**, 494–502.
67. H. G. Sun, M. W. Ruszczycky, W.-C. Chang, C. J. Thibodeaux and H.-w. Liu, *J. Biol. Chem.*, 2012, **287**, 4602–4608.
68. R. Lee, D. Monsey, A. Weston, K. Duncan, C. Rithner and M. R. McNeil, *Annal. Biochem.*, 1996, **242**, 1–7.
69. (a) R. E. Lee, M. D. Smith, R. J. Nash, R. C. Griffiths, M. McNeil, R. K. Grewal, W. Yan, G. S. Besra, P. J. Brennan and G. W. J Fleet, *Tetrahedron Lett.*, 1997, **38**, 6733–6736; (b) K. Itoh, Z. Huang and H.-w. Liu, *Org. Lett.*, 2007, **9**, 879–882; (c) N. Veerapen, Y. Yuan, D. a. R. Sanders and B. M. Pinto, *Carbohydr. Res.*, 2004, **339**, 2205–2217; (d) S. Desvergnes, V. Desvergnes, O. R. Martin, K. Itoh, H.-w. Liu and S. Py, *Bioorg. Med. Chem.*, 2007, **15**, 6443–6449; (e) S. K. Partha, A. Sadeghi-Khomami, S. Cren, R. I. Robinson, S. Woodward, K. Slowski, L. Berast, B. Zheng, N. R. Thomas and D. A. R. Sanders, *Mol. Inf.*, 2011, **30**, 873–883; (f) A. Sadeghi-Khomami, A. J. Blake, C. Wilson and N. R. Thomas, *Org. Lett.*, 2005, 7, 4891–4894.
70. M. S. Scherman, K. A. Winans, R. J. Stern, V. Jones, C. R. Bertozzi and M. R. McNeil, *Antimicrob. Agents Chemother.*, 2003, **47**, 378–382.
71. M. Soltero-Higgin, E. E. Carlson, J. H. Phillips and L. L. Kiessling, *J. Am. Chem. Soc.*, 2004, **126**, 10532–10533.
72. V. Liautard, V. Desvergnes, K. Itoh, H.-w. Liu and O. R. Martin, *J. Org. Chem.*, 2008, **73**, 3103–3115.
73. W. Pan, C. Ansiaux and S. P. Vincent, *Tetrahedron Lett.*, 2007, **48**, 4353–4356.
74. C. Ansiaux, I. N'go and S. P. Vincent, *Chem. Eur. J.*, 2012, **18**, 14860–14866.
75. E. E. Carlson, M. L. Soltero-Higgin, L. L. Kiessling and J. H. Phillips, *Biochemistry*, 2003, **42**, 8630.
76. E. E. Carlson, J. F. May and L. L. Kiessling, *Chem. Biol.*, 2006, **13**, 825–837.
77. E. C. Dykhuizen and L. L. Kiessling, *Org. Lett.*, 2009, **11**, 193–196.
78. (a) G. Rauch, H. Ehammer, S. Bornemann and P. Macheroux, *Biochemistry*, 2007, **46**, 3768–3774; (b) C. J. Thibodeaux, W.-c. Chang and H.-w. Liu, *J. Am. Chem. Soc.*, 2010, **132**, 9994–9996; (c) S. Nenci, V. Piano, S. Rosati, A. Aliverti, V. Pandini, M. W. Fraaije, A. J. Heck, D. E. Edmondson and A. Mattevi, *Proc. Natl. Acad. Sci. U. S. A.*, 2012, **109**, 18791–18796.

79. C. Wing, J. C. Errey, B. Mukhopadhyay, J. S. Blanchard and R. A. Field, *Org. Biomol. Chem.*, 2006, **4**, 3945–3950.
80. Y. Komachi, S. Hatakeyama, H. Motomatsu, T. Futagami, K. Kizjakina, P. Sobrado, K. Ekino, K. Takegawa, M. Goto, Y. Nomura and T. Oka, *Mol. Microbiol.*, 2013, **90**, 1054–1073.
81. P. H. Stoco, C. Aresi, D. D. Luckemeyer, M. M. Sperandio, T. C. Sincero, M. Steindel, L. C. Miletti and E. C. Grisard, *Exp. Parasitol.*, 2012, **130**, 246–252.
82. (a) K. Mikušová, M. Belánová, J. Korduláková, K. Honda, M. R. McNeil, S. Mahapatra, D. C. Crick and P. J. Brennan, *J. Bacteriol.*, 2006, **188**, 6592–6598; (b) M. Belánová, P. Dianisková, P. J. Brennan, G. C. Completo, N. L. Rose, T. L. Lowary and K. Mikusová, *J. Bacteriol.*, 2008, **190**, 1141–1145; (c) L. J. Alderwick, L. G. Dover, N. Veerapen, S. S. Gurcha, L. Kremer, D. L. Roper, A. K. Pathak, R. C. Reynolds and G. S. Besra, *Protein Express. Purif.*, 2008, **58**, 332–341.
83. L. Cuthbertson, J. Powers and C. Whitfield, *J. Biol. Chem.*, 2005, **280**, 30310–30319.
84. M. R. Levengood, R. A. Splain and L. L. Kiessling, *J. Am. Chem. Soc.*, 2011, **133**, 12758–12766.
85. A. Hagopian and E. H. Eylar, *Arch. Biochem. Biophys.*, 1968, **128**, 422–433.
86. J. Y. Lee and A. P. Spicer, *Curr. Opin. Cell Biol.*, 2000, **12**, 581–586.
87. P. L. De Angelis, *Glycobiology*, 2002, **12**, 9R–16R.
88. (a) M. G. Szczepina, R. B. Zheng, G. C. Completo, T. L. Lowary and B. M. Pinto, *ChemBioChem*, 2009, **10**, 2052–2059; (b) M. G. Szczepina, R. B. Zheng, G. C. Completo, T. L. Lowary and B. M. Pinto, *Bioorg. Med. Chem.*, 2010, **18**, 5123–5128.
89. J. F. May, M. R. Levengood, R. A. Splain, C. D. Brown and L. L. Kiessling, *Biochemistry*, 2012, **51**, 1148–1159.
90. (a) G. C. Completo and T. L. Lowary, *J. Org. Chem.*, 2008, **73**, 4513–4525; (b) R. A. Splain and L. L. Kiessling, *Bioorg. Med. Chem.*, 2010, **18**, 3753–3759; (c) C. D. Brown, M. S. Rusek and L. L. Kiessling, *J. Am. Chem. Soc.*, 2012, **134**, 6552–6555; (d) J. F. May, R. A. Splain, C. Brotschi and L. L. Kiessling, *Proc. Natl. Acad. Sci. U. S. A.*, 2009, **106**, 11851–11856; (e) N. L. Rose, G. C. Completo, S.-J. Lin, M. McNeil, M. M. Palcic and T. L. Lowary, *J. Am. Chem. Soc.*, 2006, **128**, 6721–6729; (f) A. K. Pathak, V. Pathak, L. Seitz, J. A. Maddry, S. S. Gurcha, G. S. Besra, W. J. Suling and R. C. Reynolds, *Bioorg. Med. Chem.*, 2001, **9**, 3129–3143; (g) A. K. Pathak, V. Pathak, W. J. Suling, S. S. Gurcha, C. B. Morehouse, G. S. Besra, J. A. Maddry and R. C. Reynolds, *Bioorg. Med. Chem.*, 2002, **10**, 923–928; (h) A. K. Pathak, V. Pathak, L. Seitz, S. S. Gurcha, G. S. Besra, J. M. Riordan and R. C. Reynolds, *Bioorg. Med. Chem.*, 2007, **15**, 5629–5650.
91. S. Cren, A. J. Blake, G. S. Besra, N. R. Thomas and S. S. Gurcha, *Org. Biomol. Chem.*, 2004, **2**, 2242–2418.
92. N. L. Rose, R. B. Zheng, J. Pearcey, R. Zhou, G. C. Completo and T. L. Lowary, *Carbohydr. Res.*, 2008, **343**, 2130–2139.

93. K. Vembaiyan, J. A. Pearcey, M. Bhasin, T. L. Lowary and W. Zou, *Bioorg. Med. Chem.*, 2011, **19**, 58–66.
94. P. Peltier, M. Beláňová, P. Dianišková, R. Zhou, R. B. Zheng, J. A. Pearcey, M. Joe, P. J. Brennan, C. Nugier-Chauvin, V. Ferrières, T. L. Lowary, R. Daniellou and K. Mikušová, *Chem. Biol.*, 2010, **17**, 1356–1366.
95. M. B. Poulin, R. Zhou and T. L. Lowary, *Org. Biomol. Chem.*, 2012, **10**, 4074–4087.
96. R. W. Wheatley, R. B. Zheng, M. R. Richards, T. L. Lowary and K. K. S. Ng, *J. Biol. Chem.*, 2012, **287**, 28132–28143.
97. L. L. Lairson, B. Henrissat, G. J. Davies and S. G. Withers, *Annu. Rev. Biochem.*, 2008, 77, 521–555.
98. L. Cattiaux, B. Sendid, M. Collot, E. Machez, D. Poulain and J. M. Mallet, *Bioorg. Med. Chem.*, 2011, **19**, 547–555.
99. L. Legentil, J. L. Audic, R. Daniellou, C. Nugier-Chauvin and V. Ferrières, *Carbohydr. Res.*, 2011, **346**, 1541–1545.
100. D. J. Owen, C. B. Davis, R. D. Hartnell, P. D. Madge, R. J. Thomson, A. K. J. Chong, R. L. Coppel and M. von Itzstein, *Bioorg. Med. Chem. Lett.*, 2007, **17**, 2274–2277.
101. M. Suleman, J. P. Gangneux, L. Legentil, S. Belaz, Y. Cabezas, C. Manuel, R. Dureau, O. Sergent, A. Burel, F. Daligault, V. Ferrières and F. Robert-Gangneux, *Antimicrob. Agents Chemother.*, 2014, **58**, 2156–2166.

CHAPTER 10

Carbohydrate-containing Matrix Metalloproteinase Inhibitors (MMPIs): A Second Childhood for Sulfonamidic Inhibitors?

CRISTINA NATIVI,*[a] BARBARA RICHICHI[a] AND STEFANO ROELENS[b]

[a] Dipartimento di Chimica, Università di Firenze, via della Lastruccia 13, Sesto F.no (Fi) 50019, Italy; [b] CNR, Istituto di Metodologie Chimiche (IMC), Dipartimento di Chimica, via della Lastruccia 13, Sesto F.no (Fi) 50019, Italy
*Email: cristina.nativi@unifi.it

10.1 Matrix Metalloproteinases

Matrix metalloproteinases (MMPs) are a widespread family of metalloenzymes deputed to exert fundamental physiological roles, which are also responsible for important diseases. These calcium-dependent zinc-containing peptidases, also known as matrixins, are responsible for tissue remodeling, wound healing, and extra cellular matrix (ECM) degradation.[1] Under normal physiological conditions, MMPs are only minimally expressed, for their expression is strictly regulated by endogenous processes and inhibitors.[2] An imbalance of these activities results in over-expression of MMPs which, in

RSC Drug Discovery Series No. 43
Carbohydrates in Drug Design and Discovery
Edited by Jesús Jiménez-Barbero, F. Javier Cañada and Sonsoles Martín-Santamaría

Published by the Royal Society of Chemistry, www.rsc.org

turn, leads to a variety of pathological disorders, including atherosclerosis, thrombosis, heart failure, and cancer.[3]

The first vertebrate MMP discovered, collagenase, was isolated and characterized in 1962, but it was only several years later, in 1985, that significant advances in the field occurred with the identification of many new members of this family of endopeptidases. At present, the mammalian MMP family includes up to 23 enzymes, the majority of which are constituted by a pro-domain, deputed to hydrolytic activity, and a hemopexin-like domain, which likely plays a role in substrate recognition.

10.2 Matrix Metalloproteinase Inhibitors: Structural Basis

Owing to their undisputed involvement in many pathologies, MMPs have been recognized as useful pharmaceutical targets,[4] so that a great deal of effort has been directed to design and develop valuable inhibitors of MMPs. A huge number of inhibitors have been reported, some of which showed affinities in the low nanomolar range, but up to now all drug candidates have failed in clinical trials. Indeed, low bioavailabilities, limited solubilities in physiological media, and low selectivities are the main reasons for failure of the efforts dedicated to synthetic MMP inhibitors (MMPIs).

The low selectivities of MMPIs endowed with very high affinities (nanomolar or sub-nanomolar) *versus* several MMPs have been matter of debate. As a matter of fact, it became clear over the years that MMPs present a common property within the family, that is, a remarkable ability of adapting their binding pocket to the inhibitor's shape,[5] which determines the observed lack of selectivity of inhibitors toward different MMPs.

The three-dimensional structures of enzyme–inhibitor adducts obtained for several ligands with various MMPs clearly show that the vast majority of inhibitors share similar features,[6] such as the ability to bind to the Zn ion, to fit into the hydrophobic pocked, named S_1', and to bind to the peptidic skeleton of the enzyme. On the other hand, approaches combining high-resolution X-ray structures and NMR data confirmed the adaptivity and the conformational heterogenicity of some loop regions in several MMPs (Figure 10.1).[7]

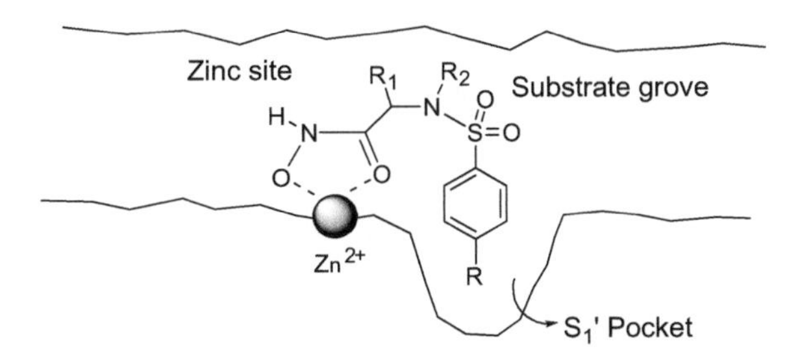

Figure 10.1 Schematic representation of the active site of MMPs. (Bertini and co-workers[7] are acknowledged.)

Small molecules (synthetic and natural), as well as macromolecular endogenous inhibitors, have been evaluated as potential drugs to treat the overexpression of MMPs.[8] In the 1990s the development of synthetic inhibitors met with great success in terms of new molecules proposed as drug candidates. The main features shared by the large majority of them were: (a) a functional group able to chelate the active-site Zn(II) ion (named zinc binding group, ZBG); (b) one side chain providing van der Waals interactions with the S_1' pocket; and (c) one or more functional groups to form hydrogen bonding interactions with the enzyme backbone. Among all, it is worth mentioning those belonging to the class of succinyl hydroxamates, like Batimastat[8b] and Marimastat,[8a] and those belonging to the class of sulfonamide hydroxamates, such as NNGH {*N*-isobutyl-*N*-[(4-methoxyphenylsulfonyl)glycyl]hydroxamic acid}[9] and CGS27023A,[10] which have been considered lead structures for at least two decades (Figure 10.2).

Batimastat, Marimastat, and CGS25966 entered in clinical trials but their low bioavailability and scarce selectivity prevented any realistic development as drugs for tumor treatment. In order to overcome these serious drawbacks, new drug discovery techniques, like X-ray analysis combined with NMR (see above), or multidimensional NMR spectroscopy techniques known as "SAR by NMR", allowed for the identification of potent MMPIs on the basis of NMR-derived structural information.[11]

10.2.1 Water-soluble MMPIs

In 1995, Bertini and co-workers[7] reported on the X-ray structure of the catalytic domain of MMP-12 in the presence of NNGH. The crystallographic

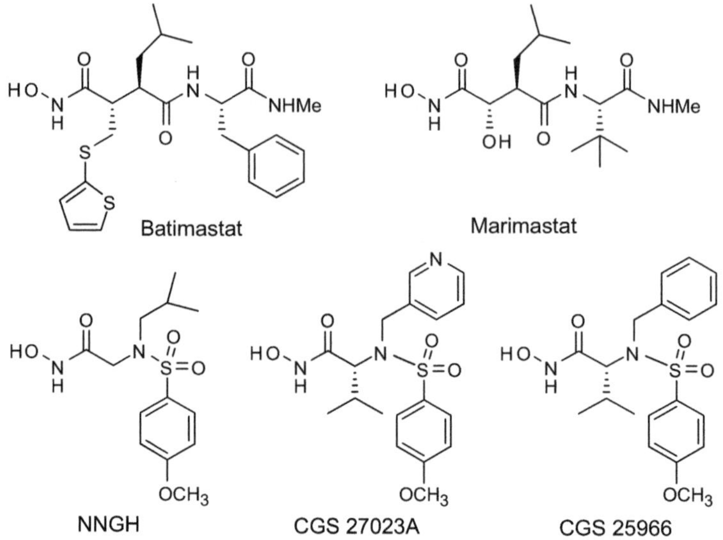

Batimastat Marimastat

NNGH CGS 27023A CGS 25966

Figure 10.2 Structures of Batimastat, Marimastat, NNGH, CGS27023A, and CSG25966.

structure confirmed the interaction of the hydroxamate moiety of the inhibitor with the catalytic zinc ion of the active site of the enzyme, as well as the binding interaction of the aromatic group of NNGH with the $S_1{}'$ subsite of the MMP-12. In addition, it was ascertained that the isopropyl group on the sulfonamide nitrogen atom points away from a second shallow pocked (named $S_2{}'$) and, notably, does not directly participate in binding.

Although the presence of a hydrophobic residue was considered a requirement in the design of high affinity MMPIs, in 2007 a new family of NNGH-related MMPIs featuring a hydrophilic residue was obtained with the aim of overcoming the limitations of NNGH and NNGH-based inhibitors.[12] These molecules, characterized by the replacement of the isopropyl moiety with polar groups of increasing complexity and polarity, although soluble in water, still showed affinities *versus* MMPs in the nanomolar range (Table 10.1).

According to the leading concept followed in designing inhibitors **1a–1h**, a residue of β-glucose was also attached to the sulfonamide nitrogen through a two-carbon spacer.[13] The synthesis of inhibitor 2 is depicted in Scheme 10.1.

The glycosyl inhibitor **2** featured higher water solubility (>30 mM) compared to NNGH, together with a remarkable affinity towards several MMPs (Table 10.2).[13]

Particularly noteworthy were the inhibition constants K_i of **2** towards MMP-7 and MMP-13, both of which are smaller than those reported for NNGH toward the same enzymes.

Direct evidence for the lack of significant participation in binding of the glucose ring came from the X-ray crystal structure of the MMP-12 · **2** complex (Figure 10.3).[13]

Table 10.1 MMP inhibition constants (K_i, nM) of compounds **1a–1h**. (Attolino and co-workers[12] are acknowledged.)

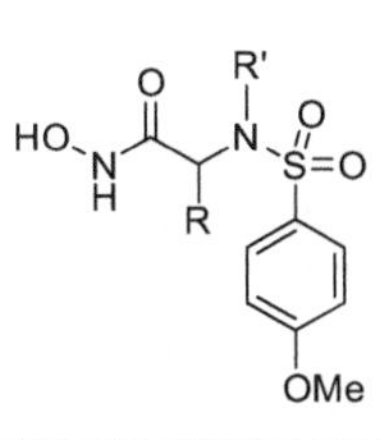

1a R = H, R' = $(CH_2)_2$-OH
1b R = H, R' = $CH_2(CHOH)CH_2OH$
1c R = CH_2OH, R' = $CH_2(CHOH)CH_2OH$
1d R = CH_2OH, R' = H
1e R = $CH(CH_3)OH$, R' = H
1f R = H, R' = $(CH_2)_2$-OH
1g R = H, R' = $(CH_2)_2$-OH
1h R = H, R' = $CH_2(CHOH)CH_2OH$

Compound	MMP-1	MMP-7	MMP-8	MMP-12	MMP-13
NNGH	174	13000	9.0	4.3	31
1a	128	1500	13	7.6	1.0
1b	160	9000	24	8.0	3.6
1c	32	344	8.0	7.0	1.7
1d	143	823	9.0	6.0	2.2
1e	287	984	10.0	11.0	2.9
1f	151	3000	173	40	46
1g	67	1600	4.6	31	2.7
1h	742	4000	48	60	18

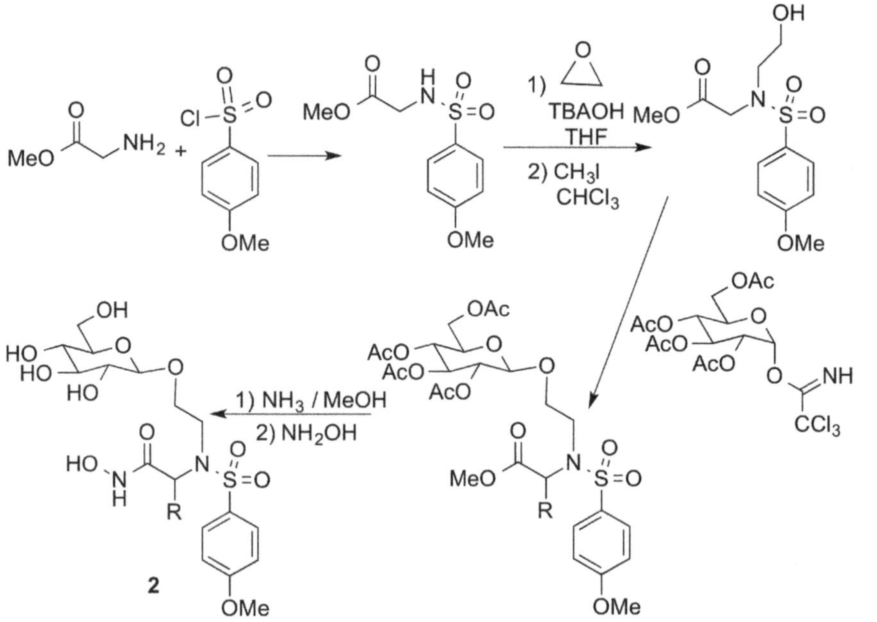

Scheme 10.1 Synthesis of inhibitor **2**.
(Calderone and co-workers[13] are acknowledged.)

Table 10.2 Inhibition constants (K_i, nM) of compound **2** towards several MMPs.[a]

	MMP-1	MMP-7	MMP-8	MMP-12	MMP-13
NNGH	174	13000	9.0	4.3	3.1
Inhibitor 2	286	2000	9.1	14.3	1.7

[a] K_i measured *in vitro* by fluorimetric assays.

In analogy to the complex of NNGH,[7] the MMP-1·**2** complex showed the hydroxamate moiety chelated to the penta-coordinated zinc and the *para*-substituted benzene ring nested inside the S_1' pocket. The glucose portion protruded out of the protein towards the solvent region. The presence of a short linker proved to be essential, as binding was nearly depleted when the glucose residue was directly linked to the sulfonamidic scaffold.[13]

Although the interaction between the glucose residue and the protein was only marginal, it appeared to play a beneficial role in terms of selectivity. As a matter of fact, compound **2** showed improved selectivity towards MMP-13, which was inhibited more effectively than MMP-8 (over fivefold) and MMP-12 (almost tenfold). In addition, it was demonstrated that inhibitor **2** does not appreciably interact with human serum albumin (HSA), whereas NNGH is strongly bound under the same experimental conditions. Overall, the glucose moiety discouraged binding to HSA but not to MMPs.

Other examples of water-soluble MMPIs have been reported in the literature;[14] one of them (compound **A**) presented a glucose moiety[15]

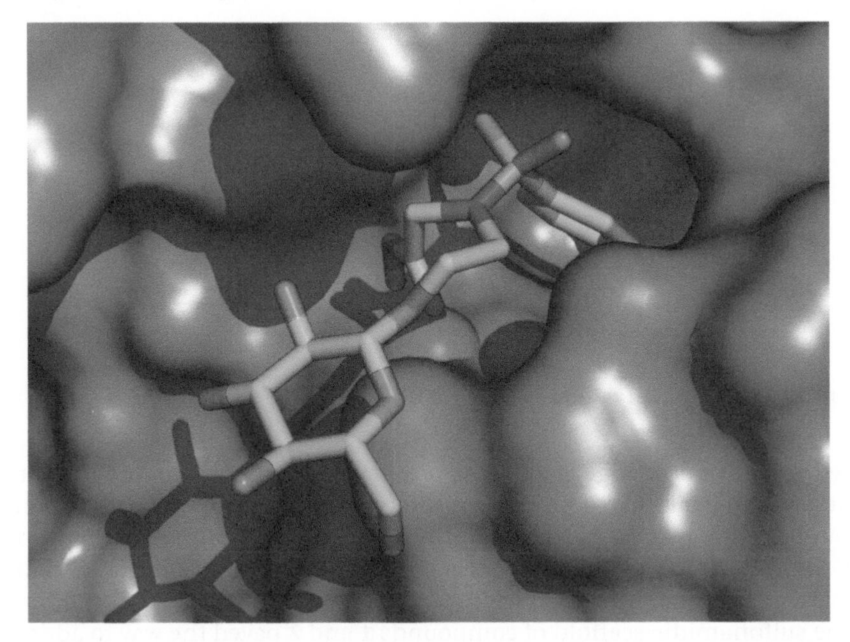

Figure 10.3 X-ray crystal structure of the MMP-12 · **2** complex (only the catalytic site of MMP-12 is showed for clarity).
(Calderone and co-authors[13] are acknowledged.)

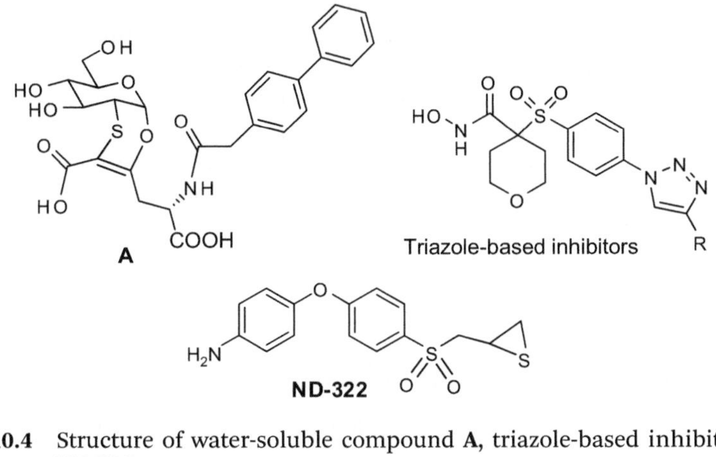

A

Triazole-based inhibitors

ND-322

Figure 10.4 Structure of water-soluble compound **A**, triazole-based inhibitors, and ND-322.

analogous to inhibitor **2**, but none was structurally related to NNGH (Figure 10.4).

10.3 Dual-specific Targeting and Therapeutic MMPIs

Along with their low solubility in water, the disappointing results reported for MMPIs as therapeutics can also be ascribed to their lack of selectivity.

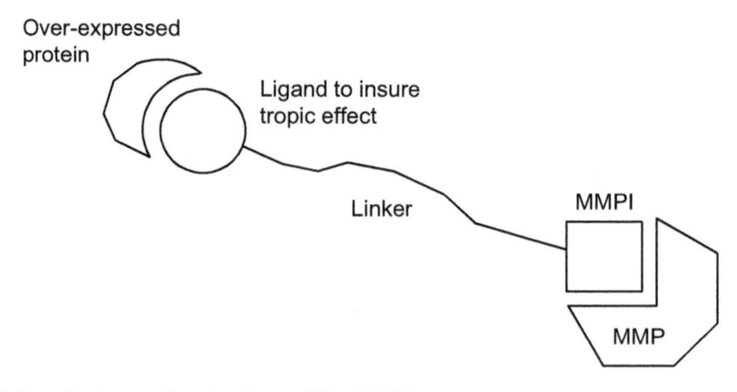

Figure 10.5 Cartoon of a dual-specific MMPI.

Based on the knowledge gained in recent years on the enzyme/inhibitor structure and capitalizing on the improvement obtained with inhibitors **1a–1h** and **2** with respect to the majority of reported MMPIs,[16] it appeared evident that there was the need for a change of perspective in the development of a new generation of efficient MMPIs. The synthetic flexibility offered by the sulfonamidic scaffold of compounds **1** and **2** paved the way to address a major issue in targeting MMPs' activity regulation, namely, interfering with MMPs at the sites of disease progression, that is, delivering the inhibitor at the specific site where the MMPs are over-expressed.[17] This could be achieved by installing on the sulfonamidic nitrogen a hydrophilic epitope recognized by specific lectins, *e.g.* galectins, to preferentially deliver the inhibitor conjugate to or near tumors, thus increasing therapeutic efficiency and reducing the extent of side effects. Following this strategy, a dual-specific targeting molecule able to probe tumor cells or stroma MMPs over-expression was obtained,[18] which used a hydrophilic residue not only to improve solubility, but also to direct the inhibitor conjugate towards MMPs over-expressed by tissues, by binding to a specific protein which is co-expressed (or co-over-expressed) with MMPs in the tumoral region (Figure 10.5).

10.3.1 Tumor Progression and Extracellular Galectin-3

A tumor is an extremely complex pathological status. A great deal of evidence now indicates that cancer cells and their microenvironment deeply influence each other. The microenvironment allows tumor cells to proliferate and spread, and it also activates the signaling cascade that regulates processes like tumor cell invasion, adhesion, and angiogenesis. Lectins are an example of cell adhesion proteins which specifically bind to carbohydrate structures playing a main role in cell recognition. Galectin-3 (Gal3) is a member of the lectin family which regulates many biological functions and signaling pathways in normal and cancer cells.[19] Gal3 is expressed in a variety of tumors and its density and localization depends on tumor aggressiveness and metastatic potentials. As it happens for MMPs, Gal3 over-expression correlates with tumor progression and with the decreased survival of patients.[20]

Extracellular Gal3 was found to be involved in cell-matrix adhesion. More recently, stimulation of MMPs' production and its cleavage by MMPs have been reported. It is noteworthy that cells presenting Gal3 resistant to MMP cleavage showed a lower attitude to promote tumor progression with respect to cleavage-sensitive Gal3. The analysis of Gal3 cleavage in cancer has also been proposed to be a useful tool for validation of the effectiveness of MMP inhibitor therapy.[19]

The presence of Gal3 in the tumor invasion zone and the role of tumor progression by up-regulating MMP production reveal a clear functional connection between these two proteins and highlights their potential for the effective delivering of an inhibitor *in situ*, with the extra benefit of blocking clinically unfavorable galectin activities.[21]

Taking into account that these adhesion/growth regulatory lectins present a marked affinity to β-galactosides, the introduction of carbohydrates, such as galactose or lactose, as hydrophilic epitopes in sulfonamidic-based inhibitors appeared to be an effective strategy to achieve the first example of dual-specific targeting and therapeutic MMPIs.

10.3.2 Synthesis of Bifunctional MMPIs for Presentation by Gal3

Gal3 is a member of the galectins, a family of soluble carbohydrate-binding proteins. They are characterized by conserved peptide sequences in their carbohydrate recognition domains (CRD)[22] which selectively bind to β-galactoside-containing glycoconjugates. Gal3 is composed of an N-terminal, non-lectin domain, and a C-terminal CRD domain, both of which are essential for its role in signal transduction and cellular adhesion.[23] Proteolytic cleavage of Gal3 results in the removal of the N-terminal domain, preventing Gal3 from exerting its functions without affecting/altering its affinity for native ligands. Thus, both full-length and truncated forms of Gal3, as well as other tumor-associated galectins, preserve their binding properties *versus* β-galactosides.

In this context, the synthesis of the bifunctional MMPI **3** was reported,[18] which is structurally related to compound **2** but features a lactose portion strategically inserted to bind to Gal3 (Figure 10.6).

The new molecule represents the first example of bifunctional architecture, which combines MMP inhibition properties with galectin-targeting and -blocking abilities, features that can be ascribed to the β-galactoside substituent. The strong, though yet poorly understood, correlation between MMPs and galectins in tumor progression makes Gal3 and the inhibitor **3** ideal tools for testing (see Figure 10.6).

Sulfonamide **3** presents a hydroxamic acid (as zinc binding group) linked to lactose through a 16-membered spacer characterized by two amidic

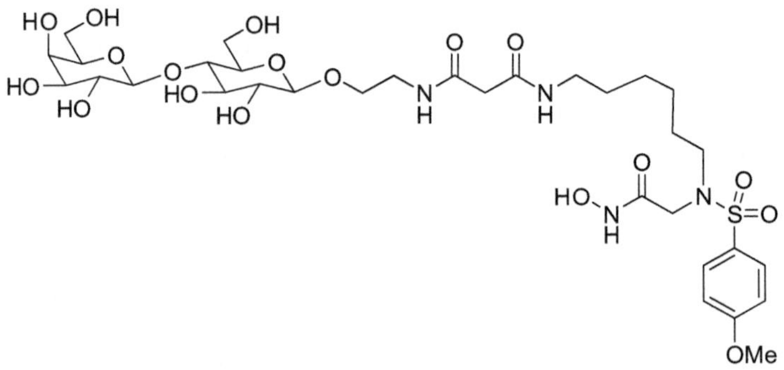

Figure 10.6 Structure of inhibitor 3.

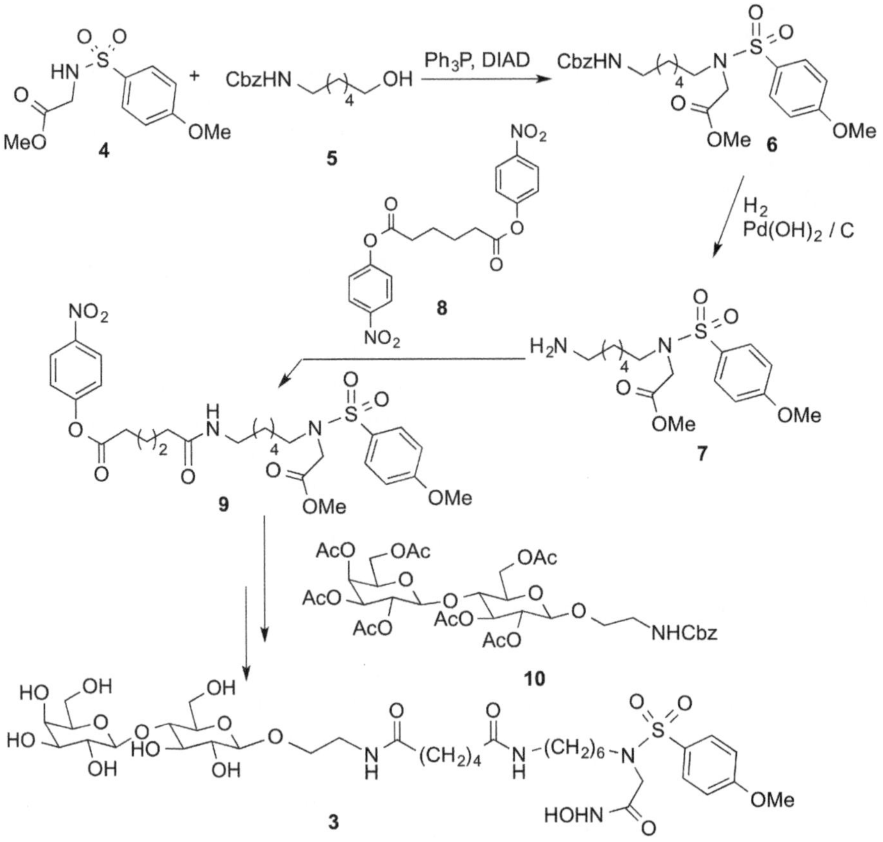

Scheme 10.2 Synthesis of inhibitor 3.

linkages. In the synthesis of compound **3**, shown in Scheme 10.2, the sulfonamide of the glycine methyl ester **4** is reacted with the benzyloxycarbonyl (Cbz)-protected aminohexanol **5** under Mitsunobu conditions to form the methyl ester **6**.[13] The elongation of the linker was obtained by treating the unprotected amino derivative **7** with the adipic ester **8**[24] and *N*-methylmorpholine (NMM) to give **9** which, in turn, was reacted with the glycosyl donor **10**. Protecting group removal afforded the divalent inhibitor **3** (Scheme 10.2).

The inhibitory potency of compound **3**, tested on a set of MMPs (*i.e.* MMP-1, -7, -9, -12, and -13), confirmed that functionalization with lactose did not impair the inhibition efficacy of the ligand. As a matter of fact, with only two exceptions (MMP-1 and MMP-7), compound **3** showed K_i values in the low nanomolar range for all the MMPs tested. The ability of **3** to bind to MMP and Gal3 simultaneously was demonstrated through STD and trNOESY NMR experiments and by HSQC NMR titrations, making use of ^{15}N-labeled proteins. The catalytic domain of MMP-12 (catMMP-12) and the CRD of chicken Gal3 (CG3) were used in these studies.[18]

The CG3↔**3** binary complex proved that molecular recognition involves CG3 and the lactose residue of **3**. Notably, ligand **3** binds to CG3 at the same site as the native ligand (lactose) does. Studies carried out on the catMMP-12↔**3** binary complex showed that the binding mode of **3** is analogous to that previously described for other MMPIs structurally related to NNGH (see above) and indicate a strong binding between the two partners.

The ternary complex CG3↔**3**↔catMMP-12 was also investigated. The bivalency of **3** was evaluated through HSQC spectra of the ^{15}N-labeled catalytic domain of MMP-12 in the presence of an equimolar amount of **3**. Under these conditions, MMP-12 is fully saturated. Upon addition of CG3, a dramatic decrease was observed in the intensity of the cross-peaks corresponding to the ^{15}N-labeled MMP-12 amino acids residues located in the region of the hydroxamate binding site and in the adjacent β-strand, while the intensity of the peaks corresponding to the amino acids in the α-helix A remained unaffected (Figure 10.7).[18]

This behavior strongly suggests that a large entity, that is, a complex with the lectin, was formed in solution. Altogether, these data indicated the formation of the ternary complex and demonstrated that ligand **3** is able to gather CG3 and MMP-12 side-by-side.

10.4 Conclusions

In conclusion, MMPIs structurally related to the golden standard NNGH, but featuring a more favorable profile in terms of bioavailability and solubility, can be strategically functionalized to target specific proteins co-overexpressed with MMPs in cancers. These dual-target molecules might represent a new approach in designing MMPIs to overcome the limitation of their scarce selectivity and likely represent a new way to address the harmful effects of uncontrolled MMP production.

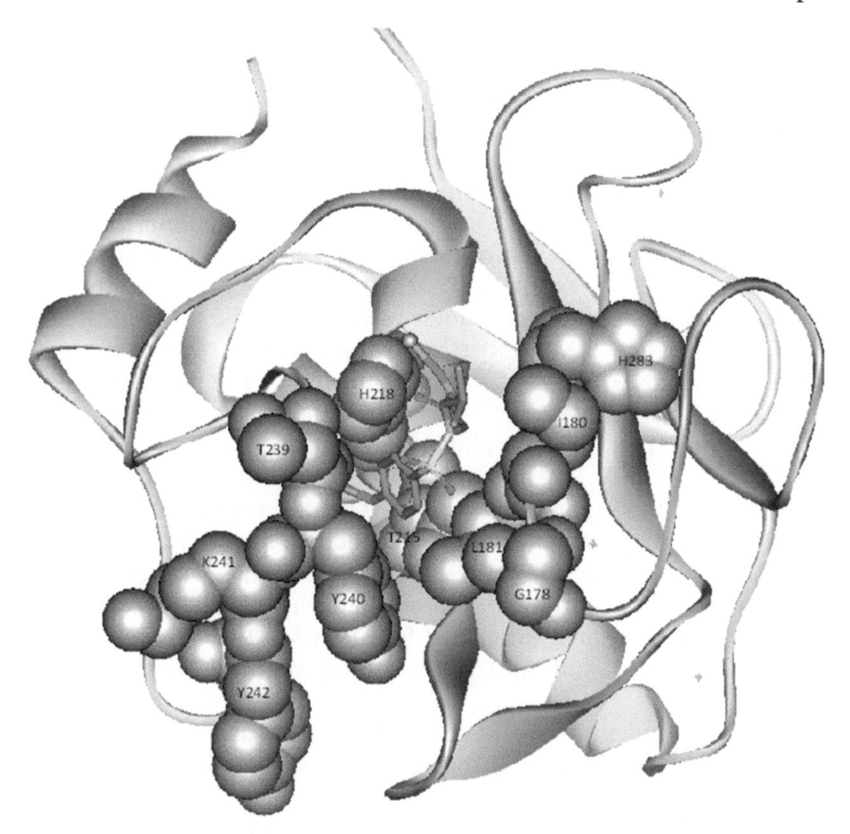

Figure 10.7 X-Ray structure of the catalytic domain of catMMP12 complexed with NNGH (pdb: 1Z3J). The residues whose backbone NH/Ha crosspeak shift the most upon ligand 3 binding are highlighted in *gray* CPK representation.
(Bartoloni and co-authors[18] are acknowledged.)

Acknowledgements

The authors are grateful to Prof. H.J. Gabius (University of München) for fruitful discussions on galectin 3, and to Prof. J. Jimenez-Barbero (CSIC, Madrid), Prof. C. Luchinat (University of Florence), and Dr M. Fragai (University of Florence) for long lasting scientific support and friendship.

References

1. A. Page-McCaw, A. J. Ewald and Z. Werb, *Nat. Rev. Mol. Cell Biol.*, 2007, **8**, 221.
2. (a) V. Knauper, H. Will, C. Lopez-Otis, B. Smith, S. J. Atkinson, H. Stanton, R. M. Hembry and G. Murphy, *J. Biol. Chem.*, 1996, **271**, 17124; (b) C. M. Overall, E. Tam, G. A. McQuibban, C. Morrison,

U. M. Wallon, H. F. Bigg, A. E. King and C. R. Roberts, *J. Biol. Chem.*, 2000, **275**, 39497.

3. (a) B. Fingleton, *Curr. Pharm. Des.*, 2007, **13**, 333; (b) R. P. Verma and C. Hansch, *Biorg. Med. Chem.*, 2007, **15**, 2223.
4. (a) M. Whittaker, C. D. Floyd, P. Brown and A. J. H. Gearing, *Chem. Rev.*, 1999, **99**, 2735; (b) J. W. Skiles, N. C. Gonnella and A. Y. Jeng, *Curr. Med. Chem.*, 2004, **11**, 2911.
5. F. J. Moy, *J. Am. Chem. Soc.*, 2002, **124**, 12658.
6. (a) B. Lovioy, A. R. Welch, S. Carr, C. Luoung, C. Broka, R. T. Hendricks, J. A. Campbell, K. A. M. Walker, R. Martin, H. Van Wart and M. F. Browner, *Nat. Struct. Biol.*, 1999, **6**, 217; (b) R. Lang, A. Kocourek, M. Braun, H. Tschesche, R. Huber, W. Bode and K. Maskos, *J. Mol. Biol.*, 2001, **312**, 731.
7. I. Bertini, V. Calderone, M. Cosenza, M. Fragai, M. Lee, C. Luchinat, S. Mangani, B. Terni and P. Turano, *Proc. Natl. Acad. Sci. U. S. A.*, 2005, **102**, 5334.
8. (a) R. P. Beckett, A. H. Davidson, A. H. Drummond, P. Huxley and M. Whittaker, *Drug Discovery Today*, 1996, **1**, 16; (b) J. Ngo, A. Graul and J. Castaner, *Drugs Future*, 1996, **21**, 1215; (c) A. J. P. Dockerty, A. Lyons, B. J. Smith, E. M. Wright, P. E. Stephens and T. J. R. Harrys, *Nature*, 1985, **318**, 66; (d) W. G. Stetler-Stevenson, H. C. Krutzsch and L. A. Liotta, *J. Biol. Chem.*, 1989, **264**, 17374.
9. A. Y. Jeng, M. Chou and D. T. Parker, *Biorg. Med. Chem.*, 1998, **8**, 897.
10. L. J. MacPherson, E. K. Bayburt, *et al.*, *J. Med. Chem.*, 1997, **40**, 2525.
11. P. J. Hajduk and J. Greer, *Nat. Rev. Drug Discovery*, 2007, **6**, 211.
12. (a) Attolino, V. Calderone, E. Dragoni, M. Fragai, B. Richichi, C. Luchinat and C. Nativi, *Eur. J. Med. Chem.*, 2010, **45**, 5919; (b) I. Bertini, V. Calderone, M. Fragai, A. Giachetti, M. Loconte, C. Luchinat, M. Maletta, C. Nativi and K.-J. Yeo, *J. Am. Chem. Soc.*, 2007, **129**, 2466.
13. V. Calderone, M. Fragai, C. Luchinat, C. Nativi, B. Richichi and S. Roelens, *ChemMedChem*, 2006, **1**, 598.
14. (a) B. Fabre, K. Filipiak, J. M. Zapico, N. Díaz, R. J. Cabajo, A. K. Schott, M. P. Martínez-Alcázar, D. Suárez, A. Pineda-Lucena, A. Ramos and B. de Pascual-Teresa, *Org. Biomol. Chem.*, 2013, **11**, 6623; (b) M. Gooyit, M. Lee, V. A. Schroeder, M. Ikejiri, M. A. Suckow, S. Mobashery and M. Chang, *J. Med. Chem.*, 2011, **54**, 6676.
15. M. Fragai, C. Nativi, B. Richichi and C. Venturi, *ChemBioChem*, 2005, **6**, 1345.
16. (a) M. Egeblad and Z. Werb, *Nat. Rev. Cancer*, 2002, **2**, 161; (b) W. C. Parks, C. L. Wilson and Y. S. Lopez-Boado, *Nat. Rev. Immunol.*, 2004, **4**, 617.
17. S. H. Park, H. S. Min, B. Min, J. Myung and S. H. Paek, *Neuropathology*, 2008, **28**, 497.
18. M. Bartoloni, B. E. Domínguez, E. Dragoni, B. Richichi, M. Fragai, S. André, H. J. Gabius, A. Ardá, C. Luchinat, J. Jimenéz-Barbero and C. Nativi, *Chem. – Eur. J.*, 2013, **19**, 1896.

19. P. Nangia-Makker, V. Bacan and A. Raz, *Cancer Microenviron.*, 2008, **1**, 43.
20. E. C. Lee, H. J. Woo, C. A. Korzelius, G. D. Steele and A. M. Mercurio Jr, *Arch. Surg.*, 1991, **126**, 1498.
21. (a) S. Rorive, N. Belot, C. Decaestecker, F. Lefranc, L. Gordower, S. Micik, C. A. Maurage, H. Kaltner, M. M. Ruchoux, A. Danguy, H. J. Gabius, I. Salmon, R. Kiss and I. Camby, *GLIA*, 2001, **33**, 241; (b) M. Demers, T. Magnaldo and Y. St-Pierre, *Cancer Res.*, 2005, **65**, 5205; (c) O. Roda, E. Ortiz-Zapater, N. Martinez-Bosch, R. Gutiérrez-Gallego, M. Vila-Perelló, C. Ampurdanés, H. J. Gabius, S. André, D. Andreu, F. X. Real and P. Navarro, *Gastroenterology*, 2009, **136**, 1379.
22. S. H. Barondes, D. N. Cooper, M. A. Gitt and H. Leffler, *J. Biol. Chem.*, 1994, **269**, 20807.
23. J. Nieminen, A. Kuno, J. Hirabayashi and S. Sato, *J. Biol. Chem.*, 2007, **282**, 1374.
24. X. Wu, C.-C. Ling and D. R. Bundle, *Org. Lett.*, 2004, 4407.

CHAPTER 11

Amphiphilic Aminoglycoside Antimicrobials in Antibacterial Discovery

BALA KISHAN GORITYALA, GOUTAM GUCHHAIT AND FRANK SCHWEIZER*

Department of Chemistry, University of Manitoba, Winnipeg, MB R3T 2N2, Canada
*Email: schweize@cc.umanitoba.ca

11.1 Introduction

During the middle of the 20th century, mankind witnessed the therapeutic efficacy of streptomycin, whose isolation was a landmark in the antibiotic era.[1] It was the first aminoglycoside to be characterized and proved to be the most effective antibiotic candidate against tuberculosis (TB), which has caused high human mortality over the past few centuries.[2] Ever since, diverse natural and semi-synthetic aminoglycosides (AGs) have been clinically deployed to alleviate fatal epidemic bacterial infectious diseases.[3] AGs constitute a large family of polycationic drugs used in the treatment of bacterial infections (Figure 11.1). AGs are natural products from soil actinomycetes, a group of Gram-positive bacteria or their semi-synthetic derivatives. AGs demonstrate broad-spectrum activity and are microbiologically and clinically effective at eradicating most Gram-negative bacteria and certain Gram-positive bacteria. The United States Food and Drug Administration has approved nine aminoglycosides for clinical use: gentamicin, tobramycin,

RSC Drug Discovery Series No. 43
Carbohydrates in Drug Design and Discovery
Edited by Jesús Jiménez-Barbero, F. Javier Cañada and Sonsoles Martín-Santamaría

Published by the Royal Society of Chemistry, www.rsc.org

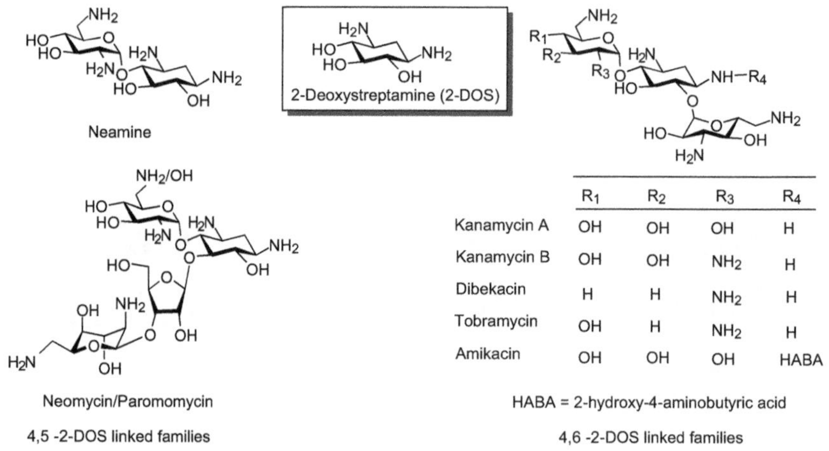

Neamine

2-Deoxystreptamine (2-DOS)

Neomycin/Paromomycin

4,5 -2-DOS linked families

	R_1	R_2	R_3	R_4
Kanamycin A	OH	OH	OH	H
Kanamycin B	OH	OH	NH_2	H
Dibekacin	H	H	NH_2	H
Tobramycin	OH	H	NH_2	H
Amikacin	OH	OH	OH	HABA

HABA = 2-hydroxy-4-aminobutyric acid

4,6 -2-DOS linked families

Figure 11.1 Structures of representative aminoglycoside families.

amikacin, streptomycin, neomycin, kanamycin, paromomycin, netilmicin and spectinomycin.[4] The use of AGs in industrialized countries has been relatively restrained in recent decades because of intrinsic toxicity and the introduction of other broad-spectrum antibiotics such as cephalosporins, carbapenems and fluoroquinolones. However, despite their drawbacks, AGs still play a key role in the fight against bacterial infections worldwide. For instance: (a) inhaled or infused tobramycin is an indispensable therapeutic approach to treat pulmonary infections in cystic fibrosis patients;[5] (b) gentamicin is widely used in the treatment of urinary infections;[6] and (c) paromomycin is an efficacious and affordable alternative to treat parasitic Leishmania infections in developing countries.[7] Furthermore, the World Health Organization[8,9] has declared that streptomycin, kanamycin and amikacin are to be used as second-line anti-tuberculosis drugs.

The antibacterial effect of AGs is afforded mainly by their binding to the 30S ribosomal subunit, leading to misreading of RNA, disruption of protein synthesis and, ultimately, accumulation of truncated and nonfunctional proteins, leading to bacterial death.[10] In addition, some AGs (including tobramycin) may also bind to other sites (including the 50S) on the bacterial ribosome.[11] The RNA interactions are relatively selective for bacterial ribosomes because structural differences lower the drugs' affinity for eukaryotic ribosomes and allow for general safe human use. Although AGs exhibit potent and fast bactericidal activity, their widespread use has been compromised by toxicity[4,12] and the worldwide emergence and spread of AG-resistant strains.[13,14] Nephrotoxicity and ototoxicity are the two most serious side effects associated with this class of antibiotic.[4] Nephrotoxicity is reversible and can be clinically managed with hydration therapy so that patients generally recover normal renal function once treatment with AGs is discontinued.[15] In contrast, AG-induced ototoxicity manifests itself in the form of irreversible hearing loss at higher frequencies. It is caused by damage to the outer and inner hair cells, with the severity of ototoxicity dependent upon the aminoglycoside used.[4] Neomycin is considered the most toxic, followed by gentamicin, kanamycin and tobramycin, while amikacin and netilmicin are considered the least toxic.[4]

A number of mechanisms are responsible for AG resistance. These include decreased uptake into cells, as a result of activation of drug efflux pumps, modified membrane potential, changes in membrane composition and covalent modification of the drug.[16–19] Enzymatic modification is the most common resistance mechanism and is achieved by aminoglycoside-resistant enzymes such as aminoglycoside *N*-acetyltransferases, aminoglycoside *O*-phosphotransferases and aminoglycoside *O*-nucleotidyltransferases.[18] AG modification results in a large decrease in binding affinity to the therapeutic target. The emergence of AG-resistant strains has instigated research efforts to develop inhibitors of AG-resistant enzymes[20–25] and modified AG analogues that lack some of the structural elements required by modifying enzymes. This has led to the development of semi-synthetic AGs such as amikacin, tobramycin and plazomicin. However, this approach is gradually becoming more and more difficult with the emergence of mechanisms giving class resistance such as 16*S* rRNA methylases, compromising nearly all aminoglycosides,[26] and up-regulation of resistance, nodulation and division (RND) efflux pumps,[27] resulting in multiple drug-resistant organisms.

11.2 Uptake of Polycationic Antibacterials

Polycationic antibacterials (PCAs) containing two or more positively charged amino and/or guanidino functions or other cationic groups define a structurally diverse class of antibacterials with broad-spectrum activity and different modes of action. This class of antibacterial agents can be further subdivided into non-amphiphilic PCAs, including the aminoglycoside antibiotics, but also amphiphilic PCAs comprised of the naturally occurring cationic antimicrobial peptides (AMPs), synthetic mimics of antimicrobial peptides (SMAMPs), synthetic polycationic lipopeptides, polycationic lipids and surfactants. The cationic charges of the PCAs ensure accumulation at polyanionic microbial cell surfaces that contain acidic polymers, such as lipopolysaccharides, and wall-associated teichoic acids in Gram-negative and Gram-positive bacteria.[28] Several PCAs, including aminoglycoside (gentamicin) and AMPs (polymyxin B, defensins, gramicidin S variants and others), transit the outer membrane by interacting at sites at which divalent cations cross-bridge adjacent polyanionic polymers. This causes a destabilization of the outer membrane that is proposed to lead to self-promoted uptake of the PCAs and/or other extracellular molecules.[28] After transit through the outer membrane, PCAs contact the anionic surface of the cytoplasmic membrane. Here, depending on the structure of the PCA, several scenarios can be envisaged. Amphiphilic PCAs can insert themselves into the cytoplasmic membrane, thereby either disrupting the physical integrity of the bilayer, *via* membrane thinning, transient poration and/or disruption of the barrier function, or translocate across the membrane and act on internal targets.[28] Non-amphiphilic PCAs, such as aminoglycoside antibiotics, must cross the membrane in order to bind to intracellular targets such as RNA, DNA and proteins. It is believed that the selective bacterial cytotoxicity of PCAs is caused by the affinity of the net negative charge found on bacterial cell membranes, in contrast to eukaryotic lipid bilayers which are typically made up of zwitterionic phospholipids.[28]

11.3 Amphiphilic Aminoglycoside Antimicrobials

Conjugation of one or more hydrophobic groups to a polycationic aminoglycoside converts a non-amphiphilic aminoglycoside antibiotic into an amphiphilic one. Compounds of this type are termed amphiphilic aminoglycoside antimicrobials (AAAs) and belong to the class of PCAs. AAAs display similar physicochemical characteristics and biological properties as seen for polycationic AMPs, AMP-mimetics, lipopeptides and other types of cationic amphiphiles. Furthermore, it has been shown that co-administration of aminoglycosides with membrane permeabilizing agents such as ionic lipids and cationic amphiphiles can result in enhanced antimicrobial activity.[29] This finding spurred initial interest to study the antibacterial properties of AAAs with the goal to develop agents that overcome bacterial resistance mechanisms.

11.3.1 Amphiphilic Neomycin Analogues

Most of the work on AAAs has focused on amphiphilic neomycin B analogues. Rinehart *et al.* synthesized and studied the antimicrobial activities of substituted hexa-*N*-benzyl neomycins **1a–e** (Scheme 11.1).[30] They believed that the low lipophilicity of neomycin might cause a poor absorption rate from the intestine;[31] hence two of the benzylated neomycin derivatives were tested for their oral activity. Among the tested analogues, hexa-*N*-(*p*-chlorobenzyl) neomycin B (**1b**) and hexa-*N*-(*p*-methoxybenzyl) neomycin B (**1d**) showed moderate activity (MIC 2 μg mL^{-1}) against *Diplococcus*

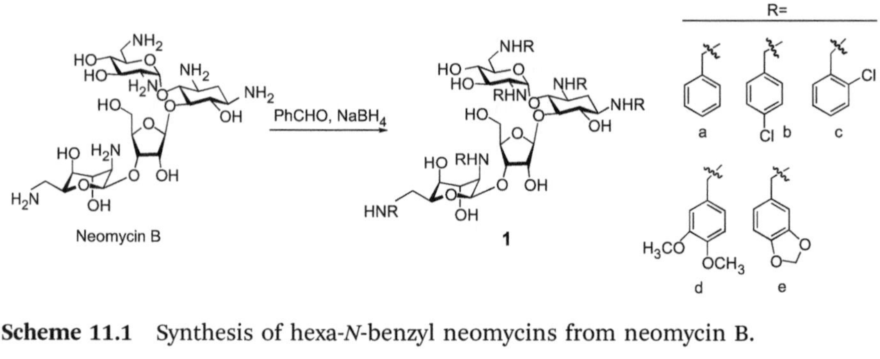

Scheme 11.1 Synthesis of hexa-*N*-benzyl neomycins from neomycin B.

Table 11.1 Minimum inhibitory concentration (MIC) of amphiphilic hexa-*N*-benzyl neomycin B derivatives.

	MIC values (μg mL^{-1})[a]							
Compound	A	B	C	D	E	F	G	H
1a	31.2	250	125	15.6	31.2	3.9	3.9	15.6
1b	250	500	250	250	31.2	31.2	2.0	31.2
1c	>500	>500	>500	>500	>500	31.2	2.0	500
1d	500	250	250	250	500	250	125	500
1e	250	250	250	62.5	250	7.8	15.6	125

[a]Bacterial strains: (A) *E. coli* UC 51; (B) *P. vulgaris* UC 93; (C) *P. aeruginosa* UC 95; (D) *K. pneumonia* UC 57; (E) *S. faecalis* UC 3235; (F) *S. aureus* UC 80; (G) *D. pneumoniae* UC 41; (H) *S. hemolyticus* UC 152.

pneumoniae UC 41 (Table 11.1). Further, hexa-*N*-benzyl neomycin B (**1a**) and hexa-*N*-(*p*-chlorobenzyl) neomycin B (**1b**) were subjected to *in vivo* studies in infected mice against *Staphylococcus aureus*. It was shown that **1a** was inactive subcutaneously at 40 mg kg^{-1} whereas analogue **1b** showed slight activity subcutaneously at 100 mg kg^{-1}. In addition, **1b** was found to be inactive orally at 800 mg kg^{-1}.

Our group has been involved in developing novel synthetic aminoglycoside–lipid conjugates through chemical modification in an attempt to delay or overcome bacterial resistance.[32]

Neomycin B was our initial choice of aminoglycoside antibiotic due to the low biological potency of neomycin toward antibiotic-resistant bacteria, including methicillin-resistant *S. aureus*. Selective tosylation of the reactive 5″-hydroxyl group in the ribose moiety, followed by subsequent chemical modifications, offered the partially protected 5″-NH_2 neomycin derivative **4**. Thereafter a variety of acids varying in chain length, saturation and hybridization, such as aliphatic C_6, C_{12}, C_{16} and C_{20} and unsaturated C_{18} acids and pyrene-containing acids, have been coupled to neomycin-amine **4** by exposure to 2-(1*H*-benzotriazol-1-yl)-1,1,3,3-tetramethyluronium tetrafluoroborate (TBTU) coupling reagent in Hünig's base/DMF and deprotection with 95% TFA, furnishing the desired neomycin–lipid conjugates **5a–g** (Scheme 11.2). All the synthetic derivatives showed significant antimicrobial activity against Gram-positive bacterial strains. The cumulative lipophilicity in the C_{16} and C_{20} lipid chain neomycin conjugates imparted greater antimicrobial activity against methicillin-resistant *Staphylococcus aureus* (MRSA) (MIC 8 μg mL^{-1}) when compared to the parent neomycin (MIC 256 μg mL^{-1}). However, in contrast, the shorter lipid chain C_6 analogue (**5a**) showed enhanced activity (MIC 8 and 4 μg mL^{-1}) against Gram-negative *E. coli* and gentamicin-resistant *E. coli* (MIC 8 and 4 μg mL^{-1}, respectively) than their long-chain neomycin counterparts (**5d** and **5e**) (MIC 64–128 μg mL^{-1}) (Table 11.2). Having aromatic and cholesterol lipophilic acid moieties did decrease the antimicrobial potencies against Gram-positive strains (Table 11.2). We have shown that the nature of the lipid influences the antimicrobial activity of neomycin–lipid acid conjugates, as evidenced by

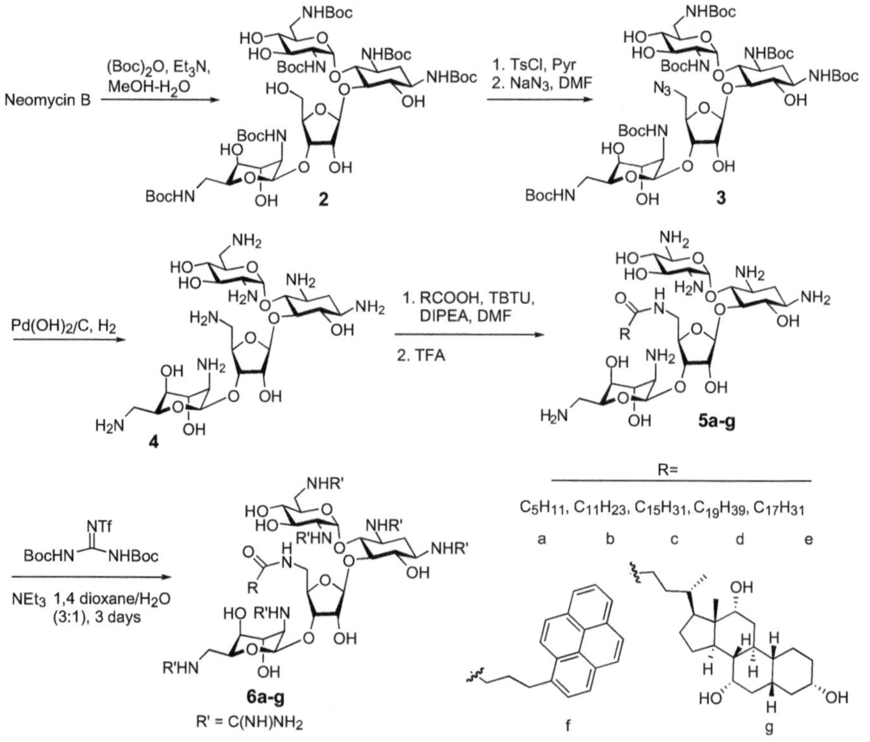

Scheme 11.2 Synthesis of C-5′′ neomycin B–lipid and guanidinylated lipid conjugates.

Table 11.2 MIC of amphiphilic neomycin and guanidinylated neomycin analogues.

	MIC values (μg mL^{-1})[a]											
Cpd.	A	B	C	D	E	F	G	H	I	J	K	L
Gen.	1	2	0.25	32	ND	ND	4	1	128	8	8	128
Neo.	1	256	0.25	0.5	ND	ND	32	4	8	ND	512	512
5a	16	>512	2	32	ND	ND	64	16	16	64	>512	>512
5b	32	>256	4	8	ND	ND	128	64	64	128	256	256
5c	4	8	2	2	ND	ND	64	32	64	ND	128	128
5d	8	8	4	4	ND	ND	64	128	128	64	128	64
5e	16	32	4	4	ND	ND	128	128	64	128	256	128
5f	16	256	2	64	ND	ND	>512	32	64	64	>512	256
5g	16	128	8	16	ND	ND	>256	32	64	64	>256	128
6b	4	8	2	2	32	16	64	16	32	16	32	32
6c	4	4	1	4	4	0.5	32	32	64	32	32	8
6f	8	8	8	8	32	16	64	32	32	32	32	16
6g	16	16	4	8	32	16	64	64	64	32	128	32

[a]Bacterial strains: (A) *S. aureus* ATCC 29213; (B) MRSA ATCC 33592; (C) *S. epidermidis* ATCC 14990; (D) MRSE CAN-ICU 61589; (E) *E. faecalis* ATCC 29212; (F) *E. faecium* ATCC 27270; (G) *S. pneumoniae* ATCC 49619; (H) *E. coli* ATCC 25922; (I) *E. coli* CAN-ICU 61714; (J) *E. coli* CAN-ICU 63074; (K) *P. aeruginosa* ATCC 27853; (L) *P. aeruginosa* CAN-ICU 62308; ND = not determined.

strong antibacterial activity of neomycin C_{16} and C_{20} lipid conjugates against MDR stains such as MRSA and MRSE.

We also investigated the influence of guanidinylation on antibacterial properties against resistant bacterial strains. The polyamine-based scaffold of neomycin B was guanidinylated[33] and conjugated with a variety of lipophilic moieties (Scheme 11.2).[34] Polyguanidinylated neomycin lipid conjugates (**6b**, **6c** and **6f**) exhibited significant antimicrobial activity against *P. aeruginosa* when compared to corresponding polyamine-headed neomycin lipid conjugates (**5b**, **5c** and **5f**) (Table 11.2). Among the compounds tested, guanidinylated neomycin C_{16} (**6c**) and pyrene-based lipid derivatives (**6f**) showed enhanced antimicrobial activities (MIC 8 and 16 μg mL^{-1}, respectively). It was also observed that polyguanidinylated neomycin B–C_{16} lipid conjugates (**6f** and **6g**) showed improved anti-MRSA activity when compared to polyamine neomycin B–C_{16} lipid conjugates (**5f** and **5g**), emphasizing the importance of guanidinylation. Furthermore, polyguanidinylated aminoglycoside-based lipid conjugates displayed reduced hemolytic activity when compared with their polyamine analogues.

Chang and co-workers synthesized 5″-modified neomycin derivatives *via* click chemistry or amide linkage formation and assayed the synthetic candidates against *E. coli* and *S. aureus* strains.[35] Carboxybenzyl-protected neomycin B (**2**) was regioselectively substituted with azide functionality at C-5″ (Scheme 11.3). In the first strategy, the neomycin B azide underwent 1,3-dipolar cyclization with various alkynes to furnish "click" neomycin B analogues. In the second strategy, the azide was reduced to an amine, which was subsequently coupled with carboxylic acids or amino acids. Catalytic hydrogenation of both the analogues furnished the desired 5″-modified neomycin B derivatives (Scheme 11.3). Click-based neomycin B derivatives (**7e** and **7g**) displayed modest antimicrobial activities, on a par with parent neomycin B. However, analogues with amide linkages manifested less antimicrobial activity than the parent aminoglycoside except the derivatives that bear long linear alkyl chain scaffolds (**8b** and **8c**) (Table 11.3). The same group has also shown that structural modifications over aminoglycosides do influence their mode of action, which in turn would govern the antimicrobial activity on the resistant bacterial strains. Binding affinity studies disclosed that unmodified pyranmycin[36] exhibits antimicrobial activity comparable to that of traditional neomycin and exerts a similar mode of action by binding toward rRNA. The modified neomycin **8c**, armed with an amphiphilic-rich hexadecanoyl group, exerts prominent antimicrobial activity against MRSA and vancomycin-resistant enterococci (VRE) strains, which usually stymies the entry of traditional aminoglycosides.[35,37] Fluorescence-based studies demonstrated that compound **8c** targets the bacterial cell membrane, leading to enhanced permeability and cellular uptake.[38]

A recent report by Baasov *et al.* described amphiphilic neomycin B analogues in the form of neomycin B–fluoroquinolone heterodimer hybrid molecules **11a–m** (Scheme 11.4). Interestingly, compounds of this type cause significant delays in bacterial resistance.[39] Neomycin B and ciprofloxacin

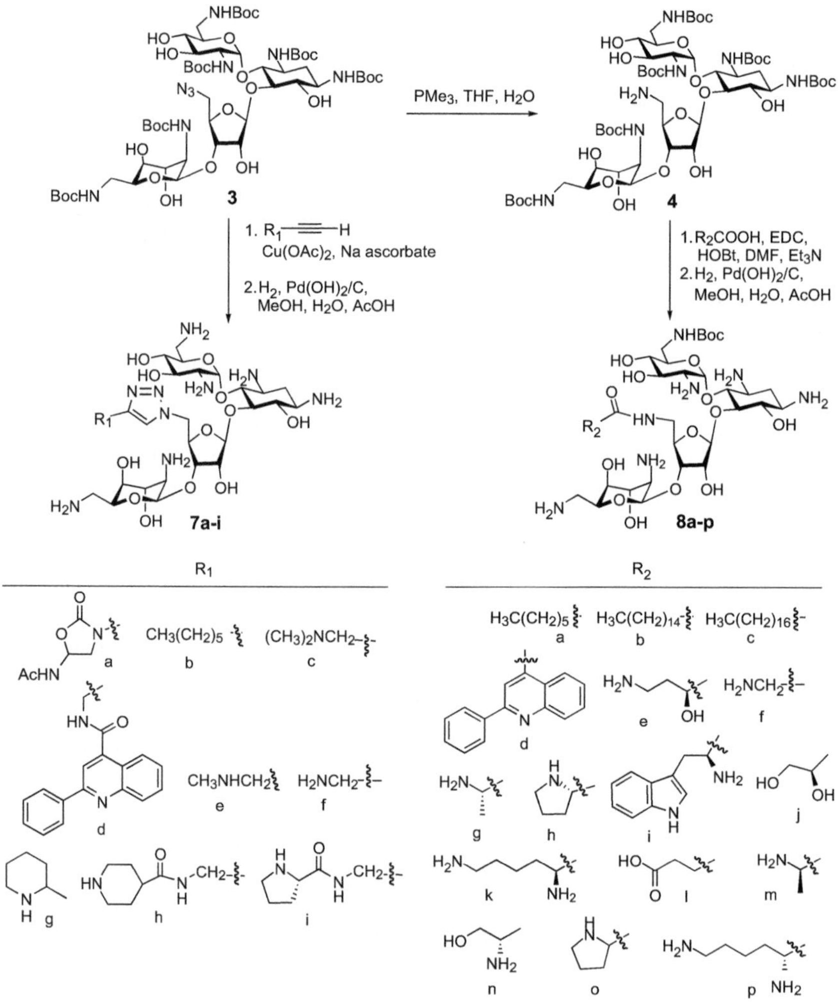

Scheme 11.3 Synthesis of C-5″ modified neomycin B–lipid analogues.

were linked *via* a series of 1,2,3-triazole functionality containing spacers, which vary in chain length. Notably, none of the lead molecules showed superior antimicrobial activity over fluoroquinolone, but significantly improved activity when compared to neomycin B (Table 11.4). In contrast, most hybrid analogues not only displayed potent antimicrobial activity when compared to parent neomycin B against Gram-positive (MSRA) and Gram-negative (neomycin-resistant and susceptible *E. coli*) strains, but also tolerated the influence of traditional aminoglycoside modifying enzymes such as APH (3′)-Ia, APH (3′)-IIIa and AAC(6′)/APH(2″).

Extensive studies have been carried out to investigate the effect of lipophilic substitutions, including polycarbamates and polyethers, on the neomycin B-based polyol scaffold.[40] Carbamates were synthesized from the

Table 11.3 MIC of C-5″ modified neomycin B–lipid analogues.

	MIC values (μg mL^{-1})[a]										
Cpd.	A	B	C	D	E	F	G	H	I	J	K
Neo. B	4	1	4–8	4–8	≥2000	4	16–32	125	64	64–125	≥250
7e	8	1–2	16	16	≥2000	ND	ND	ND	ND	ND	ND
7g	8–16	1–2	16	16	1000	ND	ND	ND	ND	ND	ND
8b	4	2	2–4	16	8	8–16	16–32	2–4	4	2–4	4–8
8c	4–8	4	2–4	32	8	8–16	32	4–8	8–16	4–8	8–16
8f	16	1	8–16	8–16	500–1000	4	8	16–32	8–16	32–64	≥250

[a]Bacterial strains: (A) *E. coli* (ATCC 25922); (B) *S. aureus* (ATCC 25923); (C) *E. coli* (TG1) (aminoglycoside susceptible strain); (D) *E. coli* (TG1) (pSF815 plasmid encoded for (AAC6′/APH(2″); (E) *E. coli* (TG1) (pTZ19U-3 plasmid encoded for APH(3′)-I); (F) *K. pneumoniae* ATCC 13883; (G) *K. pneumoniae* ATCC 700603; (H) *S. aureus* (ATCC 33591) MRSA; (I) *P. aeruginosa* ATCC 27853; (J) *E. faecalis* ATCC 29212; (K) *E. faecalis* ATCC 51299 (VRE); ND = not determined.

Comp	X	Y	Comp	X	Y	Comp	X	Y
a	-$(CH_2)_2$-	-C_6H_4NHCO-	f	-CH_2-mC_6H_4-CH_2-	-C_6H_4NHCO-	j	-$(CH_2)_2$-O-$(CH_2)_2$-	-CH_2-NHCO-
b	-$(CH_2)_4$-	-C_6H_4NHCO-	g	-CH_2-pC_6H_4-CH_2-	-C_6H_4NHCO-	k	-CH_2-mC_6H_4-CH_2-	-CH_2-NHCO-
c	-$(CH_2)_6$-	-C_6H_4NHCO-	h	-$(CH_2)_5$-	-CH_2-NHCO-	l	-CH_2-pC_6H_4-CH_2-	-CH_2-NHCO-
d	-$CH_2CH(OH)CH_2$-	-C_6H_4NHCO-	i	-$CH_2CH(OH)CH_2$-	-CH_2-NHCO-	m	-$(CH_2)_2$-	-CH_2-O-CH_2-
e	-$(CH_2)_2$-O-$(CH_2)_2$-	-C_6H_4NHCO-						

Scheme 11.4 Synthesis of ciprofloxacin–neomycin B hybrid analogues.

Table 11.4 MIC of ciprofloxacin–neomycin B hybrid analogues.

	MIC values (μg mL^{-1})[a]							MIC values (μg mL^{-1})[a]					
Cpd.	A	B	C	D	E	F	Cpd.	A	B	C	D	E	F
Cipro.	0.02	0.02	0.05	<0.005	0.02	0.20	**11f**	6	12	12	1.5	6	12
Neo. B	24	48	384	96	1.5	384	**11g**	1.5	3	3	0.75	1.5	3
11a	12	6	24	0.75	6	48	**11h**	24	6	24	1.5	12	48
11b	12	6	24	1.5	6	24	**11i**	6	3	12	0.75	3	12
11c	6	6	12	1.5	12	48	**11j**	12	6	12	1.5	12	24
11d	6	3	6	0.75	6	24	**11k**	6	6	12	0.75	3	6
11e	24	12	12	1.5	12	24	**11m**	3	3	12	0.38	0.75	6

[a]Bacterial strains: (A) *E. coli* R477-100; (B) *E. coli* ATCC 25922; (C) *E. coli* AG 100B; (D) *E. coli* AG 100A; (E) *B. subtilis* ATCC 6633; (F) MRSA ATCC 43300.

tert-butyl carbamate (Boc)-protected neomycin B analogue **2** *via* cabamoylation of the hydroxy groups with various commercially available hydrophobic isocyanates; subsequent deblocking with TFA afforded amphiphilic polycarbamates (**12a–d**). Amphiphilic polyether analogues **13a–d** were prepared from the Boc-protected neomycin B **2** *via* *O*-alkylation followed by

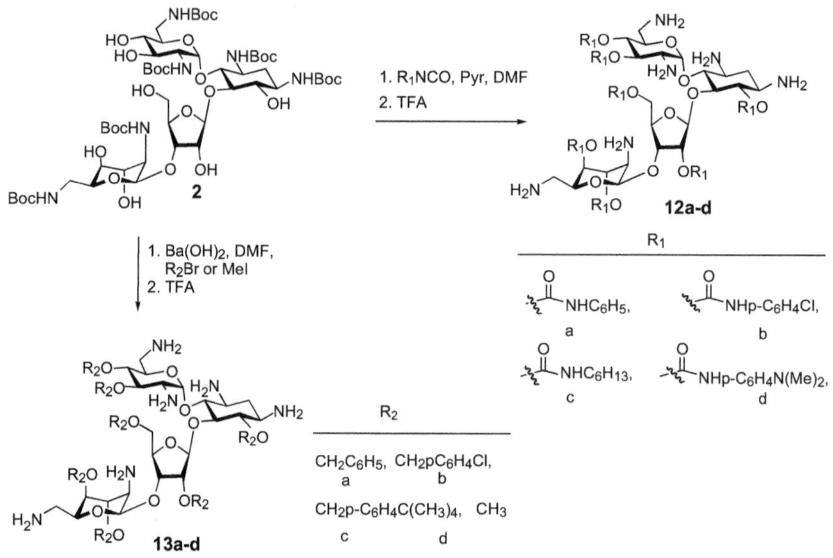

1. R_1NCO, Pyr, DMF
2. TFA

1. $Ba(OH)_2$, DMF, R_2Br or MeI
2. TFA

R_1: a NHC_6H_5, b $NHp\text{-}C_6H_4Cl$, c NHC_6H_{13}, d $NHp\text{-}C_6H_4N(Me)_2$

R_2: a $CH_2C_6H_5$, b $CH_2pC_6H_4Cl$, c $CH_2p\text{-}C_6H_4C(CH_3)_4$, d CH_3

Scheme 11.5 Synthesis of neomycin B-derived polycarbamate and polyether analogues.

Table 11.5 MIC of neomycin B-derived polycarbamate and polyether analogues.

	MIC values (μg mL^{-1})[a]												
Cpd.	A	B	C	D	E	F	G	H	I	J	K	L	M
Gen.	1	2	0.25	32	4	1	128	8	8	128	>512	128	0.25
Neo.	1	256	0.25	0.5	32	4	8	ND	512	512	>512	64	1
12a	1	1	0.5	0.5	16	32	16	16	256	128	4	32	256
12c	2	256	1	2	8	8	8	32	>256	256	>256	>256	4
13a	4	4	2	2	16	8	16	4	32	64	8	16	8
13b	64	64	32	64	128	128	>128	64	>128	>128	>128	>128	128

[a]Bacterial strains: (A) *S. aureus* ATCC 29213; (B) MRSA ATCC 33592; (C) *S. epidermidis* ATCC 1490; (D) MRSE ATCC 14990; (E) *S. pneumoniae* ATCC 49619; (F) *E. coli* ATCC 25922; (G) *E. coli* ATCC 6174; (H) *E. coli* CAN-ICU 63074; (I) *P. aeruginosa* ATCC 27853; (J) *P. aeruginosa* CAN-ICU 62308; (K) *S. maltophilia* CAN-ICU 62584; (L) *A. baumannii* CAN-ICU 63169; (M) *S. pneumoniae* ATCC 13883; ND = not determined.

deprotection with TFA (Scheme 11.5). The heptaphenyl carbamate **12a** of neomycin B manifested superior antimicrobial activity against MRSA (256-fold increment), *S. aureus*, *S. epidermidis* and MRSE (MIC < 1 μg mL^{-1}). However, significant loss of Gram-positive activity was observed with the introduction of *p*-chloro and *p*-dimethylamino substituents in the phenyl ring of the polycarbamates. The most potent cationic polyether **13a** exhibits broad spectrum activity against MRSE, MRSA and *P. aeruginosa* (Table 11.5).

Cationic antimicrobial peptides (AMPs) and host defense peptides display broad-spectrum antibacterial activity, especially against antibiotic-resistant bacteria, and are currently being investigated for clinical use.[41] Often the antibacterial effect of cationic antimicrobial peptides could be attributed to their physiochemical properties rather than to the precise amino acid sequence. Previously, it was reported that ultrashort amphiphilic peptide sequences, as short as tripeptides, display potent antibacterial activity.[42] Inspired by these short amphiphilic peptide sequences, we have prepared a different class of amphiphilic aminoglycoside derivative termed "aminoglycoside–peptide triazole candidates" (APTCs).[43] The physicochemical properties of AMPs and APTCs suggest a membranolytic mode of action against bacterial strains.[28] We ratiocinated that conjugating aminoglycoside antibiotics with amphiphilic short peptides would enhance antimicrobial activity against resistant bacterial strains. Copper(I)-catalyzed click-based cycloaddition of neomycin B-based azide to small peptide sequences containing alkyne moieties furnished the desired APTCs **14a–d** (Scheme 11.6). Amino groups in tryptophan–dipeptide containing neomycin B (**14c**) were further guanidinylated (**14d**) to explore basicity effects in APTCs. Activities of **14c** and **14d** APTCs were found to be much higher and can surpass the acquired resistance of MRSA against neomycin B. It has also been demonstrated that more amphiphilic conjugates (**14c**) exert greater antimicrobial activity (16 μg mL^{-1} against MRSA) compared to protected less hydrophobic conjugates **(14b)** (128 μg mL^{-1} against MRSA), thus realizing the fact that increase in the amphiphilicity boosts antibacterial activity (Table 11.6).

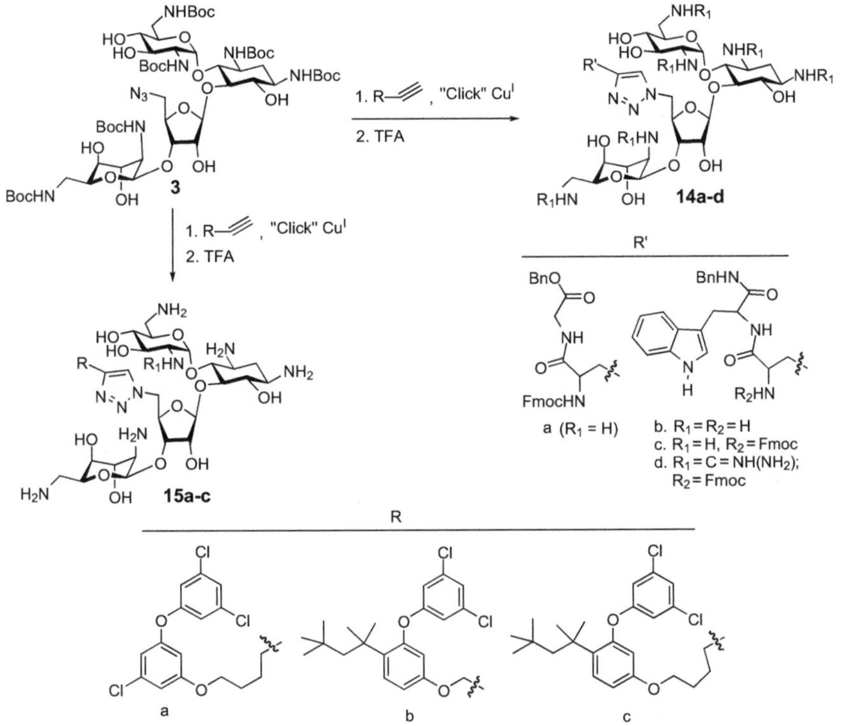

Scheme 11.6 Synthesis of neomycin B–peptide and phenolic conjugates.

Table 11.6 MIC of neomycin B–peptide and phenolic conjugates.

	MIC values (μg mL^{-1})[a]									
Cpd.	A	B	C	D	E	F	G	H	I	J
Neo.	1	256	0.25	0.25	8	2	4	32	512	512
14a	16	64	4	8	>512	32	32	64	512	512
14b	32	128	8	8	>512	32	32	ND	512	256
14c	8	16	4	8	64	16	32	ND	128	64
14d	4	8	1	2	64	32	64	64	128	128
15a	4	8	ND	2	64	16	16	64	128	64
15b	8	8	ND	1	64	16	64	64	128	64

[a]Bacterial strains: (A) *S. aureus* ATCC 29213; (B) MRSA ATCC 33592; (C) *S. epidermidis* ATCC 14990; (D) MRSE CAN-ICU 61589; (E) *S. pneumoniae* ATCC 49619; (F) *E. coli* ATCC 25922; (G) *E. coli* CAN-ICU 61714; (H) *E. coli* CAN-ICU 63074; (I) *P. aeruginosa* ATCC 27853; (J) *P. aeruginosa* CAN-ICU 62308; ND = not determined.

Our group also reported the synthesis and antibacterial evaluation of neomycin B–phenolic conjugates **15a–c**. Phenolic scaffolds were prepared *via* a copper(I)-catalyzed 1,3-dipolar cycloaddition reaction between the phenol-modified alkyne and neomycin-based azide positioned at C-5″ of the ribose moiety (Scheme 11.6).[44] When compared to the parent neomycin B, the phenolic conjugates **15a** and **15b** were found to be potent and exhibited comparable activity against Gram-positive bacteria and significant activity against multidrug-resistant strains such as MRSA and *P. aeruginosa* (Table 11.6).

11.3.2 Amphiphilic Paromomycin Analogues

Intricate molecular structural aspects associated with aminoglycoside interactions at the ribosomal level have prompted Hanessian and co-workers to design new types of paromomycin analogues.[45] They have introduced a plethora of aminoalkyl ethers at the C-2 hydroxyl group in ring III of the paromomycin framework and tested antimicrobial activities against *E. coli* (ATCC 25922) and *S. aureus* (ATCC 13709). Most of the lipophilic 2″-ethylamino substituents (Scheme 11.7), comprising aliphatic and aromatic derivatives circumscribing the 2″-position, have shown more or less similar activity as paromomycin against *S. aureus*, whereas smaller and polar functionalities displayed better activities against *E. coli* (Table 11.7). Analogues **17b** and **17d** possess significant antibacterial activities against *S. aureus*. Understanding the ribosomal A structural pattern through X-ray studies has disclosed some critical points pertaining to the structural requirements of aminoglycosides and its pharmacokinetic properties. It was shown that removing ring IV from the unmodified antibiotic completely diminishes its antimicrobial activity, while the corresponding 2″-ether-substituted analogues retrieve the antimicrobial activity.

In recent years, one of the remarkable synthetic modifications on aminoglycosides is the development of amikacin, an N^1-acylated γ-amino-α-hydroxybutyryl (HABA) derivative of kanamycin.[46] It was proven to be

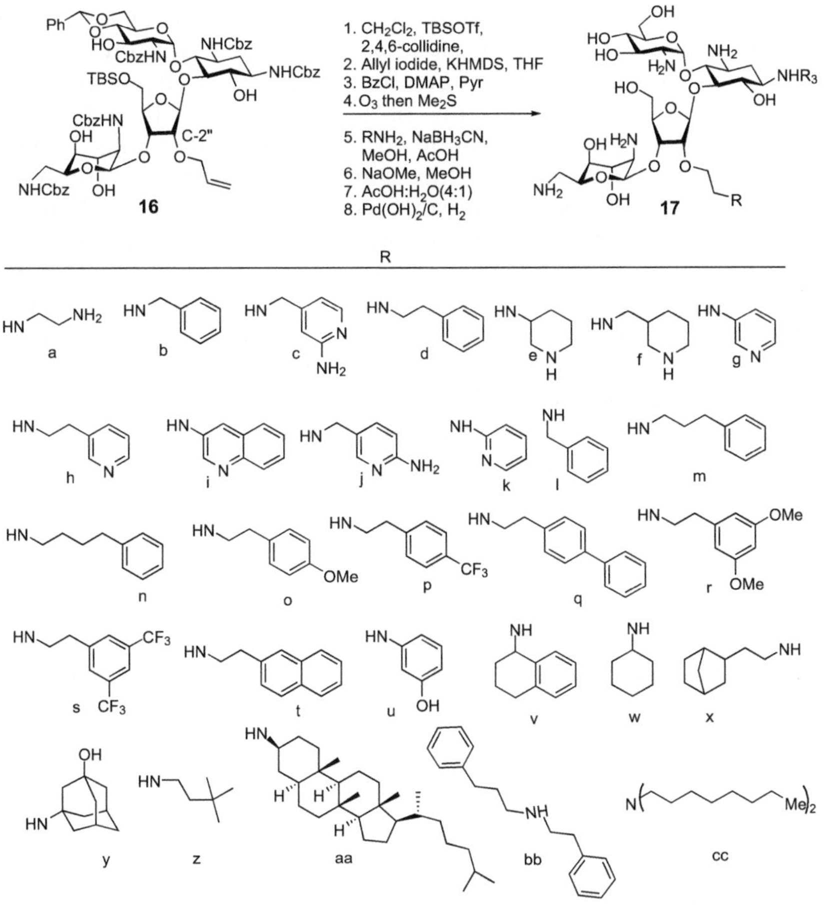

1. CH_2Cl_2, TBSOTf, 2,4,6-collidine,
2. Allyl iodide, KHMDS, THF
3. BzCl, DMAP, Pyr
4. O_3 then Me_2S
5. RNH_2, $NaBH_3CN$, MeOH, AcOH
6. NaOMe, MeOH
7. AcOH:H_2O(4:1)
8. $Pd(OH)_2$/C, H_2

Scheme 11.7 Synthesis of C-2″ ether analogues of paromomycin derivatives.

Table 11.7 MIC values of C-2″ ether analogues of paromomycin derivatives.

	MIC values (μg mL^{-1})							
Cpd.	*E. coli*	*S. aureus*	Cpd.	*E. coli*	*S. aureus*	Cpd.	*E. coli*	*S. aureus*
Paro.	3–6	1–2	**17j**	3–6	0.3–0.6	**17t**	10–20	1–3
17a	6–12	2–3	**17k**	12–25	2–3	**17u**	12–25	2–3
17b	12–25	<1.5	**17l**	3–6	3–5	**17v**	>10	1.25–2.5
17c	6–12	0.6–1	**17m**	3–5	0.6–1.2	**17w**	3–6	3–5
17d	3–6	0.3–0.6	**17n**	10–20	3–5	**17x**	>10	5–10
17e	1.5–3	3–6	**17o**	5–10	<0.6	**17y**	20–40	1–3
17f	1.5–3	3–6	**17p**	5–10	0.6–1.2	**17z**	10–20	0.6–1
17g	12–25	<1.5	**17q**	20–40	5–10	**17aa**	>10	2.5–5
17h	12–25	<1.5	**17r**	10–20	5–10	**17bb**	10–20	0.6–1.2
17i	12–25	1–2	**17s**	>10	2.5–5	**17cc**	5–10	1.2–2.5

highly potent against certain strains of *Pseudomonas aeruginosa*. Earlier reports were known on the synthesis of paromomycin and neomycin HABA derivatives.[47,48] It was also known that aminoalkyl ether chains at the C-2″ position in the ribofuranosyl ring of paromomycin deliver promising antibacterial effects.[45] Hanessian and co-workers put effort into designing new analogues of paromomycin, which combine N^1-HABA and hydrophobic appendages at C-2″ in one framework. Both 2″-*O*-(phenethylamino) ethyl ether and N^1-HABA scaffolds were incorporated in paromomycin and the 3′,4′-dideoxy-paromomycin ring system (**22**) *via* key intermediates **16** and **21** (Scheme 11.8).[49] It was shown that lipophilic N^1-HABA paromomycin (**18**) manifested impressive antibacterial activity against *E. coli* and *S. aureus* (Table 11.8), whereas the corresponding 3′,4′-dideoxygenated paromomycin analogue (**22**) displayed miniscule reduced inhibitory activity (Table 11.8).

There has also been a patent claim that describes the use of *N*-alkylated paromomycin analogues as inhibitors of efflux pumps, based on studies with *Haemophilus* influenza.[50]

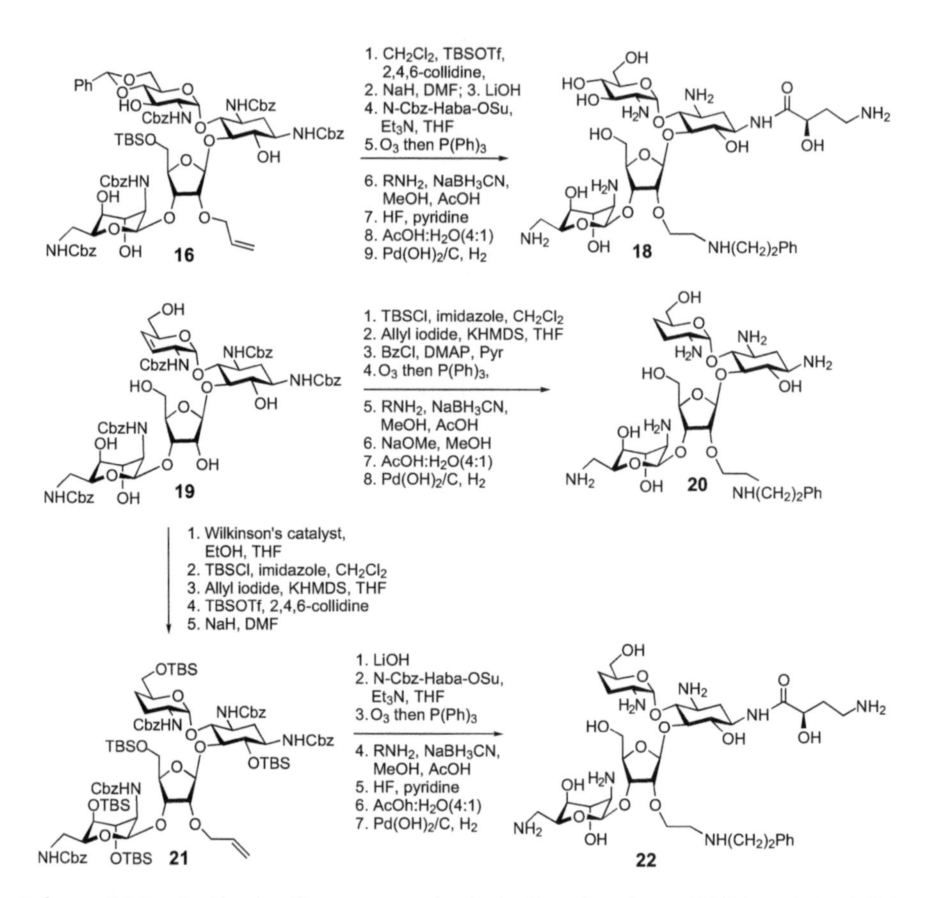

Scheme 11.8 Synthesis of paromomycin derivatives bearing a HABA moiety at N-1.

Table 11.8 MIC values of paromomycin derivatives with N-1 HABA.

Cpd.	*E. coli* (ATCC 25922)	*S. aureus* (ATCC 13709)
Paro.	3–6	1–2
Neo. B	3–6	1–2
17d	3–6	0.3–0.6
18	0.6–1.3	0.6–1.3
20	40	1–3
22	2.5–5.0	0.6–1.3

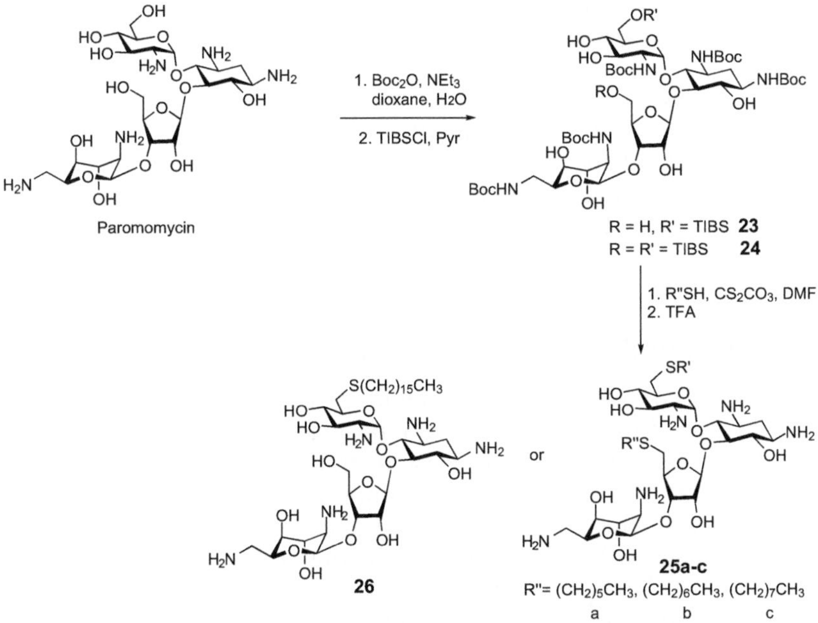

Scheme 11.9 Synthesis of amphiphilic mono- and dialkylated paromomycin analogues.

Recently, Fridman has reported the synthesis of a series of cationic amphiphilic paromomycin analogues (Scheme 11.9) for potential use to treat skin infections.[51] They accessed the readily available two primary hydroxyl groups at the C-6′ and C-5″ positions in paromomycin to achieve chemoselective dialkylation. Dialkylated analogues bearing C_6–C_8 (**25a–c**) aliphatic thioether tethers and monoalkylated C_{16} (**26**) aminoglycosides were assayed against a range of skin infections caused by bacterial strains. In general, the antibacterial activity of paromomycin analogues containing C_6, C_7 and C_8 dialkylated chain analogues displayed augmented antibacterial activity, with C_8 being the most active, compared to the parent paromomycin, against a variety of *Streptococcus pyogenes* strains (Table 11.9). The monoalkylated paromomycin analogue (**26**) caused extensive hemolysis of rat RBCs (33.8 ± 5.2% at a concentration of 16 μg mL^{-1}) when compared to the

Table 11.9 MIC values of mono- and dialkylated paromomycin analogues compared to unmodified paromomycin and gramicidin D.

	MIC values (μg mL^{-1})[a]													
Antibiotic	A	B	C	D	E	F	G	H	I	J	K	L	M	N
Gram. D	32	64	8	2	>64	>64	<0.5	<0.5	<0.5	<0.5	<0.5	<0.5	16	2
Paro.	>64	>64	2	>64	16	>64	64	64	32	32	64	64	64	16
25b	8	16	4	4	16	4	4	2	2	4	4	4	4	4
25c	4	8	2	2	4	4	2	2	2	2	2	2	2	2

[a]Bacterial strains: (A) *S. aureus* Oxford NCTC6571; (B) MRSA; (C) *S. epidermidis* ATCC12228 (biofilm negative); (D) *S. epidermidis* ATCC35984/RP62A (biofilm positive); (E) *S. aureus* Cowan ATCC12598; (F) *S. pyogenes* serotype M12 (strain MGAS9429); (G) *S. pyogenes* serotype M1T1; (H) *S. pyogenes* serotype M2; (I) *S. pyogenes* serotype M3; (J) *S. pyogenes* serotype M5; (K) *S. pyogenes* serotype M24; (L) *S. pyogenes* JRS75; (M) *S. pyogenes* glossy; (N) *S. pyogenes* serotype T5.

dialkylated derivatives (**25a–c**), which showed no perceptible hemolysis at the same concentration. Subsequently, alkylated derivative antimicrobial activities were compared to that of the membrane-targeting antibiotic gramicidin D. Interestingly, compounds **25b** and **25c** demonstrated greater antibacterial activity than gramicidin D against certain *Staphylococci* strains (Table 11.9).

11.3.3 Amphiphilic Tobramycin Analogues

Hanessian and co-workers have also reported modifications on tobramycin.[53] Lack of hydroxy functionality at the 3′-position in the 2-deoxystreptamine ring makes this aminoglycoside resistant to the aminoglycoside-modifying enzyme APH (3′).[52] In an attempt to mimic paramomycin rings III and IV that interact with the lower stem of the A-site RNA, tobramycin C-5 analogues appended with various basic functionalities have been prepared (Scheme 11.10).[53] Some of the derivatives were as active as the parent tobramycin (**29d–f**, MIC 0.5–2.5 μg mL^{-1}) (Table 11.10). Two guanidino derivatives (**29d** and **29e**) displayed potent antibacterial activity against *P. aeruginosa* (ATCC 27853) (MIC 12.5 μg mL^{-1}).

Previous work in our group emphasized the importance of amphiphilicity in restoring the antimicrobial activities of kanamycin, neomycin and neamine against Gram-positive and Gram-negative bacteria.[32,40] As a logical extension, we delved into tobramycin by placing lipid scaffolds at the C-6″ position in ring III to analyze the lipid tail effects on antimicrobial properties.[54] A series of tobramycin-based triazole conjugates (**32a,b**) and polycationic lipids (**33a,b**) were synthesized (Scheme 11.11) and tested against a variety of bacterial strains, including Canadian hospital isolates. Our results revealed that amphiphilic tobramycin analogues like amide **33a** exhibited good activity against Gram-positive strains (MIC $\leq$ 4–8 μg mL^{-1}) and displayed potent activity (MIC $<$ 0.25 μg mL^{-1}) against methicillin-resistant *Staphylococcus epidermidis.* Interestingly, replacement of the C_{16} lipid tail by a partially fluorinated lipid tail (**33b**) imparted better activities (MIC 16 μg mL^{-1}) against two *Pseudomonas aeruginosa* strains (Table 11.11).

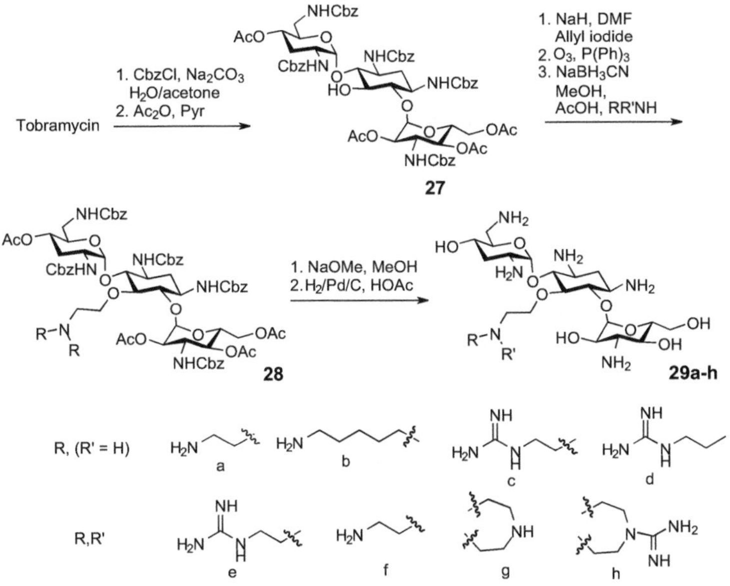

Scheme 11.10 Synthesis of C-5 aminoalkyl ether tobramycin analogues.

Table 11.10 MIC values of C-5 aminoalkyl ether tobramycin analogues.

Compound	*E. coli* (ATCC 25922)	*S. aureus* (ATCC 13709)
Kanamycin A	2.5–5	1.2–2.5
Kanamycin B	1.2–2.5	0.3–0.6
Tobramycin	0.6–1.2	0.3–0.6
Amikacin	1.2–2.5	1.2–2.5
Paromomycin	2.5–5	1.2–2.5
Neamine	>10	>10
29d	2.5–5	>10
29e	0.6–1.2	2.5–5
29f	1–2	5–10

Fridman and co-workers also reported the synthesis and antibacterial properties of 6″-modified tobramycin analogues. The synthesised amphiphilic tobramycin analogues containing a repertoire of linear as well as branched aliphatic thioether alkyl chains varying in carbon atoms from 6 to 22, including cyclic and substituted aromatic rings (**37a–r**, Scheme 11.12).[55] The compounds were evaluated for their antibacterial activity against Gram-positive and Gram-negative bacterial strains that are highly resistant to tobramycin, such as MRSA, VRE (tobramycin MIC ≥ 150 μg mL^{-1}), organisms expressing aminoglycoside-modifying enzyme (AME) cloned *E. coli* BL21 (DE3) strains, *i.e.* AAC(6′)/APH(2″), AAC(3)-IV and multi-acylating enzymes (Table 11.12). In general, the antibacterial activities of the derivatives increased with the extension of the aliphatic arm from C_6 to C_{16}. However,

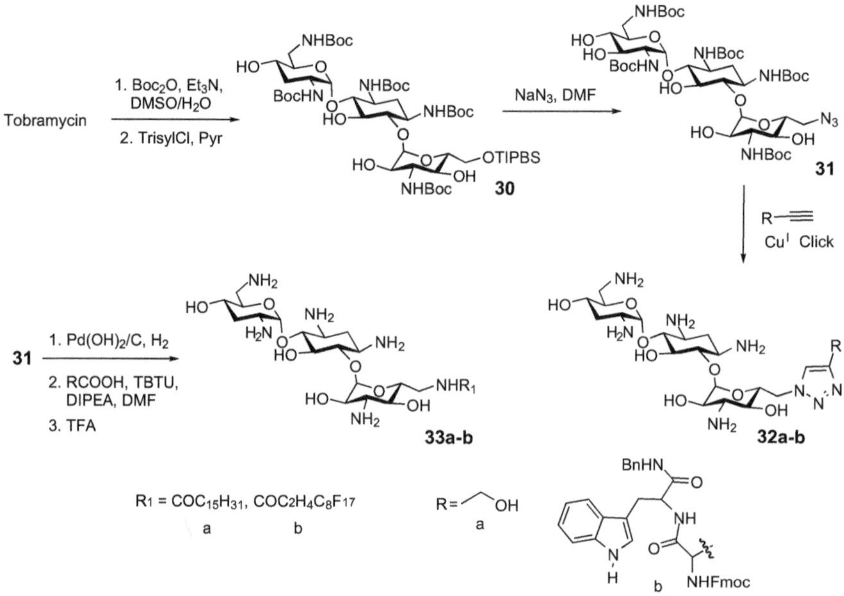

Scheme 11.11 Synthesis of triazole- and amide-linked tobramycin derivatives.

Table 11.11 MIC values of triazole- and amide-linked tobramycin derivatives.

	MIC values ($\mu g\ mL^{-1}$)[a]														
Cpd.	A	B	C	D	E	F	G	H	I	J	K	L	M	N	O
Tob.	0.5	0.5	≤0.25	2	8	16	2	0.5	8	8	0.5	16	>512	16	>0.25
32b	8	16	8	16	32	16	64	16	32	32	32	16	256	256	256
33a	8	8	4	<0.25	8	4	128	32	32	8	128	16	256	256	32
33b	16	32	8	≤0.25	16	0.5	128	64	64	64	16	16	>512	>512	128

[a]Bacterial strains: (A) *S. aureus* ATCC 29213; (B) MRSA ATCC 33592; (C) *S. epidermidis* ATCC 14990; (D) MRSE CAN-ICU 61589; (E) *E. faecalis* ATCC 29212; (F) *E. faecium* ATCC 27270; (G) *S. pneumoniae* ATCC 49619; (H) *E. coli* ATCC 25922; (I) *E. coli* CAN-ICU 61714; (J) *E. coli* CAN-ICU 63074; (K) *P. aeruginosa* ATCC 27853; (L) *P. aeruginosa* CAN-ICU 62308; (M) *S. maltophilia* CAN-ICU 62584; (N) *A. baumannii* CAN-ICU 63169; (O) *S. pneumoniae* ATCC 13883.

activities subsided for the chain lengths C_{18} and C_{22}. Antibacterial activity was dramatically increased from C_{10} to C_{16} with the C_{14} (**34e**) derivative being the remarkable one, with superior MIC values ranging between 0.3 to 18.8 $\mu g\ mL^{-1}$ against 19 strains (Table 11.12). However, the sulfoxides/sulfones (**35a–r**/**36a–r**) of corresponding C_{12} and C_{14} derivatives rendered diminished activity when compared to the parent derivatives against most of the strains. It has been demonstrated that unlike the parent antibiotic, the C_{14} analogue (**34e**) does not bind to ribosomal RNA, instead targeting the bacterial membrane. It was also shown that AMEs had trifling deactivating impact on these cationic amphiphilic AGs.

The same group also prepared 6″-modified amphiphilic tobramycin analogues bearing a triazole ring (**38a–c**) or an amide linkage (**39a–c**)

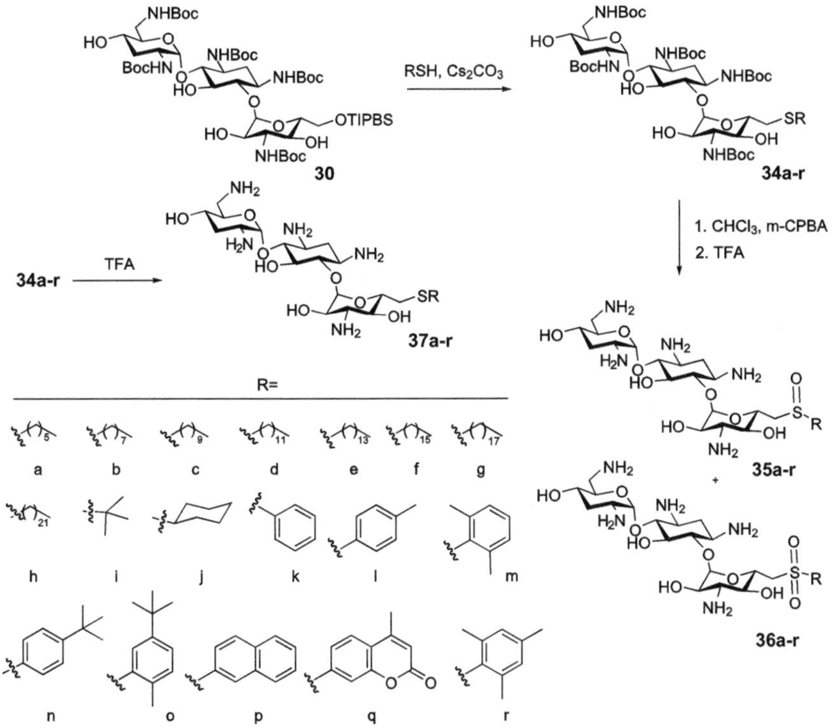

Scheme 11.12 Synthesis of C-6″ thioether tobramycin analogues.

Table 11.12 MIC values of 6″-thioether tobramycin analogues.

	MIC values (μg mL^{-1})[a]									
Cpd.	A	B	C	D	E	F	G	H	I	J
Tob.	9.4	>150	18.8	75	18.8	2.3	>150	>150	150	18.8
37c	37.5	75	150	9.4	18.8	4.7	>150	>150	>150	>150
37e	9.4	9.4	2.3	2.3	1.2	0.3	18.8	4.7	2.3	4.7
37f	4.7	9.4	9.4	9.4	2.3	1.2	18.8	4.7	4.7	18.8

[a]Bacterial strains: (A) MRSA; (B) *S. pyogenes* serotype M12 (strain MGAS9429); (C) *S. mutans* UA159; (D) *B. subtilis* 168; (E) *B. anthracis* 34F2 Sterne strain; (F) VRE; (G) *E. faecalis* ATCC 29212; (H) *E. coli* BL21 (DE3) with AAC(3)-IV-Int-pET19b-pps; (I) *E. coli* BL21 (DL3) with Eis; (J) *E. coli* TOlC.

(Scheme 11.13).[56] However, when compared to the corresponding thiol ether derivative **37e**, the antibacterial activity for most of the analogues could not be further improved (Table 11.13). However, some of the thioether (**37e**), triazole (**38b**) and amide analogues (**39b**) exhibited greater potency against tobramycin-resistant *S. pyogenes* (Tables 11.12 and 11.13). It was earlier shown that low micromolar concentrations of aliphatic chain carboxylic acids ranging from C_{12} to C_{14} preferentially inhibited the biofilm formation by *S. aureus* and *Listeria monocytogene* strains.[57] Minimal biofilm inhibition

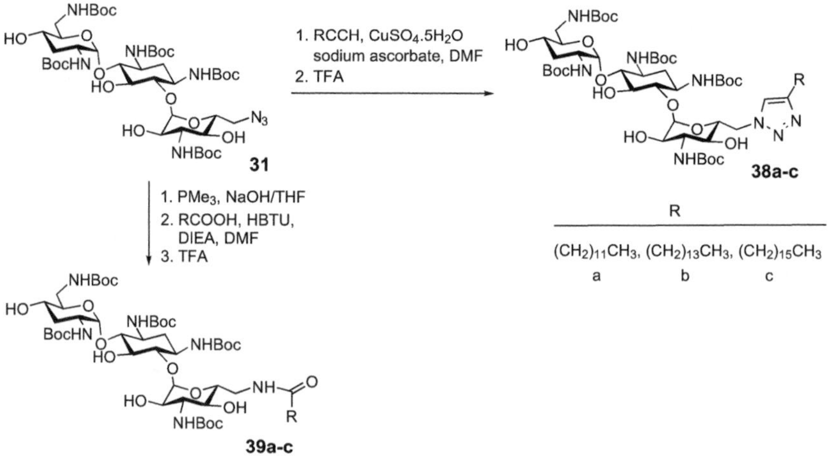

Scheme 11.13 Synthesis of triazole- and amide-linked tobramycin derivatives.

Table 11.13 MIC values of triazole- and amide-linked tobramycin derivatives.

Cpd.	MIC values ($\mu g\ mL^{-1}$)^a A	B	C	D	E	F	G	H	I	J
Tob.	64	>128	128	>128	>128	16	128	<1	>128	16
38b	4	64	4	32	128	8	8	4	128	32
39b	4	32	4	32	64	8	4	4	128	16

^{a}Bacterial strains: (A) *S. pyogenes* serotype M12 (strain MGAS9429); (B) MRSA; (C) *S. mutans* UA159; (D) VRE; (E) *E. faecalis* ATCC 29212; (F) *S. aureus* ATCC 9144; (G) *S. epidermidis* ATCC 35984; (H) *S. epidermidis* ATCC 33347; (J) *S. sonnei* clinical isolate 6831.

concentration (MBIC) values were evaluated against *S. mutants* UA159 and *S. epidermidis* ATCC 35984 to demonstrate the efficiency of the C_{12} and C_{14} aliphatic tobramycin analogues for their biofilm growth properties. Although there was a clear improvement in MBIC values (ranging 4–32 $\mu g\ mL^{-1}$) when compared to the parent tobramycin (MBIC range of 64–128 $\mu g\ mL^{-1}$), it becomes evident that this increment could be due to the self-prompted antibacterial activity and rules out the validity of these candidates as biofilm growth inhibitors. It was also shown that the C_{12} aliphatic chain 6″-amide analogue (**39a**) exhibited lower hemolytic activity ($10.2 \pm 0.8\%$ at 128 $\mu g\ mL^{-1}$) on rat RBCs than corresponding thioether ($71.6 \pm 8.3\%$) and triazole ($89.1 \pm 1.6\%$) analogues.

Tobramycin derivatives bearing 3″,6″- and 4′,6″-dithioether alkyl chain scaffolds (**41a–c, 43a–c**) were synthesized (Scheme 11.14) and tested for their antibacterial activity against a range of bacterial strains.[58] As shown in Table 11.14, the 4′,6″-dithioether analogues **41b**, **41c** and **43a** were found to be potent against MRSA. In addition, **43a** displayed both a low percentage of hemolysis and demonstrated potent antimicrobial activity against seven of the eight bacterial strains tested (Table 11.14).

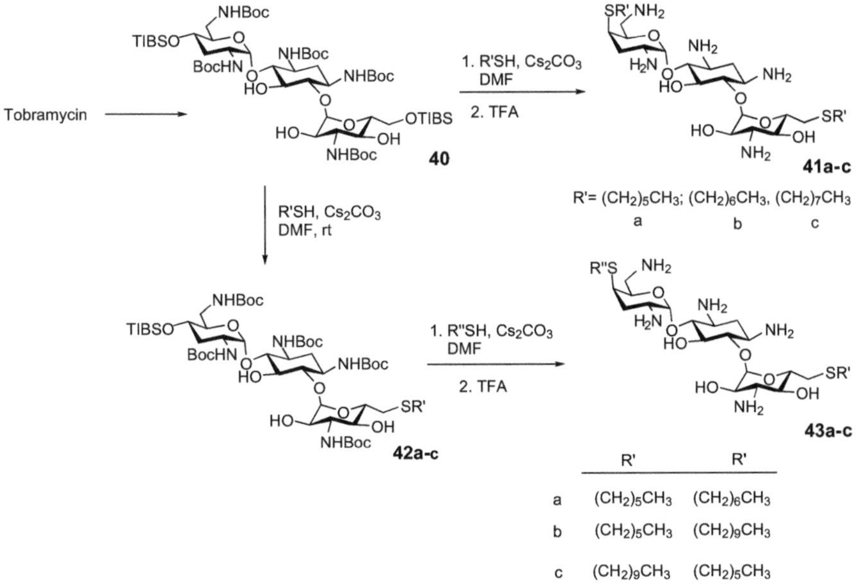

Scheme 11.14 Synthesis of tobramycin-based amphiphilic dithioether analogues.

Table 11.14 MIC values of tobramycin-based amphiphilic dithioether analogues.

	MIC values (μg mL^{-1})^{a}							
Cpd.	A	B	C	D	E	F	G	H
Tob.	16	>32	0.3	>32	16	32	>32	8
41b	4	8	2	4	>32	>32	1	4
41c	4	4	4	8	>32	32	1	4
43a	4	8	2	4	8	32	4	2

^{a}Bacterial stains: (A) *S. aureus* Oxford NCTC6571; (B) MRSA; (C) *S. epidermidis* ATCC 12228 (biofilm negative); (D) *S. epidermidis* ATCC 35984/RP62A (biofilm positive); (E) *S. aureus* Cowan, ATCC 12598; (F) *S. pyogenes* serotype M1T1; (G) *S. pyogenes* JRS75; (H) *S. pyogenes* serotype T5.

11.3.4 Amphiphilic Kanamycin Analogues

Umezawa reported the synthesis of tetra-*N*-(phenylalkyl)kanamycin derivatives **44a–p** and evaluated their antimicrobial properties against Gram-positive, Gram-negative and mycobacteria.[59] Tetra-*N*-(phenylalkyl)-kanamycins were synthesized by initial formation of Schiff's bases derived from appropriate aldehyde and kanamycin A and subsequent reduction with sodium borohydride (Scheme 11.15). Various *N*-substituted kanamycin derivatives such as **44b, 44c, 44d, 44j** and **44m** have shown small improvement in the antimicrobial activity when compared to the parent antibiotic. However, only a diminutive increment in the antimicrobial activity for tetra-*N*-(4-chlorobenzyl)kanamycin (**44k**) against *P. aeruginosa* has been observed.

Our group has extended the aminoglycoside carbamate chemistry to a tetracationic kanamycin A scaffold (Scheme 11.16), where the

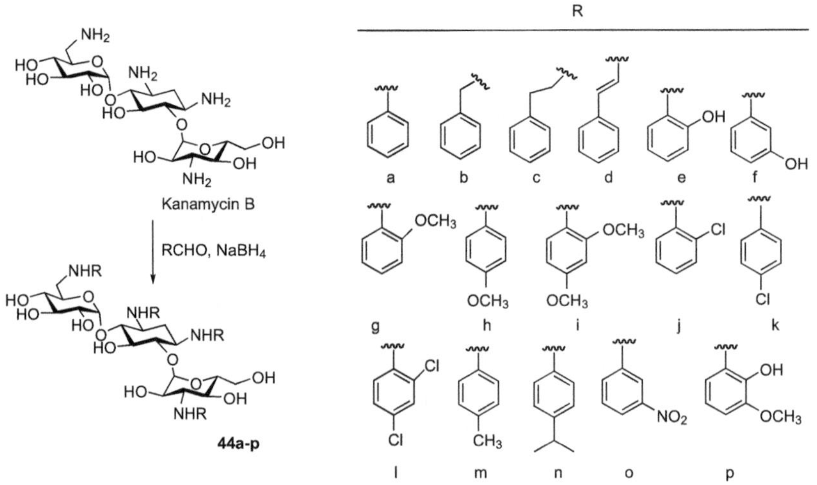

Scheme 11.15 Synthesis of tetra-*N*-(phenylalkyl)kanamycin analogues.

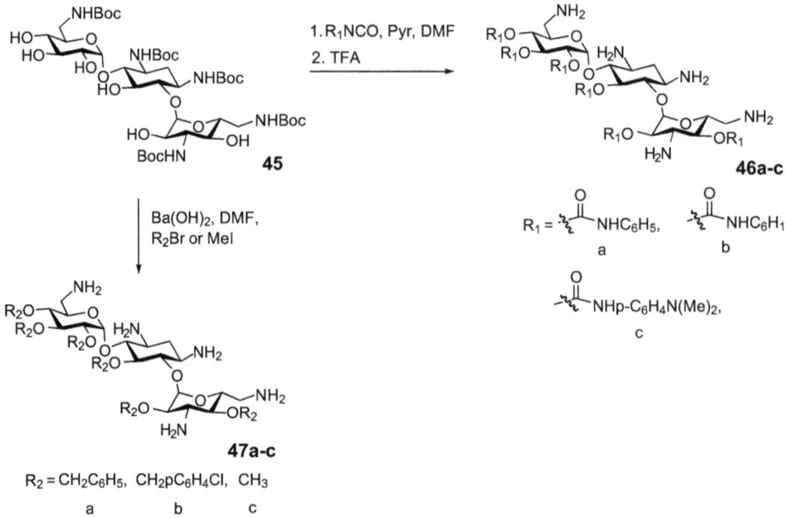

Scheme 11.16 Synthesis of kanamycin A-derived polycarbamate and polyether analogues.

heptacarbamate analogues **46a–c** exhibited reduced Gram-positive activity (up to 32-fold) and weak Gram-negative activity (MIC > 64 μg mL^{-1}) (Table 11.15).[40] The most potent kanamycin polyether analogue **47b** exhibited at least 16-fold upsurge in activity against MRSA and MRSE.

Kanamycin-based peptide triazole conjugates (**48a,b**) have also been reported (Scheme 11.17).[43] It has been shown that the kanamycin A-dipeptide conjugate **48b** manifested a 16-fold lower MIC value against

Table 11.15 MIC values of kanamycin A-derived polycarbamate and polyether analogues.

	MIC values (μg mL^{-1})^a												
Cpd.	A	B	C	D	E	F	G	H	I	J	K	L	M
Gen.	1	2	0.25	32	4	1	128	8	8	128	>512	128	0.25
Neo.	1	256	0.25	0.5	32	4	8	ND	512	512	>512	64	1
Kan. A	4	>512	2	128	8	8	16	32	>512	>512	>512	32	1
46a	32	32	4	8	64	>256	>256	>256	>256	>256	>256	>256	>256
47b	8	8	4	8	32	32	32	16	>64	>64	>64	>64	>64

^{a}Bacterial stains: (A) *S. aureus* ATCC 29213; (B) MRSA ATCC 33592; (C) *S. epidermidis* ATCC 1490; (D) MRSE ATCC 14990; (E) *S. pneumoniae* ATCC 49619; (F) *E. coli* ATCC 25922; (G) *E. coli* ATCC 6174; (H) *E. coli* CAN-ICU 63074; (I) *P. aeruginosa* ATCC 27853; (J) *P. aeruginosa* CAN-ICU 62308; (K) *S. maltophilia* CAN-ICU 62584; (L) *A. baumannii* CAN-ICU 63169; (M) *S. pneumoniae* ATCC 13883; ND = not determined.

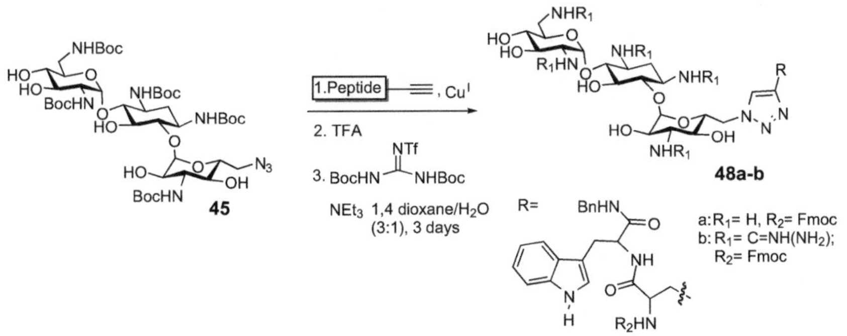

Scheme 11.17 Synthesis of triazole and guanidinylated kanamycin analogues.

Table 11.16 MIC values of triazole and guanidinylated kanamycin analogues.

	MIC values (μg mL^{-1})^a									
Cpd.	A	B	C	D	E	F	G	H	I	J
Kan.	4	>512	2	128	8	8	16	32	>512	>512
48a	16	32	8	16	64	32	32	32	128	16
48b	4	8	2	8	64	64	64	64	128	128

^{a}Bacterial stains: (A) *S. aureus* ATCC 29213; (B) MRSA ATCC 33592; (C) *S. epidermidis* ATCC 14990; (D) MRSE CAN-ICU 61589; (E) *S. pneumoniae* ATCC 49619; (F) *E. coli* ATCC 25922; (G) *E. coli* CAN-ICU 61714; (H) *E. coli* CAN-ICU 63074; (I) *P. aeruginosa* ATCC 27853; (J) *P. aeruginosa* CAN-ICU 62308.

MRSE, whereas another dipeptide conjugate (**48a**) exhibited a ≥32-fold lower MIC against kanamycin A-resistant *Pseudomonas aeruginosa* (Table 11.16).

Chang and co-workers have developed a novel class of kanamycin-based aminoglycosides (**52a–c**) derived through stereoselective glycosylation between the neamine core and an α-glycosyl donor (Scheme 11.18).[60,61] Incorporation of a C_8 alkyl aliphatic chain at the 4″-O-position of ring III furnished compound **52b** with a broad spectrum of antifungal activity (Table 11.17).

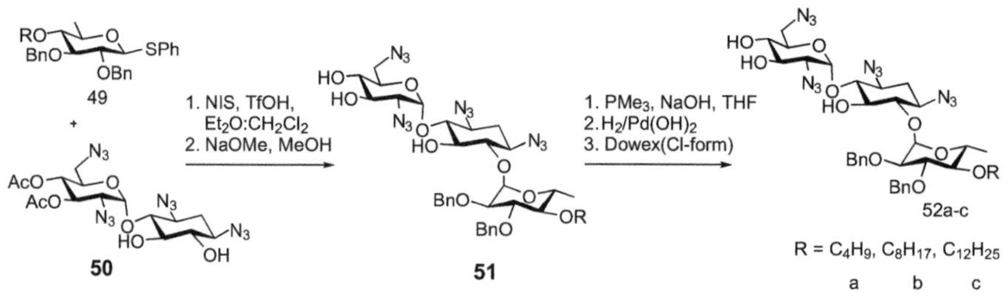

Scheme 11.18 Synthesis alkyl-modified kanamycin analogues.

Table 11.17 MIC values of alkyl-modified kanamycin analogues.

	MIC values (μg mL^{-1})^a													
Cpd.	A	B	C	D	E	F	G	H	I	J	K	L	M	N
Kan. B	>250	>250	>250	>250	>250	ND	ND	ND	ND	ND	ND	ND	ND	ND
52b	7.8	31.3	3.9	31.3	7.8	7.8	15.6	15.6	31.3	31.3	31.3	31.3	31.3	31.3

^{a}Fungi: (A) *R. pilimanae* (ATCC 26423); (B) *C. albicans* (ATCC 10231); (C) *S. cerevisiae* W303; (D) *F. graminearum* B-4-5A; (E) *F. oxysporum*; (G) *P. irregulare*; (H) *P. ultimum* (I) *P. parasitica*; (J) *R. stolonifer*; (K) *C. cladosporioides*; (L) *C. brachyspora*; (M) *B. cinerea*; (N) *P. spp.*; ND = not determined.

Ye and co-workers have synthesized C-4′-modified kanamycin B derivatives encompassing dimethylamino (**56a**), acetamido (**56b**), 4-aminobutanoylamino (**56c**), cyclohexylcarbonylamino (**56d**), 2-(2-methylbenzamido)acetamido (**56e**) and piperidin-4-ylcarbonylamino (**56f**) scaffolds (Scheme 11.19).[62] In general, except for **56a** the other kanamycin analogues retained or furnished enhanced antibacterial activity against some bacterial strains. An intriguing 2- to 16-fold enhancement was observed for compound **56c** against four reference strains, as shown in Table 11.18.

Amikacin-based amphiphilic cationic polycarbamates and polyethers were synthesized by employing a previously described protocol (Scheme 11.20).[40] The pentacationic octaphenyl carbamate (**58a**) and octahexyl carbamate (**58b**) rendered 2- to 16-fold increased Gram-positive activity. In addition, the tetracationic octahexyl carbamate (**58b**) exhibited an 8- to 32-fold increment in antibacterial activity against *E. coli* and *S. maltophilia* strains (Table 11.19). Amikacin-derived pentacationic unsubstituted heptabenzyl ethers perpetuated the antibacterial activity (Table 11.19).

11.3.5 Amphiphilic Neamine Analogues

Based on the affinity of ethylenediamine units harboured with naphthyl, anthryl or acridyl scaffolds toward polyA.polyU (RNA model) and poly(dA).poly(dT) (DNA model) for groove binding,[63] Décout's group designed and synthesized neamine derivatives equipped with phenyl, naphthyl, pyridyl or quinolyl rings (Scheme 11.21).[64,65] The 3′,4′-(**61f**), 3′,6-(**61g**) and 3′,4′,6-(**61j**) 2-naphthylmethylamine derivatives displayed good antibacterial

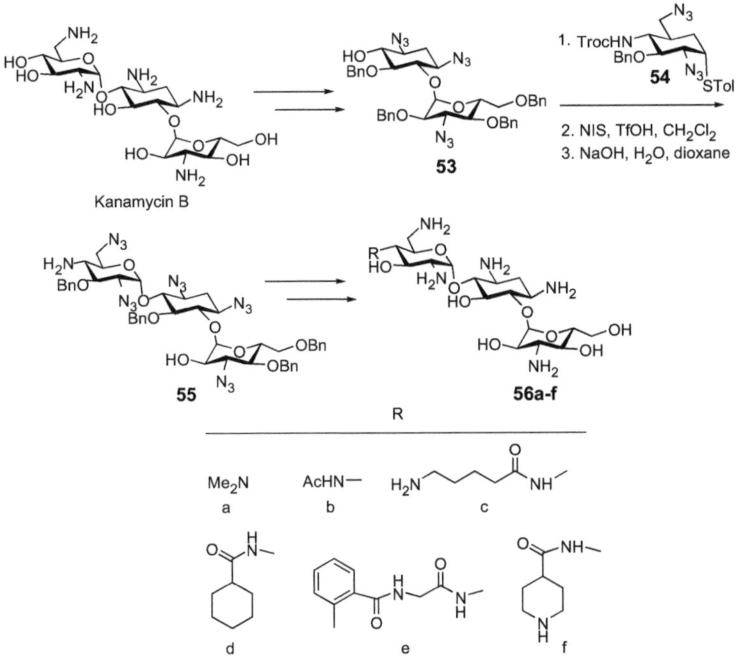

Scheme 11.19 Synthesis of C-4′-modified kanamycin B analogues.

Table 11.18 MIC values of C-4′-modified kanamycin B analogues.

	MIC values (μg mL^{-1})[a]									
Cpd.	A	B	C	D	E	F	G	H	I	J
Kan. B	4	4	1	>128	0.25	32	>128	16	64	64
56a	128	64	32	>128	16	32	>128	64	>128	>128
56b	2	8	2	16	0.5	2	>128	2	4	>128
56c	1	1	0.25	2	<0.06	0.25	32	1	0.5	32
56d	2	2	1	4	0.25	1	>128	32	8	64
56e	4	4	2	64	0.5	8	>128	64	32	64
56f	2	2	1	2	0.25	0.25	128	8	2	128

[a]Bacterial strains: (A) *E. coli* ATCC 25922; (B) *E. coli* ATCC 32518; (C) *S. aureus* ATCC 29213; (D) *S. aureus* ATCC 33591; (E) *S. epidermidis* ATCC 12228; (F) *S. epidermidis* f: 08-18 MR clinical isolate; (G) *S. epidermidis* f: 07-9, clinical isolate expressing APH(3′)-IIIa; (H) *K. pneumoniae* ATCC 700603; (I) *P. aeruginosa* ATCC 27853; (J) *E. faecalis* ATCC 29212.

activities against *Staphylococcus aureus.* The neamine derivative **61j** also displayed prominent antimicrobial activity against Gram-negative bacterial strains (MIC *P. aeruginosa*, 4–8 μg mL^{-1}; *E. coli*, 4–16 μg mL^{-1}; *A. lwoffi*, 4–32 μg mL^{-1}) and strains over-expressing efflux pumps in *P. aeruginosa* (4–16 μg mL^{-1}) (Table 11.20). To optimize the antibacterial activity of the neamine derivatives, the same group fine-tuned the lipophilicity of the naphthyl moiety. They have synthesized 3′,6-di- and 3′,4′,6-tri(2-naphthylalkyl)neamine derivatives.[66] They have also prepared neamine derivatives

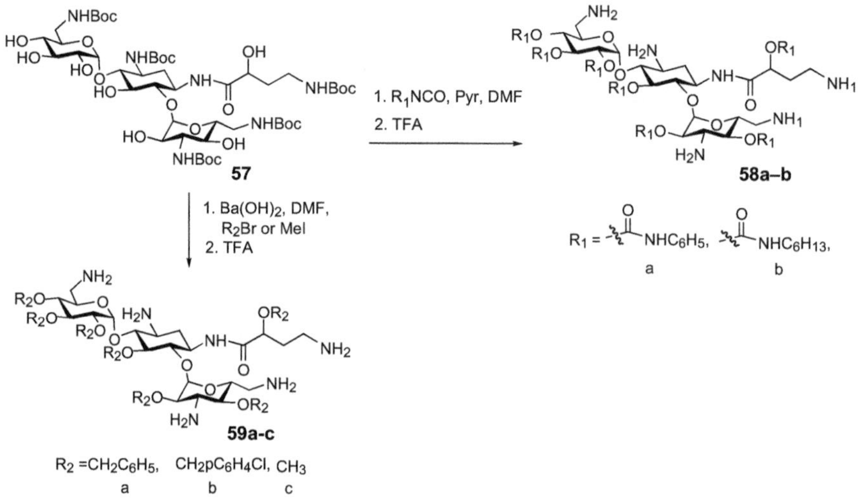

Scheme 11.20 Synthesis of amikacin-derived polycarbamate and polyether analogues.

Table 11.19 MIC values of amikacin-derived polycarbamate and polyether analogues.

Cpd.	MIC values ($\mu g\ mL^{-1}$)[a] A	B	C	D	E	F	G	H	I	J	K	L	M
Gen.	1	2	0.25	32	4	1	128	8	8	128	>512	128	0.25
Neo.	1	256	0.25	0.5	32	4	8	ND	512	512	>512	64	1
Ami.	4	8	1	2	8	4	2	32	4	128	>512	128	0.5
58a	8	16	2	2	32	256	256	256	>256	256	128	256	256
58b	4	4	4	4	32	4	8	4	64	16	16	>128	>128
59a	4	4	1	1	32	16	16	16	128	64	32	32	16

[a]Bacterial strains: (A) *S. aureus* ATCC 29213; (B) MRSA ATCC 33592; (C) *S. epidermidis* ATCC 1490; (D) MRSE ATCC 14990; (E) *S. pneumoniae* ATCC 49619; (F) *E. coli* ATCC 25922; (G) *E. coli* ATCC 6174; (H) *E. coli* CAN-ICU 63074; (I) *P. aeruginosa* ATCC 27853; (J) *P. aeruginosa* CAN-ICU 62308; (K) *S. maltophilia* CAN-ICU 62584; (L) *A. baumannii* CAN-ICU 63169; (M) *S. pneumoniae* ATCC 13883; ND = not determined.

bearing simple linear alkyl chains devoid of naphthyl scaffolds. Increase in the lipophilicity in the dialkyl series led to broad-spectrum antibacterial activities. For instance, the dialkyl derivatives 3′,6-di-2NP (**61s**), 3′,6-di-2NB (**61t**) and 3′,6-diNn (**61x**) were shown to be active against susceptible and resistant Gram-positive and Gram-negative bacteria (Table 11.20).

11.3.6 Amphiphilic Paromamine Analogues

To investigate the role of the 6′-amino functionality on the antibacterial activity of small amphiphilic aminoglycosides with a neamine core, Décount's group synthesized 3′,6-di-2NM (**63a**), 3′,4′,6-tri-2NM (**63b**),

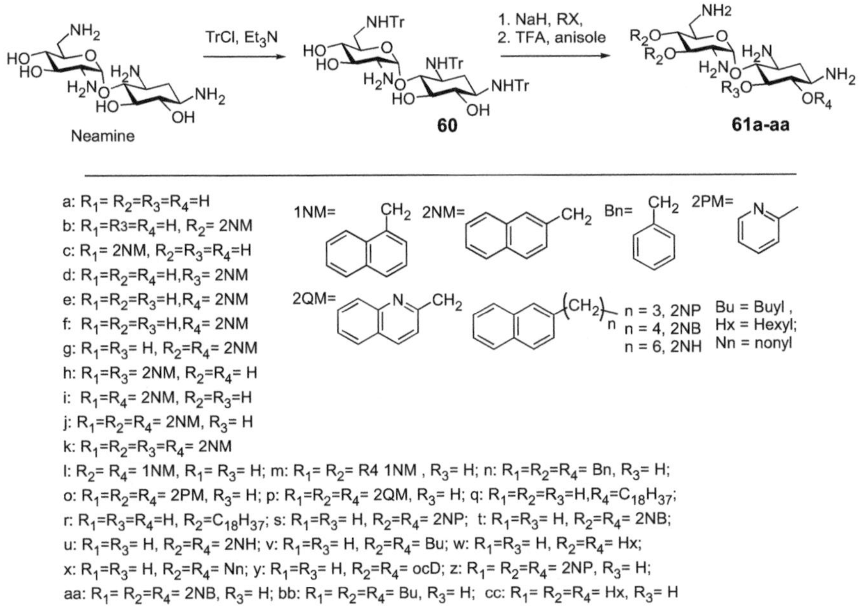

Neamine → (TrCl, Et_3N) → **60** → (1. NaH, RX, 2. TFA, anisole) → **61a-aa**

a: R_1= R_2=R_3=R_4=H
b: R_1=R3=R_4=H, R_2= 2NM
c: R_1= 2NM, R_2=R_3=R_4=H
d: R_1=R_2=R_4=H,R_3= 2NM
e: R_1=R_2=R_3=H,R_4= 2NM
f: R_1=R_2=R_3=H,R_4= 2NM
g: R_1=R_3= H, R_2=R_4= 2NM
h: R_1=R_3= 2NM, R_2=R_4= H
i: R_1=R_4= 2NM, R_2=R_3=H
j: R_1=R_2=R_4= 2NM, R_3= H
k: R_1=R_2=R_3=R_4= 2NM
l: R_2= R_4= 1NM, R_1= R_3= H; m: R_1= R_2= R4 1NM , R_3= H; n: R_1=R_2=R_4= Bn, R_3= H;
o: R_1=R_2=R_4= 2PM, R_3= H; p: R_1=R_2=R_4= 2QM, R_3= H; q: R_1=R_2=R_3=H,R_4=$C_{18}H_{37}$;
r: R_1=R_3=R_4=H, R_2=$C_{18}H_{37}$; s: R_1=R_3= H, R_2=R_4= 2NP; t: R_1=R_3= H, R_2=R_4= 2NB;
u: R_1=R_3= H, R_2=R_4= 2NH; v: R_1=R_3= H, R_2=R_4= Bu; w: R_1=R_3= H, R_2=R_4= Hx;
x: R_1=R_3= H, R_2=R_4= Nn; y: R_1=R_3= H, R_2=R_4= ocD; z: R_1= R_2=R_4= 2NP, R_3= H;
aa: R_1= R_2=R_4= 2NB, R_3= H; bb: R_1= R_2=R_4= Bu, R_3= H; cc: R_1= R_2=R_4= Hx, R_3= H

1NM= ; 2NM= ; Bn= ; 2PM= ; 2QM= ; n = 3, 2NP; n = 4, 2NB; n = 6, 2NH; Bu = Buyl , Hx = Hexyl; Nn = nonyl

Scheme 11.21 Synthesis of amphiphilic neamine derivatives.

Table 11.20 MIC values of amphiphilic neamine derivatives.

	MIC values (μg mL^{-1})[a]									
Cpd.	A	B	C	D	E	F	G	H	I	J
Neo. B	2	0.5	>128	64	128	32	16–32	2	4	32
Nea.	32	2	>128	>128	>128	>128	32–64	32	>128	32
61f	4	8	>128	32	>128	>128	>128	32	16	16
61g	8	64	>128	128	128	>128	>128	64	64	64
61j	4	4	32	8	8	4	16	16	4	4
61m	1	2	128	8	8	8	128	8	2	16
61s	2	ND	ND	4	16	16–32	ND	16	8	16
61t	2	4	64	4	8	8	32	8	4	8
61x	2	ND	ND	4	4	4	ND	4–8	2–4	4
61cc	4	4	64	8	8	8	8–16	4	4	4

[a]Bacterial strains: (A) *S. aureus* ATCC 25923; (B) *A. lwoffi* ATCC 17925; (C) *A. lwoffi* AI.88-483 APH3′-VIA; (D) *P. aeruginosa* ATCC 27853; (E) *P. aeruginosa* Psa.F03 AAC6′-IIA; (F) *P. aeruginosa* PA22 (PT629) surexp MexXY; (G) *K. pneumoniae* ATCC 700603; (H) *E. coli* ATCC 25922; (I) *E. coli* PAZ505H8101 AAC6′-IB; (J) *E. coli* L58058.1 ANT2″-IA; ND = not determined.

3′,6-di-1NM (**63c**) and 3′,4′,6-tri-1NM (**63d**) naphthyl paromamine derivatives (Scheme 11.22).[66] The tri-NM derivatives appeared to be two- to fourfold more active than di-NM paromamine derivatives against Gram-positive and Gram-negative bacteria. In addition, the 1NM and 2NM isomers showed similar antibacterial activities against most of the Gram-positive and Gram-negative strains in the paromamine series (Table 11.21).

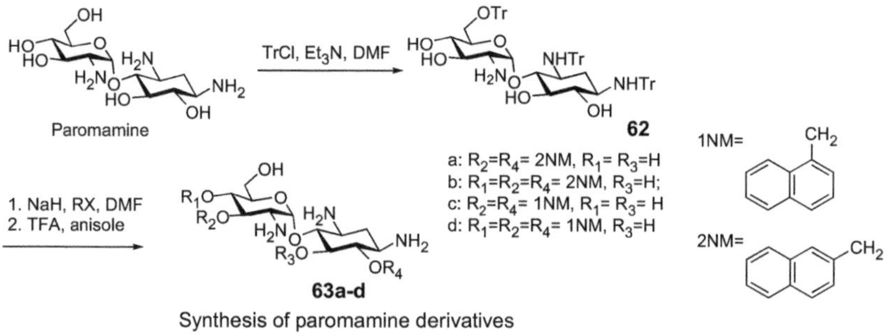

Paromamine

TrCl, Et_3N, DMF

62

a: R_2=R_4= 2NM, R_1= R_3=H
b: R_1=R_2=R_4= 2NM, R_3=H;
c: R_2=R_4= 1NM, R_1= R_3= H
d: R_1=R_2=R_4= 1NM, R_3=H

1. NaH, RX, DMF
2. TFA, anisole

63a-d

Synthesis of paromamine derivatives

Scheme 11.22 Synthesis of amphiphilic paromamine derivatives.

Table 11.21 MIC values of amphiphilic paromamine derivatives.

	MIC values (μg mL^{-1})[a]									
Cpd.	A	B	C	D	E	F	G	H	I	J
Nea.	32	2	>128	>128	>128	>128	32–64	32	>128	32
Par.	32–64	ND	ND	>128	>128	>128	ND	>128	>128	>128
63a	8	64	>128	128	64–128	128	>128	128	64	64
63b	2	1	>128	32	32	32	32	64	16	32
63c	32	128	>128	64	128	>128	128	128	64	128
63d	2	1–2	128	32	32	32	128	16	16	16

[a]Bacterial strains: (A) *S. aureus* ATCC 25923; (B) *A. lwoffi* ATCC 17925; (C) *A. lwoffi* AI.88-483 APH3′-VIA; (D) *P. aeruginosa* ATCC 27853; (E) *P. aeruginosa* Psa.F03 AAC6’-IIA; (F) *P. aeruginosa* PA22 (PT629) surexp MexXY; (G) *K. pneumoniae* ATCC 700603; (H) *E. coli* ATCC 25922; (I) *E. coli* PAZ505H8101 AAC6′-IB; (J) *E. coli* L58058.1 ANT2″-IA; ND = not determined.

11.4 Concluding Remarks

Interest in the development of AAAs is growing because of the emergence of bacteria which are resistant to all classes of antibiotics. The development of membrane-targeting antibacterials like AAAs appears to be a promising route to combat bacterial resistance, as this pathway appears to be less prone to resistance development. Many MDR bacteria which are resistant to aminoglycosides, β-lactams, fluoroquinolones and tetraccyclines are still susceptible to membrane-targeting antibacterials like colistin. However, the development of AAA-based lead structures faces a major challenge: systemic toxicity. The majority of AAAs are not able to differentiate between bacterial and host cells, which leads to significant toxicity. Moreover, many AAAs are strongly protein bound and their antibacterial activity is reduced in the presence of hydrophobic proteins or serum. The examples discussed in this chapter illustrate that AAAs possess great potential to combat MDR infections and as such may provide a new source of antibacterial leads in the years to come.

Acknowledgements

The authors thank the Canadian Institutes of Health Research (CIHR) for financial support (MOP 119335).

References

1. A. Schatz, E. Bugie and S. A. Waksman, *Proc. Soc. Exp. Biol. Med.*, 1944, **55**, 66.
2. S. A. Waksman, *Antimicrob. Agents Chemother.*, 1965, **5**, 9.
3. S. Umezawa and T. Tsuchiya, Total synthesis and chemical modification of the aminoglycoside antibiotics, ed. H. Umezawa and I. R. Hooper, in *Aminoglycoside Antibiotics: Hand Book of Experimental Pharmacology*, Springer-Verleg, New York, 1982, vol. 32, pp. 37–110.
4. J. Xie, A. E. Talaska and J. Schacht, *Hear Res.*, 2011, **281**, 28.
5. B. Frederiksen, S. Lanng, C. Koch and N. Hoiby, *Pediatr. Pulmonol.*, 1996, **21**, 153.
6. L. G. Greer, S. W. Roberts, J. S. Sheffield, V. L. Rogers, J. B. Hill, D. D. Mcintire and G. G. Wendel Jr., *Infect. Dis. Obstet. Gynecol.*, 2008, 891426.
7. S. Sundar, T. K. Jha, C. P. Thakur, P. K. Sinha and S. K. Bhattacharya, *N. Engl. J. Med.*, 2007, **356**, 2571.
8. S. E. Haydel, *Pharmaceuticals*, 2010, **3**, 2268.
9. J. A. Caminero, G. Sotgiu, A. Zumla and G. B. Migliori, *Lancet Infect. Dis.*, 2010, **10**, 621.
10. M. Kaul, C. M. Barbieri and D. S. Pilch, *J. Am. Chem. Soc.*, 2006, **128**, 1261.
11. W. S. Champney, *Bact. Curr. Drug. Targets: Infect. Disord.*, 2001, **1**, 19.
12. R. A. Giuliano, G. J. Paulus, G. A. Verpooten, V. M. Pattijin, D. E. Pollet, E. J. Nouwen and M. E. De Broe, *Kidney Int.*, 1984, **26**, 838.
13. J. D. Hayes and C. R. Wolf, *Biochem. J.*, 1990, **272**, 281.
14. G. A. Jacoby and G. C. Archer, *N. Engl. J. Med.*, 1991, **324**, 601.
15. J. Heller, *Intl. Urol. Nephrol.*, 1984, **16**, 243.
16. H. W. Taber, J. P. Mueller, P. F. Miller and A. S. Arrow, *Microbiol. Rev.*, 1987, 429.
17. G. D. Wright, A. M. Berghuis and S. Mobashery, *Adv. Exp. Med. Biol.*, 1998, **456**, 27.
18. T. Shakya and G. D. Wright, in *Mechanisms of Aminoglycoside Antibiotic Resistance in Aminoglycoside Antibiotics*, ed. D. P. Arya, Wiley, 2007, pp. 119–140.
19. G. D. Wright, *Chem. Commun.*, 2011, **47**, 4055.
20. F. Gao, X. Yan, O. M. Baettig, A. M. Berghuis and K. Auclair, *Angew. Chem., Int. Ed.*, 2005, **44**, 6859.
21. F. Gao, X. Yan, T. Shakaya, O. M. Baettig, S. Ait-Mohand-Brunet, A. M. Berghuis, G. D. Wright and K. Auclair, *J. Med. Chem.*, 2006, **49**, 5273.
22. M. L. Magalhães, M. W. Vetting, F. Gao, L. Freiburger, K. Auclair and J. S. Blanchard, *Biochemistry*, 2008, **47**, 579.
23. Y. Gao, X. Yan, O. Zahr, A. Larsen, K. Yong and K. Auclair, *Bioorg. Med. Chem. Lett.*, 2008, **18**, 5518.
24. F. Gao, X. Yan and K. Auclair, *Chemistry*, 2009, **15**, 2064.

25. L. A. Freiburger, O. M. Baettig, T. Sprules, A. M. Berghuis, K. Auclair and A. K. Mittermaier, *Nat. Struct. Mol. Biol.*, 2011, 288.
26. Y. Zhou, H. Yu, Q. Guo, *et al.*, *Eur. J. Clin. Microbiol. Infect. Dis.*, 2010, **29**, 1349.
27. H. Nikaido and Y. Takatsuka, *Biochim. Biophys Acta*, 2009, **1794**, 769.
28. R. E. W. Hancock and H. Sahl, *Nat. Biotechnol.*, 2006, **24**, 1551.
29. S. A. Shelburne, D. M. Musher, K. Hulten, H. Ceasar, Y. M. Lu, I. Bhaila and J. R. Hamill, *Antimicrob. Agents Chemother.*, 2004, **48**, 4016.
30. W. T. Shier and K. L. Rinehart, Jr., *J. Antibiot.*, 1973, **26**, 547.
31. J. E. Hawkins, in *Pharmacology of Neomycin, in Neomycin, its Structure and Practical Application*, ed. S. A. Waksman, The Williams and Wilkins Co., Baltimore, Md., 1958, pp. 130–146.
32. S. Bera, G. G. Zhanel and F. Schweizer, *J. Med. Chem.*, 2008, **51**, 6160.
33. T. J. Baker, N. W. Luedke, Y. Tor and M. Goodman, *J. Org. Chem.*, 2000, **65**, 9054.
34. S. Bera, G. G. Zhanel and S. J. Schweizer, *Antimicrob. Chemother.*, 2010, **65**, 1224.
35. J. Zhang, F. I. Chiang, L. Wu, P. G. Czyryca, D. Li and C.-W. T. Chang, *J. Med. Chem.*, 2008, **51**, 7563.
36. C.-W. T. Chang, Y. Hui, B. Elchert, J. Wang, J. Li and R. Rai, *Org. Lett.*, 2002, **4**, 4603.
37. J. Zhang, K. Keller, J. Y. Takemoto, M. Bensaci, A. Litke, P. G. Czyryca and C.-W. T. Chang, *J. Antibiot.*, 2009, **62**, 539.
38. V. Udumula, Y. W. Ham, M. Y. Fosso, K. Y. Chan, R. Rai, J. Zhang, J. Li and C.-W. Chang, *Bioorg. Med. Chem. Lett.*, 2013, **23**, 1671.
39. V. Pokrovskaya, V. Belakhov, M. Hainrichson, S. Yaron and T. Baasov, *J. Med. Chem.*, 2009, **52**, 2243.
40. S. Bera, G. G. Zhanel and F. Schweizer, *J. Med. Chem.*, 2010, **53**, 3626.
41. M. L. Mayer, D. M. Easton and R. E. W. Hancock, Fine tuning host responses in the face of infection: Emerging roles and clinical applications of host defence peptides, in *Antimicrobial Peptides: Discover, Design and Novel Therapeutic Strategies*, ed. G. Wang, CABI, Oxfordshire, 2010, ch. 12, pp. 195–220.
42. S. E. Blondelle, E. Takahashi, K. T. Dinh and R. A. Houghten, *J. Appl. Bacteriol.*, 1995, **78**, 39.
43. S. Bera, G. G. Zhanel and S. Schweizer, *Bioorg. Med. Chem. Lett.*, 2010, **20**, 3031.
44. B. Findlay, G. G. Zhanel and F. Schweizer, *Bioorg. Med. Chem. Lett.*, 2012, **22**, 1499.
45. S. Hanessian, J. Szychowski, S. S. Adhikari, G. Vasquez, P. Kandasamy, E. E. Swayze, M. T. Migawa, R. Ranken, B. Francüois, J. W. Bartoschek, J. Kondo and E. Westhof, *J. Med. Chem.*, 2007, **50**, 2352.
46. H. Kawaguchi, S. Nakagawa and K. Fijistawa, *J. Antibiot.*, 1972, **25**, 695.
47. N. Takayuki, N. Susumu and T. Soichiro, *German Pat* 2,322,576, 1973.
48. C. Battistini, G. Franceschi, F. Zarini, G. Cassinelli, F. Arcamone and A. Sanfilippo, *J. Antibiot.*, 1982, **35**, 98.

49. S. Hanessian, P. Kandasamy, J. Szychowski, A. Giguère, E. E. Swayze, M. T. Migawa, B. Francüois, J. Kondo and E. Westhof, *Bioorg. Med. Chem. Lett.*, 2010, **20**, 7097.
50. M. L. Nelson and M. N. Alekhsun, WO2004062674. (2004).
51. Y. B. Zrihen, M. I. Herzog, M. Feldman, A. S. Segev, Y. Roichman and M. Fridman, *Bioorg. Med. Chem.*, 2013, **21**, 3624.
52. D. H. Fong and A. M. Berghuis, *Antimicrob. Agents. Chemother.*, 2009, **53**, 3049.
53. S. Hanessian, M. Tremblay and E. E. Swayze, *Tetrahedron*, 2003, **59**, 983.
54. R. Dhondikubeer, S. Bera, G. G. Zhanel and S. Schweizer, *J. Antibiot.*, 2012, **65**, 495.
55. I. M. Herzog, K. D. Green, Y. B. Zrihen, M. Feldman, R. R. Vidavski, A. E. Boock, R. S. Fainaro, A. Eldar, S. G. Tsodikova and M. Fridman, *Angew. Chem., Int. Ed.*, 2012, **51**, 5652.
56. I. M. Herzog, M. Feldman, A. E. Boock, R. S. Fainaro and M. Fridman, *Med. Chem. Commun.*, 2013, **4**, 120.
57. U. T. Nguyen, I. B. Wenderska, M. A. Chong, K. Koteva, G. D. Wright and L. L. Burrows, *Appl. Environ. Microbiol.*, 2012, **78**, 1454.
58. Y. B. Zrihen, M. Herzog, M. Feldman and M. Fridman, *Org. Lett.*, 2013, **15**, 6144.
59. A. Fujii, K. Maeda and H. Umezawa, *J. Antibiot.*, 1968, **21**, 340.
60. J. Li, H.-N. Chen, H. Chang, J. Wang and C.-W. T. Chang, *Org. Lett.*, 2005, 7, 3061.
61. C.-W. T. Chang, M. Fosso, Y. Kawasaki, S. Shrestha, M. F. Bensaci, J. Wang, C. K. Evans and J. Y. Takemoto, *J. Antibiot.*, 2010, **63**, 667.
62. R. B. Yan, M. Yuan, Y. Wu, X. You and X.-S. Ye, *Bioorg. Med. Chem.*, 2011, **19**, 30.
63. N. Lomadze, H.-J. Schneider, M. T. Albelda, E. Garcia-Espana and B. Verdejo, *Org. Biomol. Chem.*, 2006, **4**, 1755.
64. I. Baussanne, A. Bussière, S. Halder, C. Ganem-Elbaz, M. Ouberai, M. Riou, J.-M. Paris, E. Ennifar, M.-P. Mingeot-Leclercq and J.-L. Dècout, *J. Med. Chem.*, 2010, **53**, 119.
65. M. Ouberai, F. E. Garch, A. Bussiere, M. Riou, D. Alsteens, L. Lins, I. Baussanne, Y. F. Dufrêne, R. Brasseur, J.-L. Dècout and M.-P. Mingeot-Leclercq, *Biochim. Biophis. Acta*, 2011, **1808**, 1716.
66. L. Zimmerman, A. Bussiere, M. Ouberai, I. Baussanne, C. Jolivalt, M.-P. Mingeot-Leclercq and J.-L. Dècout, *J. Med. Chem.*, 2013, **56**, 7691.

CHAPTER 12

From the Capsular Polysaccharide to a Conjugate Vaccine Containing Haemophilus influenzae *Type b Synthetic Oligosaccharide*

MARIA C. RODRÍGUEZ MONTERO, JOSÉ A. RUÍZ GARCÍA, YURY VALDÉS BALBÍN AND VICENTE VÉREZ BENCOMO*

Center for Biomolecular Chemistry, calle 21 y 200, Siboney, Playa, La Habana, Cuba
*Email: vicente.verez@cqb.cu

12.1 Introduction

Conjugate vaccines were first developed for *Haemophilus influenzae* type b (Hib) and have now become a formidable tool in the battle against Hib infections. They have been further extended to prevent diseases by other encapsulated bacteria such as *Streptococcus pneumoniae* and *Neisseria meningitides*. Prompted by the strong stimulus provided by the necessity of near 500 million doses per year around the globe, a conjugate vaccine against Hib was also the first vaccine to be developed using a synthetic antigen. This discovery should not be oversimplified as the mere advent of alternatives for production methods for glycoconjugate-based vaccines.

RSC Drug Discovery Series No. 43
Carbohydrates in Drug Design and Discovery
Edited by Jesús Jiménez-Barbero, F. Javier Cañada and Sonsoles Martín-Santamaría

Published by the Royal Society of Chemistry, www.rsc.org

In fact, it also became a potent incentive for the development of other synthetic oligosaccharides to be used as glycoconjugate vaccines.

In the present chapter we will provide a short overview of all these aspects.

12.2 Bacteria and Disease

Hib is a cause of bacterial infections that are particularly severe among infants. Estimates at the beginning of 21st century considered that Hib caused some 10 million serious illnesses worldwide, as well as 371 000 deaths in children aged 1–59 months.[1]

The bacterium was first described by Pfeiffer in 1892, who isolated it during an outbreak of influenza. In the 1930s the viral origin of influenza was established and Hib was then reassigned to play a role in secondary bacterial infection. Almost at the same time, Pittman identified six capsular types (a–f) and observed that virtually all isolates from cerebrospinal fluid and blood were of the capsular type b.[2]

Before the introduction of effective vaccines, Hib was the leading cause of bacterial meningitis and other invasive bacterial disease among children below five years old; approximately one in 200 children in this age group developed an invasive Hib disease. Nearly all Hib infections occurred among these children, while approximately two-thirds of the cases occurred among children below 18 months old.[3]

For the majority of individuals, Hib is a mere commensal of the nasopharynx. Only a minority of those exposed (or those that are carriers of the organism) suffer invasive disease, due to a naturally acquired immunity.[3] In fact, it is no coincidence that most of the bacteria responsible for invasive bacterial disease in childhood display a polysaccharide capsule. This capsule may provide a survival advantage for these organisms during transmission and colonization. In the pathogenesis of invasive diseases, it also facilitates survival in the blood, through resistance to complement-mediated killing and phagocytosis.[4,5] It has been suggested that the polysaccharide capsule may confer a survival advantage by allowing evasion of mucosal immune responses or by facilitating transmission between hosts by reducing desiccation.[6] In terms of invasive disease, the polysaccharide capsule has also been shown to inhibit serum bactericidal activity and complement-mediated phagocytosis.[7,8]

12.3 Capsular Polysaccharide as the Main Antigen

Considering all the benefits that bacteria obtain from their capsular polysaccharides, it is not surprising that anticapsular polysaccharide antibodies are protective against Hib invasive disease. The first evidence was already obtained 80 years ago,[3] as shown in Figure 12.1. As can be seen from the figure (dashed line), the highest incidence of disease occurs for young children, between 6–12 months old; after two years of life, the invasive disease is rarely seen. In contrast, "the bactericidal power of the blood"

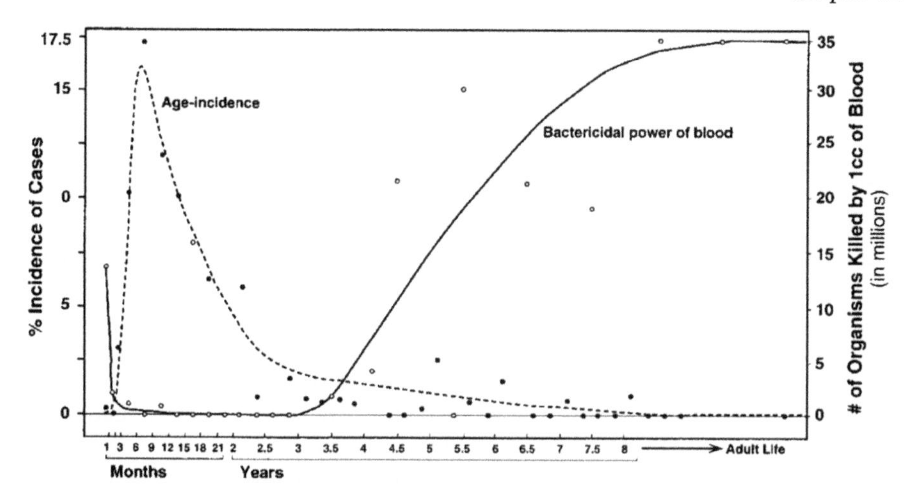

Figure 12.1 *Dashed line*: incidence of cases of invasive Hib disease from infancy to adult life. *Solid line*: bactericidal power of the blood as measured by the ability of the serum to opsonize and kill Hib organisms in the presence of phagocytes.
(Reproduced from Fothergill and Wright,[3] with permission.)

follows an inverse pattern. It is represented by the solid lines and reflects the ability of the serum to opsonize and kill Hib organisms in the presence of phagocytes.

Relatively high levels of transplacentally acquired anti-Hib antibodies fall over the first months of life to very low levels by around six months old. Subsequently, antibody titers rise again during the second year of life and peak by eight years of age, presumably as a result of exposure to Hib in the nasopharynx or to other organisms with cross-reactive antigens.

The age-specific incidence of invasive disease is inversely related to the presence of "bactericidal power in the blood" or titer of anti-Hib antibodies. Indeed, the highest incidence of disease in an unvaccinated population takes place in the time interval between the loss of the maternal antibodies and the generation of antibodies by the child's own B cells.

The protective properties of anti-Hib antibodies were also demonstrated to be specific for the capsular polysaccharide for rabbit serum.[9,10] These properties are ideal for using the Hib capsular polysaccharide[11] as a vaccine candidate.

The structure of the Hib capsular polysaccharide (CPS) was first correctly elucidated by using a combination of periodic oxidation, paper chromatography, and ^{13}C NMR spectroscopy.[12] Further ^{1}H NMR studies of the polysaccharide[13] confirmed the structure as the polyribosylribitol phosphate (PRP) depicted in Figure 12.2.

The PRP is attractive as a vaccine antigen since invasive disease is almost exclusively restricted to type b organisms. Different evidence has demonstrated that anti-PRP antibodies are a key part of the protective natural

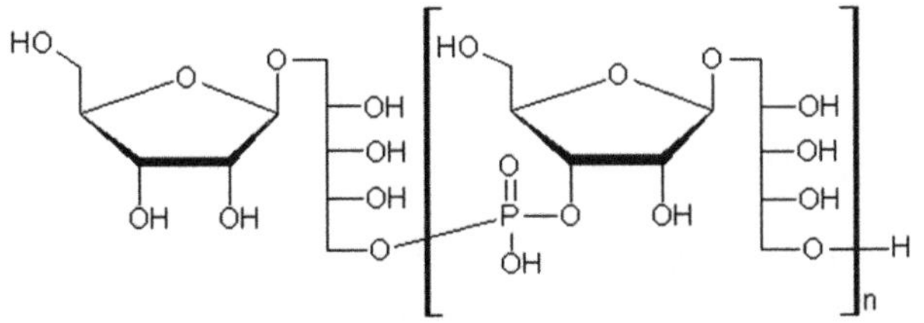

Figure 12.2 Structure of the Hib capsular polysaccharide.

immunity. Thus, in order to protect those individuals at high risk of acquiring Hib disease, the induction of anti-PRP antibodies for young enough humans has been the goal of the vaccine development process.

However, early observations with Hib demonstrated the limitations of the polysaccharide as a vaccine antigen. When given during the first two years of life, the purified PRP induces relatively low titers of serum antibodies. These are usually insufficient to protect against invasive disease.[14,15] In terms of immunological memory, the antibody induced by PRP wanes quickly. Moreover, subsequent immunization shows no evidence of immunological priming in any age group.[16] As a general pattern, the majority of invasive diseases occurred during the first two years of life.[17] In other populations, particularly in the developing world, the majority of disease occurs even earlier during infancy.[18,19]

The polysaccharide, although recognized by the B-cell receptors, cannot be presented to the T cells in conjunction with the major histocompatibility complex class II (MHC-II) molecules. Therefore, B cells lack the ability to directly recruit cognate T-cell helpers when stimulated with the polysaccharide. The polysaccharide's interaction with the B cell is thus termed T-independent. Thus, the lack of specific T-cell interactions in the immune response limits the immunogenicity of PRP, since the development of memory B cells with class-switched antibodies and subsequent avidity maturation cannot occur.

12.4 The Challenge of Making a More Immunogenic CPS

The concept of a hapten and a hapten–protein conjugate was introduced by Landsteiner to provide a chemically defined system for studying the binding of an individual antibody to a unique epitope on a protein-containing antigen.[20] According to this concept, a small organic molecule (hapten) is chemically coupled to a large protein (carrier) to provide a hapten–carrier conjugate. Animals immunized with such a conjugate produce specific antibodies not only against the carrier protein (unaltered epitopes), but also for the hapten determinant.

Later, Avery and Goebel applied this concept to cellobiouronic acid[21,22] and to the *Streptococcus pneumoniae* (Sp) serotype 3 CPS. Both conjugates to horse serum globulin were able to induce polysaccharide-specific antibodies. Remarkably, they conferred immunity to challenge by live Sp serotype 3.

In this context, Hib PRP was shown to be more immunogenic when covalently linked to a protein carrier and to show boosted responses, characteristic of T-dependent memory.[23,24] Hib was the first example of the development of a conjugate vaccine as a pharmaceutical,[25] thus initiating the era of glycoconjugate-based vaccines.

The covalent linkage of the protein and the polysaccharide is essential for achieving the enhanced immunogenicity of the glycoconjugate vaccines.[26] In addition, *in vitro* studies, using peripheral blood mononuclear cells from adults treated with the Hib conjugate vaccine, indicated that the maximum antibody production and T-cell proliferation required direct contact between the T and B cells.[27] These data are indeed consistent with the classical mode of presentation of the carrier protein in conjunction with the MHC-II to the T cell (Figure 12.3). This leads to a germinal center formation with antibody class switching, avidity maturation, and memory B-cell production.[28] Such cognate T- and B-cell interactions require the involvement of co-stimulatory molecules, such as CD40–CD40L and CD27/CD70.[29,30] In fact, CD4 + T cells specific for the carrier protein have been detected following glycoconjugate vaccination, secreting both T helper-1 and T helper-2 cytokines.[31] Studies of

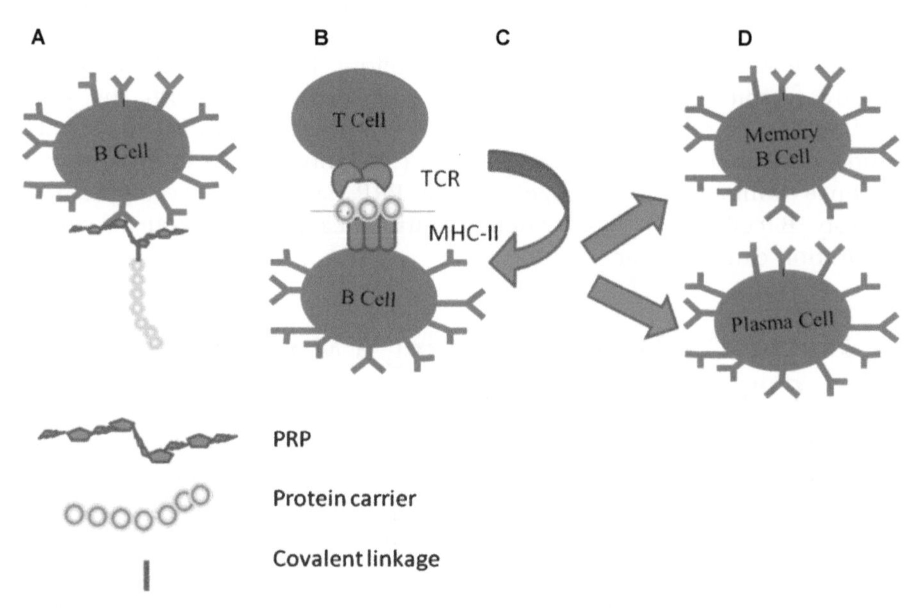

Figure 12.3 Schematic representation of (A) recognition of a PRP in a glycoconjugate by BCR in a B-cell; (B) after processing, presentation of a peptide fragment in MHC-II to a T-cell that recognizes it through TCR; (C) T-cell help to a B-cell for affinity maturation; and (D) transformation of a B-cell under T-cell cytokines either in a plasmatic or memory cell.

other glycoconjugate vaccines in children have shown an increase in CD40L mRNA expression in peripheral blood mononuclear cells following vaccination, further suggesting the importance of cognate T- and B-cell interactions in the response to conjugate vaccines.[32]

While there still appears to be some variability in the immune response with age, conjugate vaccines bypass the relative unresponsiveness of the infant immune system for the plain polysaccharide and provide the basis for the generation of memory B cells as well as priming of the immune system.

Conjugates that are very different by their composition have become successful vaccines despite small differences in their effect (see below). Different molecular weight PRPs have been used in conjugate vaccines, with distinct covalent linkers to the protein carrier. Useful protein carriers are as diverse as tetanus toxoid (TT), modified diphtheria toxoid CRM197 (CRM197), or the Neisseria meningitis outer membrane protein (NmOMP). The observed differences were more pronounced when the results for the diphtheria toxoid (DT) conjugate (a weak response was induced) and for the NmOMP conjugate (immunogenic in infants after a single dose) were compared.[33] The latter appear to additionally engage Toll-like receptor 2 on dendritic cells, perhaps altering the regulation of T-cell responses to the vaccine, thus contributing to the improved response after a single dose.

12.5 Conjugate Vaccines Using a CPS

The development of a successful conjugate vaccine has been a scientific and technological challenge since the initial efforts for Hib. Basically, the success in achieving such a pharmaceutical[34] can be defined by the proper selection of (i) the polysaccharide modification strategy, (ii) the chosen carrier protein and its modification, (iii) the employed conjugation methods and (iv) the use of robust analytical methods for the evaluation of the final glycoconjugate. In fact, different successful strategies have been employed to solve the previous steps.

High molecular weight Hib CPS has directly been activated, leading to randomly active sites positioned along the polysaccharide chain. For example, CNBr activation of **1** at basic pH takes place at all possible hydroxyl groups, giving O–CN bonds randomly distributed all along the chain.[25] Although highly efficient, this method uses the very toxic CNBr. Thus, 1-cyano-4-(dimethylamino)pyridinium tetrafluoroborate (CDAP) was developed as alternative reagent for the synthesis of **2**.[35] Transformation of **2** using adipic acid dihydrazide (ADH) as spacer leads to the activated polysaccharide **3**, ready for conjugation (Figure 12.4).

The coupling of **3** to the TT was performed using a TT-COOH derivative, *via* water-soluble carbodiimide (EDAC) condensation (Figure 12.4). The glycoconjugate vaccine **4** prepared by this method is therefore a high molecular weight aggregate, with several CPS and protein molecules linked together. Therefore, special analytical methods are required to ensure batch to batch consistency and stability. Several Hib commercially available vaccines are

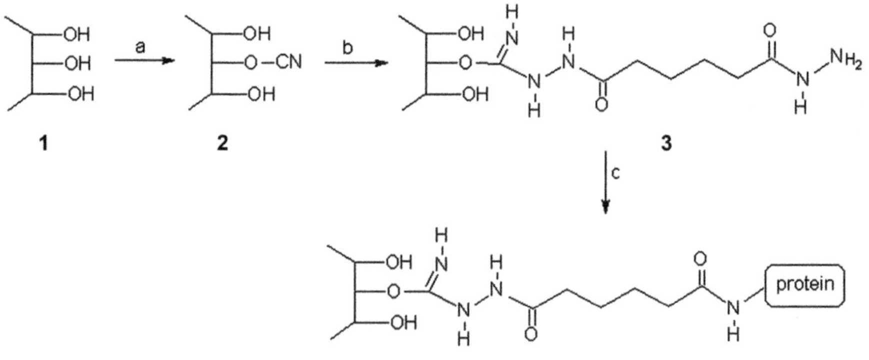

Figure 12.4 Synthesis of macromolecular glycoconjugates by activation with the CNBr–ADH method. (a) CNBr or CDAP; (b) ADH; (c) protein, EDAC.

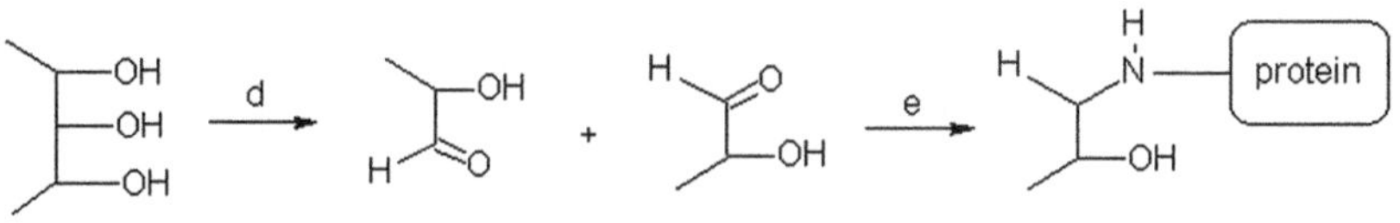

Figure 12.5 Synthesis of a glycoprotein-like glycoconjugate by using activation with the periodic oxidation–reductive amination method. (d) $NaIO_4$; (e) protein, $NaCNBH_3$.

based on these methods. For example, ProHIBiT is a glycoconjugate with DT,[36] while ActHIB, OmniHIB, Hiberix, and HibPRO are other commercially available TT conjugate vaccines.[37]

In a radically different approach, Hib CPS **1** was directly oxidized by sodium periodate (Figure 12.5). The oxidation proceeds through the diol, which is only present in the ribitol moiety. As a result, the low molecular weight activated polysaccharide **5** displays the aldehyde function at both ends.

The conjugation of **5** to CRM197[38] by reductive amination using sodium borohydride is very effective and affords **6**, which was used in the development of the commercially available HibTITER vaccine.

The Hib CPS **1** could be also reduced by acid hydrolysis to give the small fragment 7, which displays 5–15 repeating units (Figure 12.6). The terminal ribose hemiacetal was then modified by reaction with ethylenediamine to afford **8**, which was further elongated with the ADH spacer. The activated oligosaccharide **9** smoothly reacts with CRM 197, in the presence of carbodiimide, to afford **10**.[39] This preparation is used in the commercially available Vaxem.

The approaches represented in Figures 12.5 and 12.6 use rather little modified Hib CPS, thus leading to the "glycoprotein-like" conjugates **6** and

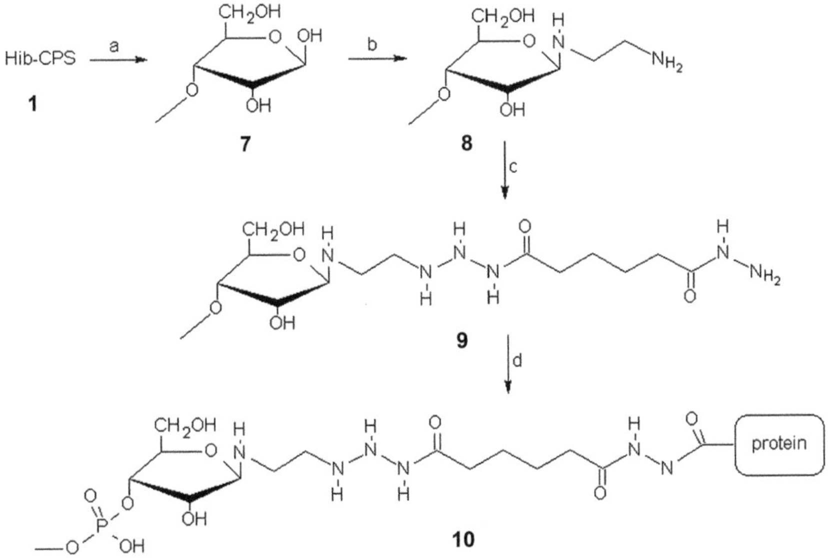

Figure 12.6 Synthesis of a "glycoprotein type" glycoconjugate by introduction of a spacer at the terminal hemiacetal. (a) Acetic acid; (b) ethylenediamine, $NaCNBH_3$; (c) ADH; (d) protein, EDAC.

10. These conjugates are rather different in their physicochemical properties to the previously described macromolecular glycoconjugate **4**. The major purification issue of the glycoconjugation process is the removal of the free unreacted polysaccharide and protein molecules. The profound difference in the properties of both glycoconjugate types implies that radically different but efficient approaches for purification should be employed.

A rather wide spectrum of molecular weights is covered by compounds **4, 6** and **10**, providing successful conjugate vaccines against Hib. Thus, one of the main conclusions of this initial glycoconjugate development is that the MW of the modified CPS is not critical provided that the B-epitope in the Hib CPS is preserved. A modified Hib CPS with intermediate molecular weight was also successfully used in a conjugate vaccine. Size-reduced Hib CPS **1** was transformed into its tetrabutylammonium salt for solubilization in dimethylformamide and dimethyl sulfoxide.[40] The process was followed by activation of the hydroxy groups with carbonyldiimidazole. A spacer was sequentially added with butanediamine (to give **11**) and bromoacetate to provide the modified Hib CPS **12**. This molecule is activated with bromoacetate at the end of the spacer. NmOMP was used as a protein carrier after thiolation (Figure 12.7).[43] Commercially available PedvaxHIB is based on compound **13**.

In 2000, after the first decade of the advent of Hib conjugate vaccines,[41] this vaccine was introduced in 22 countries, with a dramatic decline in incidence. Few vaccines in history have induced a similar impact over such a short period of time. It can be estimated that around 78% of the cases of

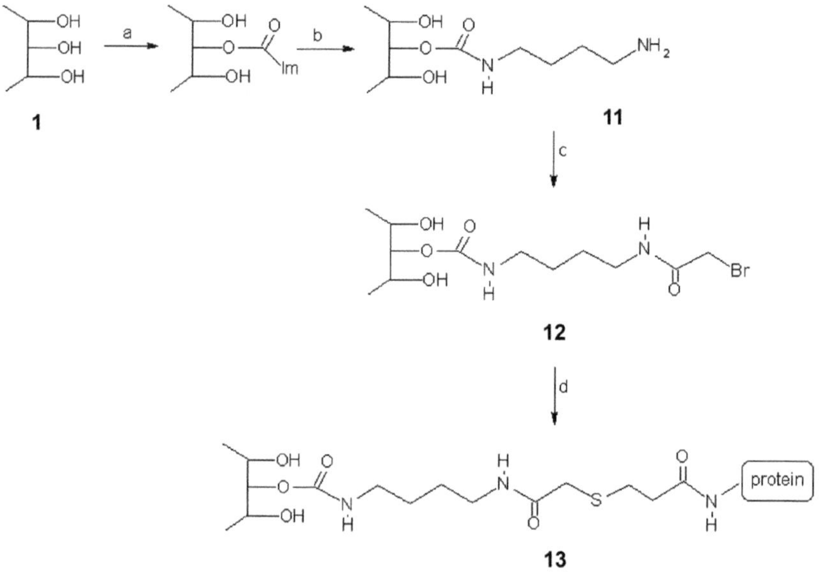

Figure 12.7 Synthesis of a glycoconjugate by a thiol–bromoacetate coupling method. (a) Carbonyldiimidazole; (b) butanediamine; (c) *p*-nitrophenyl bromoacetate; (d) protein-SH.

meningitis (21 000 of 27 000) in children aged 0 to 4 years, and around 50% of the cases of Hib disease in all age groups (38 000 of 70 000), are prevented annually by vaccination in developed regions. Associated with this impact, it is the unique glycoconjugate vaccine that allows the prevention of nasopharyngeal Hib colonization. Additionally, short-term and long-term protection against Hib infections are achieved by anti-PRP concentrations of 0.15 and 1.0 μg mL^{-1}, respectively.[42] In a non-vaccinated population, the limit of 0.15 μg mL^{-1} showed a good inverse relationship with the incidence of the disease, thus suggesting the presence of a sufficient amount of anti-polysaccharide antibodies.

A strong immunogenicity of the conjugates is observed by a good booster response of a vaccine to a challenge by plain Hib CPS, a situation that mimics the naturally occurring Hib infection.[43,44] The worldwide figures at that time were affected by the lack of precise information. However, the estimated numbers were less impressive than those previously discussed for the developed countries. In fact, only 5.9% of the cases of meningitis (21 000 of 357 000) or 8.5% of the cases of the classical Hib manifestations (38 000 of 445 000) were prevented worldwide. Clearly, the enormous impact of the initial vaccine in the developed world prompts us to perform more studies to define the real burden of the disease in the world.

In the year 2000 it was estimated that there were around 8.13 million serious illnesses worldwide (uncertainty range 7.33–13.2 million), with 371 000 deaths (247 000–527 000) in children aged 1–59 months.[45] These circumstances prompted renewed interest in developing alternatives for production of glycoconjugate vaccines, including the chemical synthesis of the Hib antigen.

12.6 The Synthetic Antigen

Small repeating unit fragments of the Hib CPS were obtained by chemical synthesis during the 1970s and 1980s.[46,47] The synthesis of larger fragments that could be useful as vaccine components were undertaken by van Boom's group,[48] and following a short interval, by Just's group.[49] The initial approaches for these syntheses already outlined the critical problems for the development of a successful vaccine. These critical problems could be summarized as follows: (i) deriving a simple route to a key disaccharide repeating unit, (ii) achieving a high-yield elongation process based on the disaccharide precursor, and (iii) employing a simple protective-group strategy, allowing a clean deprotection step.

12.6.1 Key Disaccharide Repeating Unit

Compound **16** is considered the key disaccharide repeating unit for further chain elongation. The benzyl moiety has been shown to be the best permanent protecting group, while R and R_1 are temporary protecting groups or free hydroxyl, for allowing the corresponding chain elongation. The retrosynthetic analysis (shown in Figure 12.8) for the synthesis of key disaccharide **16** leads to derivatives **14** and **15** of D-ribose and D-ribitol. As can be seen, R_2 and R_5 are the corresponding precursors of benzyl protecting groups in **16**. However, the existence of a non-participating benzyl group at position 2 does not assure the presence of the required β stereochemistry for **16**. The different solutions for achieving these processes underscore the differences between the published approaches.

The R and R_1 temporary protecting groups permit the introduction of active phosphorus derivatives. The choice should be in accordance with the chain elongation strategy, either 5→3 or 3→5. Usually, the latter one is the most favored, since it allows the use of the easily accessible primary 5-hydroxyl group of ribitol in the crucial elongation step.

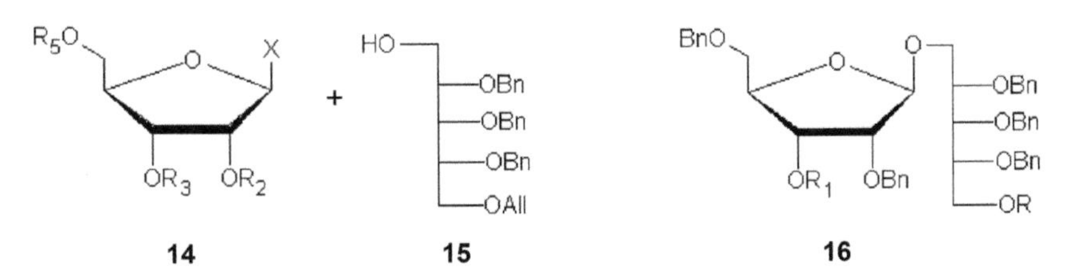

Figure 12.8 Retrosynthesis of the disaccharide key repeating unit **16**.

The β stereochemistry in the ribofuranosylation reaction is warranted by an acyl participating group at the 2-position of the D-ribose. There are two different strategies: (a) undifferentiated donor strategy, using a peracylated ribofuranosyl donor followed by deacylation and a multistep introduction of the required protective groups, or (b) a differentiated donors strategy, using 2-acylated 3- and 5-differentially protected donors. The 2-acyl substituent will warrant the required β stereochemistry, but must be exchanged to a permanent protective group in a single-step process. The first synthesis of a disaccharide for chain elongation is an example of the use of the undifferentiated strategy. The ribofuranosyl donors **17** or **18** (see Scheme 12.1) are used with the ribitol derivative **15**. Donors **17** and **15** smoothly react in the presence of TMSOTf to give the product in 88% yield.[50,51]

Donor **18**[52] was also able to very efficiently ribosylate **15**, giving the disaccharide in 97% yield (acetonitrile at −40 °C in the presence of silver perchlorate and 4 Å molecular sieves). Both compounds, after deacylation, were transformed using similar four-step complex procedures[55,56] for the introduction of protecting groups, to afford the disaccharide **19** in overall 34% yield (Scheme 12.2). The synthesis is based on the substitution of a 3,5-diol with the 1,1,3,3-tetraisopropyldisiloxane-1,3-diyl group (TIPDS). The chemical sequence is continued by the introduction of a benzyloxymethyl (BOM) group at position 2. Next, the TIPDS group is removed and the primary hydroxyl at position 5 is substituted by *tert*-butyldiphenylsilyl. The disaccharide **19** is further used as an elongation unit.

In the particular case of the ribofuranose donor **20**, a higher tendency to give the orthoester side product is usually observed when TMS triflate is used as promoter. In our laboratory, we reduce this side reaction by employing boron trifluoride etherate in dichloroethane as promoter, thus achieving a good yield of the target disaccharide.[53]

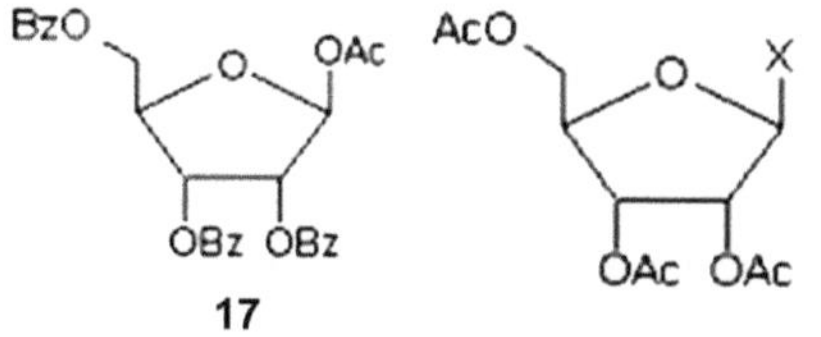

Scheme 12.1

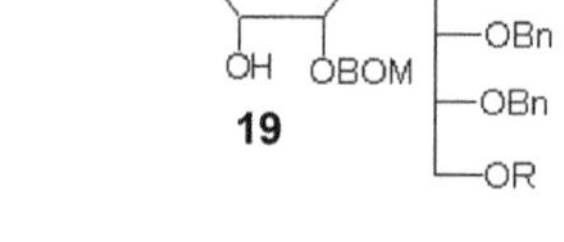

Scheme 12.2

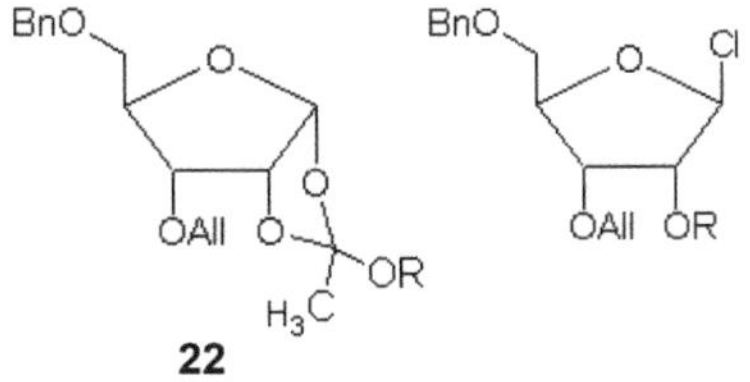

BnO O O O OAll O OR H3C 22

BnO O Cl OAll OR 21 R = Ac 23 R = Bz

Scheme 12.3

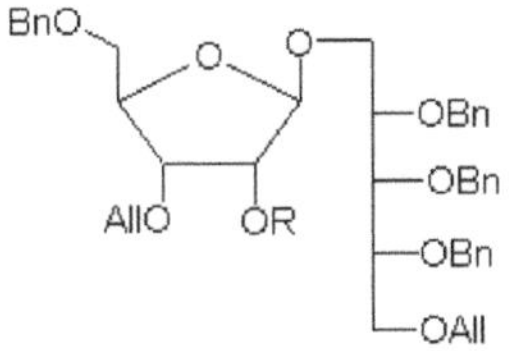

BnO O O OBn OBn OBn OAll AllO OR 24 R = Ac 25 R = Bz

Scheme 12.4

A second differentiated strategy is based on the use of a single 2-acylated donor that has another protecting groups at positions 3 and 5. Although this synthesis is much more complex than that for the peracylated donor, its transformation to the disaccharide is straightforward.

The preferred donor for this strategy is 2-*O*-acetyl-3-*O*-allyl-5-*O*-benzyl α-D-ribofuranosyl chloride **21** (see Scheme 12.3). The key intermediate is the corresponding 1,2-orthoacetate **22**, obtained from D-ribose in nine steps.[54] The chloride derivative **21** was employed for ribosylation of the ribitol derivative **15** in the presence of silver perchlorate, in 80% yield.[55] A similar chloride donor **23** obtained from the corresponding orthobenzoate also reacted with the ribitol derivative, without any catalyst, providing the corresponding disaccharide in 84% yield.[56]

The transformation of disaccharides **24** and **25** (see Scheme 12.4) into the key disaccharide **16** is very simple, since the deacetylation and benzylation steps at position 2 are straightforward. The new situation implies that R and R_1 are allyl moieties. Therefore, deallylation at both positions is needed, followed by introduction of mono- (MMT) or dimethoxytrityl (DMT) groups for differential protection of position 5 at the ribitol unit. Generally speaking, the required four-step procedure is accomplished in 65% yield.

We discovered a transformation of disaccharide **27** that is almost ideal for the synthesis of **29** (see Figure 12.9). The reaction is based on the initial formation of a dibutylstannylidene derivative, followed by selective benzylation in the presence of sodium hydride and Bu_4NI in toluene. The process allows the selective introduction of benzyl groups at positions 2 and 5 in a single step, directly obtaining a free O-3. Compared to previous procedures using undifferentiated donors, this process allows the desired key disaccharide in a single step.[57] On the other hand, the process is much simpler than that employing partially differentiated donors, and avoids the

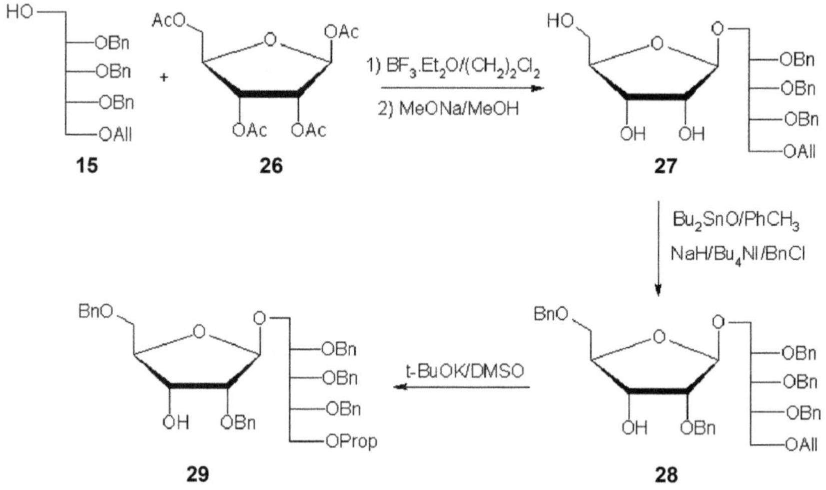

Figure 12.9 Proposed strategy for the synthesis of the key repeating unit.

multistep preparation. Additionally, the direct access to a free O-3 position avoids the complex di-deallylation process.

12.6.2 High-yielding Elongation Process on a Disaccharide Base

The elongation process is based on the formation of a phosphodiester between ribose O-3 and ribitol O-5. This concept was illustrated with the first solution synthesis of dimer **31** and trimer **32**, reported by Van Boom's group (Scheme 12.5). The coupling between the elongation unit **30** and the terminal unit was performed by activation with *N*-methylimidazole/pyridine to afford dimer **31**, in 83% yield. This synthesis is designed in such a way that permits the concomitant elongation of the saccharide chain. This process is achieved by selective deprotection of the 5-*O*-propenyl group to provide the 5-hydroxylated **31**, followed by a new addition of **30**. Thus, the protected trimer **32** is obtained, in 78% yield.[58]

In a similar approach, Just and co-workers used phosphoramidite **35**. The coupling of the monomers in the presence of tetrazole yielded 84% of dimer **33**.[59] A 2 + 2 strategy also provided the corresponding tetramer **34**.

For the chain elongation process, different approaches were employed that included the use of solution and solid-phase chemistry, as well as different types of phosphorus activation employing either phosphotriester, phosphoramidite or H-phosphonate moieties for creation of the phosphodiester bridges.

Although the solution chemistry approach is fairly simple, it affords very low yields. The repetitive elongation process requires many purification steps, with the corresponding weight losses. Thus, the use of a matrix-supported synthesis approach is recommended.

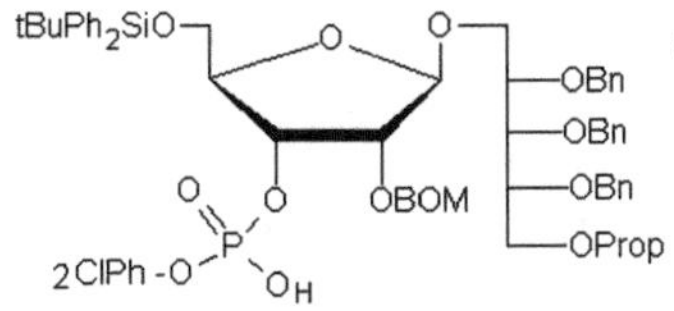

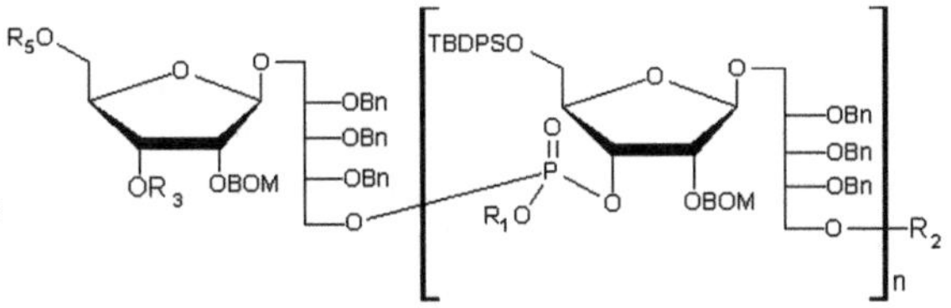

31	n=1	R_1=2-ClPh	R_2=Prop	R_3,R_5= TIPDS	
32	n=2	R_1=2-ClPh	R_2=Prop	R_3,R_5= TIPDS	
33	n=1	R_1=Me	R_2=MMTr	R_3=Lev	R_5=TBDPS
34	n=3	R_1=Me	R_2=MMTr	R_3=Lev	R_5=TBDPS

Scheme 12.5

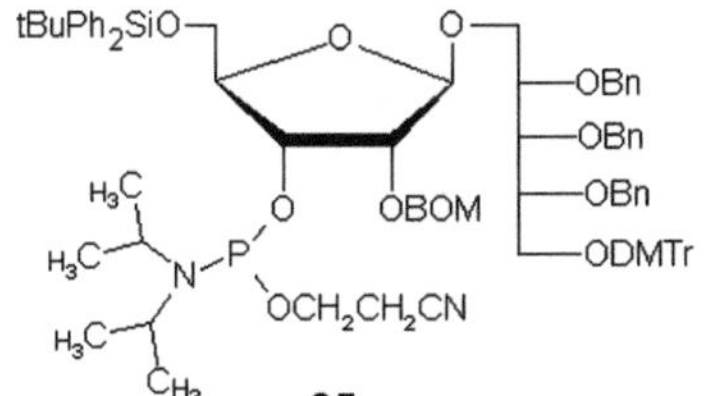

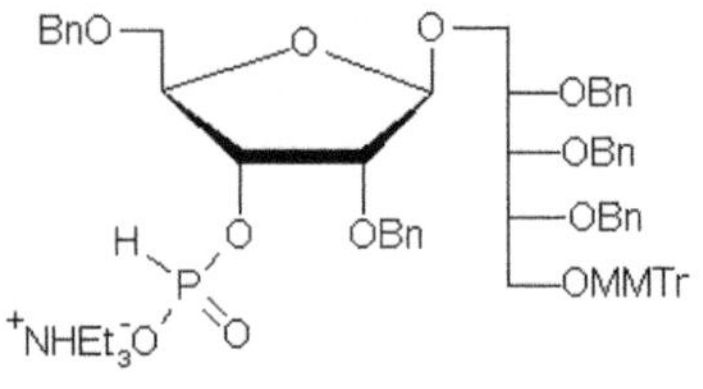

Scheme 12.6

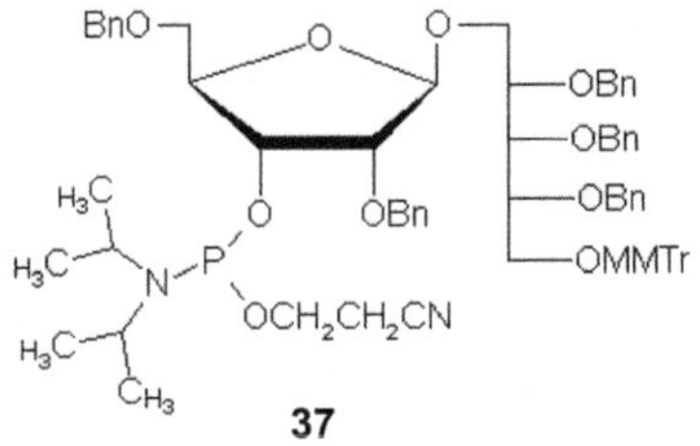

In this case, every coupling step should reach at least a 90% yield. In order to comply with this requirement, the Hib spacer-armed hexamer was obtained using a controlled-pore glass as the solid support.[59] The elongation unit selected for the solid-phase chemistry was phosphoramidite **35**. Once the disaccharide unit was attached to the solid support, the typical elongation cycle included DMT cleavage, followed by coupling with the activated disaccharide phosphoramidite **35** in the presence of 1*H*-tetrazole. Every cycle is finished by oxidation of the phosphite to a phosphate group, and capping of the unreacted compound by acetylation.

The use of solid-phase chemistry avoids the purification step after each coupling step. Moreover, the use of large excess of the coupling reagent and the key disaccharide elongation unit **35** permits driving the reaction nearly to completion. These are important advantages, besides the simple and fast sequential multistep process. In fact, it is possible to carry out all the reactions in the same vessel and the unreacted materials and the formed impurities can be easily removed by washing the support with the appropriate solvent.

A similar principle was adopted in the synthesis of the pentamer developed by Norberg's group, using H-phosphonate chemistry for the coupling process. The triethylammonium phosphonate **36** (see Scheme 12.6) was used as a the simple elongation unit in a cross-linked polystyrene solid support.[60] Once the disaccharide unit was attached to the matrix, the typical elongation cycle included MMT cleavage followed by coupling with 5 equivalents of the disaccharide phosphonate **36** in the presence of pivaloyl chloride. The capping of the unreacted material was achieved by acetylation. Oxidation of phosphonate to phosphate was performed at the final deprotection step.

Solid-phase synthesis allows access to a large CPS fragment with a high overall yield for vaccine production. However, its main disadvantage is the need of a large excess of the elongation unit, usually 5–10 molar equivalents. As a consequence the yield is rather low, *ca.* 10%, based on the precious disaccharide elongation unit.

In a different approach,[60] a soluble polymer was used as a support for the coupling reactions. This process combines the advantages from both the solution and solid-phase syntheses. On the one hand, the coupling proceeds under homogenous solution; therefore, the required excess reactant is not as big as in the solid-phase coupling approach. On the other hand, the matrix containing the compound is precipitated during the treatment, thus avoiding the complex purification steps. In fact, the efficiency reaches 90–95% per cycle, with an overall yield of 70%.

Once the disaccharide unit is fixed in a monomethoxy poly(ethylene glycol) soluble support, the typical elongation cycle consists of DMT cleavage followed by coupling with the activated disaccharide phosphoramidite **37** in the presence of 1*H*-tetrazole. Again, the capping of unreacted compound requires acetylation. The oxidation of the phosphoramidite moiety was conducted with *tert*-butyl hydroperoxide. An efficient synthesis of armed PRP_6 has been reported by this method.[64]

Another alternative that combines the advantages of both solution and solid-phase chemistry is polycondensation. As originally described,[61] a 1–6 linked mannopyranosyl phosphate oligomer, with an average of seven repeating units, was obtained in 16% yield, using a single polycondensation reaction from the partially deprotected mannopyranosyl phosphonate **38** (see Scheme 12.7) and pivaloyl chloride in pyridine. Further development of this approach with disaccharide phosphonate **39** afforded a decamer in 85% yield.[62] The reaction is rather complex and also provided the cyclic oligomer

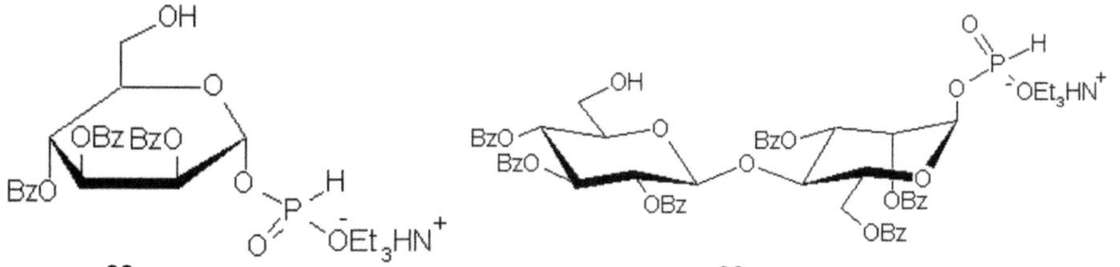

Scheme 12.7

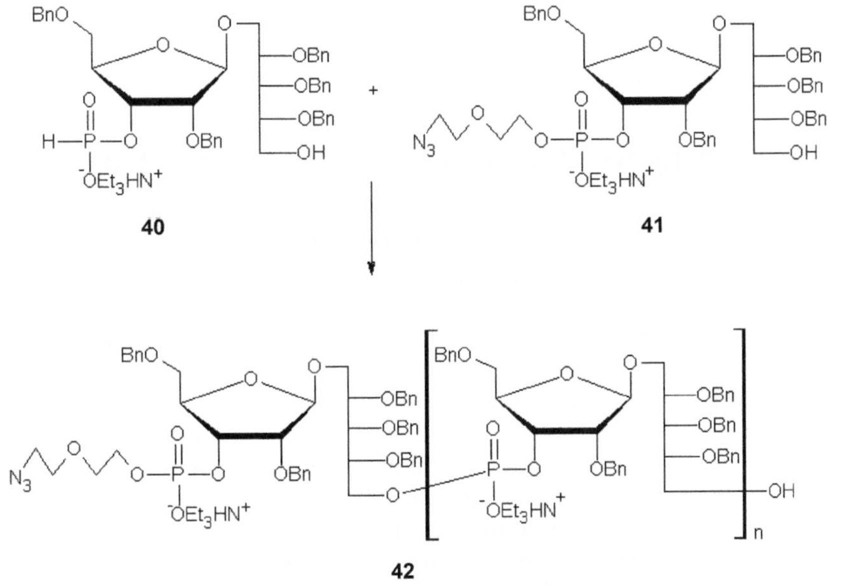

Figure 12.10 Polycondensation reaction leading to the PRP_8 spacer entity.

as the main side product. Nevertheless, this side reaction was reduced by performing the reaction at much higher concentration (10-fold).

Thus, we anticipated that the polycondensation approach could be used to provide a spacer-armed PRP oligomer, in a single step, for use in the conjugate vaccine. Our design targeted a synthetic antigen conceptually similar to **10** (see above), obtained from a CPS. Indeed, a well-defined mixture of oligomers with an activated spacer at the hemiacetal terminus was obtained (commercially available Vaxem-Hib).[63]

With this aim, we selected phosphonate **40** as the key elongation unit (see Figure 12.10). After extensive experimentation, we concluded that the spacer-armed derivative **41** should be employed as the quenching unit. The polycondensation reaction between **40** and **41** takes place, even at high scale, to yield 80% of **42**, after I_2-mediated oxidation of phosphonate to phosphate and gel filtration. This crucial discovery highly simplifies the accessibility of synthetic oligosaccharides for vaccine production.

12.6.3 Simple Protective-group Strategy Allowing Clean Deprotection

Deprotection of the oligomer is also a critical process. For those compounds having a combination of silyl protecting groups (TIPDS and *t*-$BuPh_2Si$), a treatment with Bu_4NF was employed. BOM and Bn groups were usually removed by catalytic hydrogenolysis. The reported yields ranged from 48–68%. In our experience, deprotection of oligomers having only benzyl protecting group is, in all cases, much simpler and affords compounds with higher purity and yields. Catalytic hydrogenolysis of the polycondensation product **42** affords the Hib spacer-armed oligomer with an average of eight repeating units in excellent yield and purity.[64]

12.7 Conjugate Vaccines Using Synthetic Antigens

In order to prepare small quantities of the synthetic PRP oligomer–protein conjugate as a proof of concept, glutaraldehyde was chosen as a classical model, even if the yield based on the oligosaccharide was very low.[65]

Then, other functional groups were introduced in the PRP spacer in order to obtain better yields in the conjugation step. They include the thioacetamide derivative **43**, which was obtained after derivatization with *N*-succinimidyl *S*-acetylmercaptoacetate followed by *in situ* treatment with hydroxylamine. Derivatives **45** or **47** were obtained after derivatization with the corresponding *N*-succinimidyl *m*-maleimidobenzoate or maleimidopropionate molecules (see Figure 12.11). Lysine-modified proteins or peptides have also been employed, either as bromoacetamide **44**[68,70] or thiopropionamide **46** and **48** derivatives.[71,72] These couplings use either thiol–bromoacetate or thiol–maleimide chemistry. Even if the yields based on the oligosaccharide have not been reported for most of these examples, in our experience, all these coupling conditions should afford a reasonable yield based on the synthetic PRP for vaccine production.

The TT-PRP_2, TT-PRP_3 and TT-PRP_4 conjugates were employed for immunization. As a general trend, in mice, rabbits and monkeys, the number of responder animals and the intensity of the response increased in the series: TT-PRP_2 < TT-PRP_3 < TT-PRP_4. Furthermore, TT-PRP_4 induced a similar response to that obtained with the commercial vaccine DT-PRP_{20}. While promising results have been obtained for TT-PRP_4, as far as we know, clinical evaluations of these conjugates have not been reported.

The work with synthetic PRPs has continued by conjugating them to synthetic peptides representing T-epitopes. Noteworthy, the PRP_3 conjugate to T-epitope peptides expressed in multiple copies on a polylysine backbone

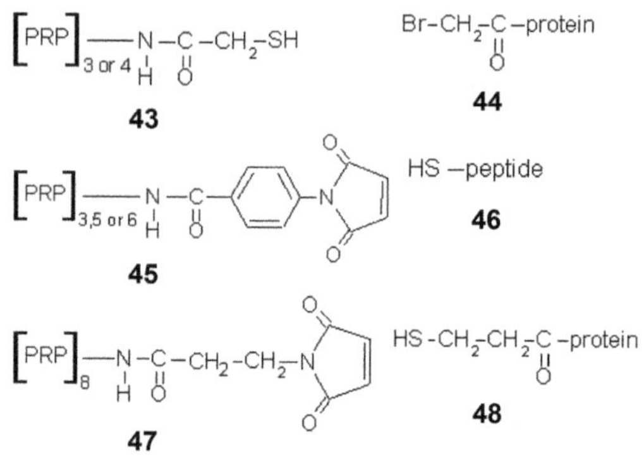

Figure 12.11 Derivatives employed for the synthesis of PRP conjugates.

induced an immune response similar to that of PRP-DT. While very promising, the proof of concept in humans for such a fully synthetic molecule has not been pursued so far.

Previous to the use of the synthetic PRPs, some experiments performed with the capsular polysaccharide fragments PRP_8 and PRP_{20} showed the real complexity of developing a conjugate vaccine.[66] These fragments were coupled to DT by reductive amination. The glycoconjugates were then tested in clinical trials with young adults and 9- to 15-month-old infants. Both vaccines consistently induced high anti-PRP Ab responses in adults. However, in infants, RRP_8-DT elicited only modest anti-PRP responses, whereas PRP_{20}-DT gave consistently high titers post second dose. As a general conclusion, the study revealed that the definition of a minimal epitope for small synthetic PRPs is a very complex task since it depends on the age of the target population. All these precedents were taken into account in our attempt at developing a low-cost process for using synthetic PRPs as Hib conjugate vaccine.

For a classical synthetic PRP_4–protein conjugate, in order to obtain $\sim$10 μg of PRP and $\sim$30 μg of TT per dose, 56 molar equivalents of PRP are required per TT equivalent. This is almost impossible in reasonable yields. Therefore, we discarded the use of PRP_4-like antigens.

One additional advantage of our polycondensation reaction is that the yield of PRP is almost independent of the PRP size. Therefore we targeted a PRP_8 target. This choice is not based on the minimal size required for an adequate immune recognition but rather on: (i) PRP_8 could be obtained in 80% yield in one step and (ii) it assured the number of moles that could be introduced with good yield into the protein.

After intense experimentation, two conjugates of PRP_8 were selected for further development, namely PRP_8-TT and PRP_8-NmOMP. The PRP_8-OMP conjugate induced a stronger immune response in mice. However, both types of conjugates induced a similar response in rabbits.[67] The wt/wt ratio of PRP to protein needed was rather different: PRP_8-TT (1 : 2.6) and PRP_8-OMP (1 : 9). The final choice was PRP_8-TT, where the PRP to TT ratio was 19 : 1 (mol/mol) or 1 : 2.6 (wt/wt). The target dose was thus 10 μg of PRP and 26 μg of TT.

The first clinical evaluation of a synthetic PRP conjugate in humans was conducted in young adults and demonstrated an excellent safety profile and an antibody response similar to that observed for the control vaccine.[68] The initial clinical trial was followed by a series of clinical trials in 4- to 5-year-old children (phase I and phase II) and infants, using three doses at 2, 4 and 6 months old (phase I and phase II). The pathway to provide clinical evidence in this case was shorter than for other vaccines, as a plasma level of 0.15 μg mL^{-1} for short-term protection and 1 μg mL^{-1} as a long-term protection[69] had previously been established as a surrogate of protection.

In November 2003, after sufficient clinical evidence, the glycoconjugate vaccine using a synthetic oligosaccharide was licensed in Cuba to be used in infants.[70] For the first time, a synthetic carbohydrate antigen became part of a commercially available vaccine. Since then, 34 million doses have been produced and used in several countries, either as Quimi-Hib or HEBERPENTA. The World Health Organization prequalified Quimi-Hib in 2009.[71] This fact means that the quality of the vaccine was inspected and considered adequate to be used by UNICEF. Furthermore, the vaccine is licensed not only in Cuba but also in several other countries. Recently, a production facility for Quimi-Hib was inaugurated in Jilin province, China, and one additional facility is under construction in Cuba.

The pentavalent vaccine containing diphtheria, tetanus, pertussis, hepatitis b, and conjugate Hib vaccine (DTP-HBV-Hib) is used in pediatric practice in more than 170 countries.[72] In the composition of the pentavalent vaccine, we found all the types of Hib glycoconjugates described in this chapter. We have promoted the inclusion of Quimi-Hib in a combined pentavalent vaccine. For that, the immunogenicity of the combined vaccine after absorption on aluminum phosphate was studied in clinical trials.[73] HEBERPENTA, a pentavalent combination vaccine containing Quimi-Hib, was licensed in 2006[74] and is used in several countries. Another important lesson learned from the Quimi-Hib story is that, with the actual trend towards the use of more and more combined vaccines, the source of antigen and the potential advantages of an individual component became less and less relevant, as far as the vaccine is stable, compatible, and affordable.

While the production and use of Quimi-Hib will continue to expand in the next years, confirming its feasibility, new conjugate vaccines with synthetic oligosaccharides are in the pipeline at different laboratories and companies.[75] Quimi-Hib precedes all of them by 10 years. In our opinion, the scientific and development process made an important contribution not only to the affordability of Hib conjugate vaccines but also to the present reflourishing of carbohydrates for drug design and discovery.

References

1. J. P. Watt, L. J. Wolfson, K. L. O'Brien, E. Henkle, M. Deloria-Knoll, N. McCall, E. Lee, O. S. Levine, R. Hajjeh, K. Mulholland and T. Cherian, *Lancet*, 2009, **374**, 903–911.
2. M. Pittman, *J. Exp. Med.*, 1931, **53**, 471.
3. L. D. Fothergill and J. Wright, *J. Immunol.*, 1933, **24**, 273–284.
4. P. Anderson, R. B. Johnston Jr. and D. H. Smith, *J. Clin. Invest.*, 1972, **51**, 31.
5. P. F. Weller, A. L. Smith, D. H. Smith and P. Anderson, *J. Infect. Dis.*, 1978, **138**, 427.
6. E. R. Moxon and J. S. Kroll, *Curr. Top. Microbiol. Immunol.*, 1990, **150**, 65.
7. A. J. Swift, E. R. Moxon, A. Zwahlen and J. A. Winkelstein, *Microb. Pathog.*, 1991, **10**, 261.
8. J. A. Winkelstein and E. R. Moxon, *J. Infect. Dis.*, 1992, **165**(Suppl. 1), S62.
9. H. Alexander, M. Heidleberger, G. Leidy and J. Yale, *Biol. Med.*, 1944, **16**, 425.

10. R. Schneerson, L. P. Rodrigues, J. C. Parke Jr. and J. B. Robbins, *J. Immunol.*, 1971, **107**, 1081.
11. P. Anderson, D. H. Smith, D. L. Ingram, J. Wilkins, P. F. Wehrle and V. M. Howie, *J. Infect. Dis.*, 1977, **136**(Suppl.), S57.
12. R. M. Crisel, R. S. Baker and D. E. Dorman, *J. Biol. Chem.*, 1975, **250**, 4926.
13. W. Egan, R. Schneerson, K. E. Werner and G. Zon, *J. Am. Chem. Soc.*, 1982, **104**, 2898.
14. J. C. Parke Jr, R. Schneerson, J. B. Robbins and J. J. Schlesselman, *J. Infect. Dis.*, 1977, **136**(Suppl.), S51.
15. H. Peltola, H. Kayhty, A. Sivonen and H. Makela, *Pediatrics*, 1977, **60**, 730.
16. H. Kayhty, V. Karanko, H. Peltola and P. H. Makela, *Pediatrics*, 1984, **74**, 857.
17. R. Booy, S. A. Hodgson, M. P. Slack, E. C. Anderson, R. T. Mayon-White and E. R. Moxon, *Arch. Dis. Child.*, 1993, **69**, 225.
18. J. V. Bennet, A. E. Platonov, M. P. E. Slack, P. Mala, A. H. Burton and S. E. Robertson, *Vaccines and Biologicals*, WHO, Geneva, 2002.
19. H. A. Bijlmer, L. van Alphen, B. M. Greenwood *et al.*, *J. Infect. Dis.*, 1990, **161**, 1210.
20. K. Landsteiner, *The Specificity of Serological Reactions*, Revised edn, Harvard University Press, Cambridge, MA, 1970.
21. W. F. Goebel, *J. Exp. Med.*, 1940, **72**, 33.
22. O. Avery and W. T. Goebel, *J. Exp. Med.*, 1929, **50**, 522.
23. R. Schneerson, O. Barrera, A. Sutton and J. B. Robbins, *J. Exp. Med.*, 1980, **152**, 361.
24. P. Anderson, *Infect. Immun.*, 1983, **39**, 233.
25. J. B. Robbins, R. Schneerson, P. Anderson and D. H. Smith, *J. Am. Med. Assoc.*, 1996, **276**, 1181.
26. P. Anderson, M. Pichichero, R. Insel, P. Farsad and M. Santosham, *J. Infect. Dis.*, 1985, **152**, 634.
27. M. A. Breukels, G. T. Rijkers, M. M. Voorhorst-Ogink and B. J. Zegers, *Infect. Immun.*, 1999, **67**, 789.
28. G. R. Siber, *Science*, 1994, **265**, 1385.
29. C. Arpin, J. Dechanet, C. Van Kooten *et al.*, *Science*, 1995, **268**, 720.
30. S. Jacquot, T. Kobata, S. Iwata, C. Morimoto and S. F. Schlossman, *J. Immunol.*, 1997, **159**, 2652.
31. K. K. Kamboj, C. L. King, N. S. Greenspan, H. L. Kirchner and J. R. Schreiber, *J. Infect. Dis.*, 2001, **184**, 931.
32. L. E. Leiva, B. Butler, J. Hempe, A. P. Ortigas and R. U. Sorensen, *Clin. Diagn. Lab. Immunol.*, 2001, **8**, 233.
33. E. Latz, J. Franko, D. T. Golenbock and J. R. Schreiber, *J. Immunol.*, 2004, **172**, 2431.
34. H. J. Jennings, R. K. Sood, Synthetic glycoconjugates as human vaccines, in *Neoglycoconjugates: Preparation and Applications*, ed. Y. C. Lee and R. T. Lee, Academic Press, San Diego, 1994, 325.

35. A. Lees, B. L. Nelson and J. J. Mondt, *Vaccine*, 1996, **14**(3), 190–198.
36. D. M. Granoff, E. G. Boies and R. S. Munson, Jr, *J. Pediatr.*, 1984, **105**, 22.
37. H. Kayhty, J. Eskola, H. Peltola, L. Saarinen and P. H. Makela, *J. Infect. Dis.*, 1992, **165**(Suppl. 1), S 165.
38. P. Anderson, M. E. Pichichero and R. A. Insel, *J. Clin. Invest.*, 1985, **76**, 52.
39. M. Porro, S. Constantino, S. Viti, F. Vannozzi, A. Naggi and G. Torri, *Mol. Immunol.*, 1985, **22**, 907.
40. S. Marburg, D. Jorn, R. L. Tolman, B. Arison, J. McCanley, P. J. Kniskern, A. Hagopian and P. P. Vella, *J. Am. Chem. Soc.*, 1986, **108**, 5282.
41. H. Peltola, *Clin. Microbiol. Rev.*, 2000, **13**, 302.
42. H. Kayhty, H. Peltola and J. Eskola, *Pediatr. Infect. Dis. J.*, 1988, 7, 574.
43. P. Anderson, M. E. Pichichero and R. A. Insel, *J. Pediatr.*, 1985, **107**, 346.
44. P. Anderson, M. Pichichero, K. Edwards, C. R. Porch and R. Insel, *J. Pediatr.*, 1987, **111**, 644.
45. J. P. Watt, L. J. Wolfson, K. L. O'Brien, E. Henkle, M. Deloria-Knoll, N. McCall, E. Lee, O. S. Levine, R. Hajjeh, K. Mulholland and T. Cherian, *Lancet*, 2009, **374**, 903.
46. J. Garegg, R. Johansson, I. Lindh and B. Samuelson, *Carbohydr. Res.*, 1986, **150**, 285.
47. J. Garegg, R. B. Lindberg and B. Samuelson, *Carbohydr. Res.*, 1977, **58**, 219.
48. P. Hoogerhout, D. Evenberg, C. A. A. van Boeckel, J. T. Poolman, E. C. Beuvery, G. A. van der Marel and J. H. van Boom, *Tetrahedron Lett.*, 1987, **28**, 1553.
49. Z. Y. Wang and G. Just, *Tetrahedron Lett.*, 1988, **29**, 1525.
50. J. P. G. Hermans, L. Poot, M. Kloosterman, G. A. van der Marel, c. A. A. van boeckel, d. Evenberg, J. T. Poolman, P. Hoogerhout and J. H. van Boom, *Recl. Trav. Chim. Pays-Bas*, 1987, **106**, 498.
51. P. Hoogerhout, J.-R. Mellema, G. N. Wagenaars, c. A. A. van boeckel, d. Evenberg, J. T. Poolman, A. W. M. Lefeber, G. A. van der Marel and J. H. van Boom, *J. Carbohydr. Chem.*, 1988, 7, 399.
52. Z. Y. Wang and G. Just, *Tetrahedron Lett.*, 1988, **29**, 1525.
53. I. Chiu, O. Madrazo and V. Verez-Bencomo, *J. Carbohydr. Chem.*, 1994, **13**, 464.
54. L. Chan and G. Just, *Tetrahedron*, 1990, **46**, 151.
55. L. Chan and G. Just, *Tetrahedron Lett.*, 1988, **29**, 4049.
56. S. Nilsson, M. Bengtsson and T. Norberg, *J. Carbohydr. Chem.*, 1992, **11**, 265.
57. V. Verez-Bencomo, R. Roy, M. C. Rodriguez, A. Villar, V. Fernandez-Santana, E. Garcia, Y. Valdes, L. Heynngnezz, I. Sosa and E. Medina, *Carbohydrate Base Vaccines*, 2008, **989**, 71.
58. P. Hoogerhout, D. Evenberg, C. A. A. van Boeckel, J. T. Poolman, E. C. Beuvery, G. A. van der Marel and J. H. van Boom, *Tetrahedron Lett.*, 1987, **28**, 1553.
59. C. J. J. Ellie, H. J. Muntendam, H. van den Elst, G. A. van der Marel, P. Hoogerhout and J. H. van Boom, *Recl. Trav. Chim. Pays-Bas*, 1989, **108**, 219.
60. A. A. Kandil, N. Chan, P. Chong and M. Klein, *Synlett*, 1992, 7, 555.

61. A. V. Nikolaev, N. S. Utkina, V. N. Shibaev, A. V. Ignatenko and B. N. Rozynov, *Bioorg. Khim.*, 1992, **18**, 272.
62. A. V. Nikolaev, J. A. Chudek and M. A. J. Fergusson, *Carbohydr. Res.*, 1995, **272**, 179.
63. M. Porro, S. Constantino, S. Viti, F. Vannozzi, A. Naggi and G. Torri, *Mol. Immunol.*, 1985, **22**, 907.
64. V. Verez-Bencomo, V. Fernández-Santana, E. Hardy, M. E. Toledo, M. C. Rodríguez, L. Heynngnezz, A. Rodríguez, A. Baly, L. Herrera, M. Izquierdo, A. Villar, Y. Valdés, K. Cosme, M. L. Deler, M. Montane, E. García, A. Ramos, A. Aguilar, E. Medina, G. Toraño, I. Sosa, I. Hernández, R. Martínez, A. Muzachio, A. Carmenates, L. Costa, F. Cardoso, C. Campa, M. Díaz and R. Roy, *Science*, 2004, **305**, 522.
65. D. Evenberg, P. Hoogerhout, C. A. A. van Boeckel, G. T. Rijkers, E. C. Beuvery, J. H. van Boom and J. T. Poolman, *J. Infect. Dis.*, 1992, **165**, S152–S155.
66. P. W. Anderson, M. E. Pichichero, E. C. Stein, S. Porcelli, R. F. Betts, D. M. Connuck, D. Korones, R. A. Insel, J. M. Zahradnik and R. Eby, *J. Immunol.*, 1989, **142**, 2464.
67. V. Fernández-Santana, F. Cardoso, A. Rodríguez, T. Carmenate, L. Penna, Y. Valdes, E. Hardy, F. Mawas, L. Heynngnezz, M. C. Rodríguez, I. Figueroa, J. Chang, M. E. Toledo, A. Musacchio, I. Hernandez, M. Izquierdo, K. Cosme, R. Roy and V. Verez-Bencomo, *Infect. Immun.*, 2004, **72**, 7115.
68. G. Toraño, M. E. Toledo, A. Baly, V. Fernandez-Santana, F. Rodriguez, Y. Alvarez, T. Serrano, A. Musachio, I. Hernandez, E. Hardy, A. Rodrıguez, H. Hernandez, A. Aguilar, R. Sanchez, M. Diaz, V. Muzio, J. Dfana, M.-C. Rodrıguez, L. Heynngnezz and V. Verez-Bencomo, *Clin. Vaccine Immunol.*, 2006, **13**, 1052.
69. H. Käyhty, H. Peltola, V. Karanko and P. H. Makela, *J. Infect. Dis.*, 1983, **147**, 1100.
70. http://www.cecmed.sld.cu/Docs/RegSan/RCP/Bio/Vacs/Quimi-Hib_InyIM.pdf, accessed November 1, 2013.
71. http://www.who.int/immunization_standards/vaccine_quality/PQ_198_Hib_1, dose_Vial_CIGB/en/index.html accessed November 1, 2013.
72. http://www.unicef.org.uk/UNICEFs-Work/What-we-do/Issues-we-work-on/UNICEFs-work-on-immunisation/pentavalent-vaccine-cold-chain/, accessed November 1, 2013.
73. A. Aguilar-Betancourt, C. A. González-Delgado, Z. Cinza-Estévez, J. Martínez-Cabrera, G. Véliz-Ríos, R. Alemán-Zaldívar, M. I. Alonso-Martínez, M. Lago-Baños, N. Puble-Alvarez, A. Delahanty-Fernandez, A. I. Juvier-Madrazo, D. Ortega-León, L. Olivera-Ruano, A. Correa-Fernández, D. Abreu-Reyes, E. Soto-Mestre, M. V. Pérez-Pérez, N. Figueroa-Baile, L. H. Pérez, A. Rodríguez-Silva, E. Martínez-Díaz, G. E. Guillén-Nieto and V. L. Muzio-González, *Hum. Vaccines*, 2008, **4**, 54.
74. http://www.cecmed.sld.cu/Docs/RegSan/RCP/Bio/Vacs/HEBERPENTA.pdf accessed November 1, 2013.
75. Y. Valdes-Balbin, M. C. Rodriguez–Montero and V. Verez-Bencomo, *Carbohydr. Chem.*, 2014, **40**, 564–595.

Subject Index